NUMERICAL ANALYSIS OF WAVELET METHODS

STUDIES IN MATHEMATICS AND ITS APPLICATIONS

VOLUME 32

ELSEVIER

AMSTERDAM – BOSTON – LONDON – NEW YORK – OXFORD – PARIS
SAN DIEGO – SAN FRANCISCO – SINGAPORE – SYDNEY – TOKYO

NUMERICAL ANALYSIS OF WAVELET METHODS

ALBERT COHEN

Laboratoire Jacques-Louis Lions
Université Pierre et Marie Curie
Paris, France

2003

ELSEVIER

AMSTERDAM – BOSTON – LONDON – NEW YORK – OXFORD – PARIS
SAN DIEGO – SAN FRANCISCO – SINGAPORE – SYDNEY – TOKYO

ELSEVIER SCIENCE B.V.
Sara Burgerhartstraat 25
P.O. Box 211, 1000 AE Amsterdam, The Netherlands

First edition 2003

Library of Congress Cataloging in Publication Data
A catalog record from the Library of Congress has been applied for.

British Library Cataloguing in Publication Data
A catalogue record from the British Library has been applied for.

ISBN: 0-444-51124-5
ISSN: 0168-2024

♾ The paper used in this publication meets the requirements of ANSI/NISO Z39.48-1992 (Permanence of Paper).
Transferred to digital printing 2005

To Thierry

Foreword

This book has initially appeared as a chapter in the Handbook of Numerical Analysis (volume VII), edited by Philippe Ciarlet and Jacques-Louis Lions. The main novelties in the present edition are contained in the last chapter. I have tried to give a synthetic survey on the main strategies which have been followed since the 1990's in the development of adaptive wavelet methods for the numerical treatment of partial differential equations, in connection with the topic of nonlinear approximation, and to point out some of the important challenges which remain in this area.

I want to express my gratitude to Jonathan Burd for his excellent proofreading job, and to all those who helped a great deal in improving this material by their valuable remarks: Martin Campos, Sidi-Mahmoud Kaber, Simon Masnou, Marie Postel and Robert Ryan. This book is also the reflection of long term interactions with Wolfgang Dahmen, Ingrid Daubechies, Ronald DeVore and Yves Meyer. Beside the beautiful inspiration that their work has been for me, I also wish to thank them for their constant support and friendship through the years.

Contents

Introduction

Since the 1960's, multiscale methods have been used in numerous areas of applied mathematics as diverse as signal analysis, statistics, computer aided geometric design, image processing and numerical analysis. The mathematical background underlying these methods was substantially reinforced with the emergence of wavelet bases in the 1980's.

The objectives of this book are to survey the theoretical results that are involved in the numerical analysis of wavelet methods, and more generally of multiscale decomposition methods, for numerical simulation problems, and to provide the most relevant examples of such mathematical tools in this particular context.

Multiscale methods are based on approximations $(f_j)_{j\geq 0}$ to the data (or the unknown) f of a given problem, at various resolution levels indexed by j. The corresponding scales of resolution h_j are usually chosen to be of order 2^{-j}: the approximation f_j should thus be viewed as a sketchy picture of f that cannot oscillate at a frequency higher than 2^j. As an example, if f is a univariate continuous function, one could choose for f_j the unique function such that $f_j(2^{-j}k) = f(2^{-j}k)$ and such that f_j is affine when restricted to $[2^{-j}k, 2^{-j}(k+1)]$ for all $k \in \mathbb{Z}$.

Formally, one obtains a *multiscale decomposition* by expanding f into the sum of its coarsest approximation and additional details:

$$f = f_0 + \sum_{j\geq 0} g_j,$$

where each $g_j = f_{j+1} - f_j$ represents the fluctuation of f between the two successive levels of resolution j and $j+1$.

In practice, these approximations and decompositions can be defined and implemented in various ways: by contrast with the one and unique Fourier transform, a "multiscale transform" can be picked out of a versatile collection of mathematical tools. Some of them can furthermore be implemented by means of fast algorithms, and are therefore more appealing for numerical applications. In the context of numerical computations, one can be interested in further decomposing each fluctuation g_j into local contributions. For specific types of multiresolution approximations, this task can be achieved using a ***wavelet basis***: one introduces an appropriate function ψ (in the case where f is a univariate function) that is well localized both in space and frequency, and oscillates in the sense that $\int \psi = 0$, and allows one to expand g_j according to

$$g_j = \sum_{k \in \mathbb{Z}} d_{j,k} \psi_{j,k},$$

where the $d_{j,k}$'s are scalar coefficients and each wavelet $\psi_{j,k} := 2^{j/2}\psi(2^j \cdot - k)$ contributes to the fluctuation of f at scale 2^{-j} in a neighborhood of size $2^{-j}|\mathrm{Supp}(\psi)|$ around the point $2^{-j}k$. Here again, there exists a versatile collection of functions ψ that are admissible for generating wavelets bases. One should thus keep in mind that, for a given problem, the efficiency of a multiscale decomposition method is strongly tied to the choice of a specific tool and associated algorithm. In the field of numerical analysis, multiscale and wavelets decompositions have been successfully used for three main tasks:

1. *Preconditioning* large systems arising from the discretization of elliptic partial differential equations,

2. *Adaptive approximation* of functions which are not uniformly smooth,

3. *Sparse representation* of initially full matrices arising in the discretization of integral equations.

It should be noted that in each of these applications, the use of multiscale and wavelet decompositions is very close in essence to other methods that have been previously introduced: multigrid methods for preconditioning, mesh refinement for adaptivity, multipole and panel clustering algorithms for integral equations.

Do wavelet-based techniques bring specific improvements, when dealing with the above tasks, in comparison to these more classical methods ?

Clearly, these techniques enlarge the "library" of tools that are available for dealing with these tasks. As it was already mentioned, an optimal choice among these tools is strongly problem dependent. More importantly, in the development and analysis of these techniques, one can benefit from a strong theoretical input from harmonic analysis and approximation theory. In particular, the possibility of characterizing various smoothness classes - e.g. Sobolev, Hölder and Besov spaces - from the numerical properties of multiscale decompositions, turns out to play a key role in the application of wavelet methods to the three previously mentioned tasks. This brings out a new and fruitful point of view, which can also be useful in the analysis of more classical, related methods.

Our goal here is not to present all the existing wavelet methods for the numerical simulation of physical processes (the amount of literature is already huge and still growing fast), but rather to give a broad account of this theoretical imput to the numerical analysis of multiscale methods.

This book is organized as follows:

In Chapter 1, we motivate the introduction of a general theory by two very simple examples of multiscale decomposition schemes. These schemes are associated with the most elementary wavelet bases: the Haar system and the Schauder hierarchical basis, that are respectively built from approximations by piecewise constant functions and piecewise affine continuous functions. We show how these schemes can be adapted to the decomposition of multivariate functions and how they can be used for adaptive approximation of functions, sparsification of integral operators, and preconditioning of elliptic equations. We discuss their inherent limitations for these applications. This chapter is descriptive in essence and the results are deliberately not organized according to the usual "lemma-theorem-corollary" sequence. The reader who is prepared to enter directly the general theory can easily skip it.

In Chapter 2, we review the general concept of multiresolution approximation, consisting in a nested sequence of approximation spaces,

$$\cdots V_{-1} \subset V_0 \subset V_1 \subset \cdots$$

which are used to define the approximation $f_j \in V_j$. This concept leads in a natural way to the construction of wavelets in terms of *details* complementing the approximation between V_j and V_{j+1}. In practice, there are two main approaches to building such tools. The first one, which is the most intuitive to the numerician, consists in constructing the spaces V_j as a nested sequence of finite element spaces, or as nested discretizations by cell averages or point values equipped with certain inter-scale operators. The second one is based on a single *scaling function* φ generating V_j in the sense that $\varphi(2^j \cdot -k)$, $k \in \mathbb{Z}$, is a Riesz basis of V_j. This last approach yields tools of a different nature than finite elements, that might seem difficult to handle for practical computations: in particular, the function φ is not given explicitly, but as a solution of an equation of the type

$$\varphi(x) = \sum_{n \in \mathbb{Z}} h_n \varphi(2x - n),$$

that expresses the embedding $V_0 \subset V_1$. A solution of such an equation is also called a *refinable function*. Since they are far less classical than finite elements in numerical analysis, and since they are a key tool in the construction of wavelet bases, we devote an important part of this chapter to the study of refinable functions. In particular, we show how they can be computed and used for numerical analysis purposes, and we analyze the relations between the properties of the coefficients h_n and properties of φ, such as smoothness, approximation power of the spaces V_j, stability, orthonormality and biorthogonality. We describe the examples of refinable functions and related multiscale decompositions that are relevant in numerical analysis, in particular biorthogonal decompositions into spline functions. We show how these tools can be adapted to multivariate domains with specific boundary conditions. We finally return to the first approach and address the construction of multiscale decompositions associated with the most commonly used discretizations in numerical analysis: point values, cell averages and finite elements.

In Chapter 3, we show how smoothness classes can be characterized from the decay properties of $(\|g_j\|_{L^p})_{j\geq 0}$ and $(\|f - f_j\|_{L^p})_{j\geq 0}$, as j goes to $+\infty$, or equivalently from the numerical properties of the wavelet coefficients $d_{j,k}$. These results turn out to be the key to the theoretical understanding of multilevel preconditioning for elliptic operators, as well as adaptive approximation. The L^p norm which is used here to measure the error is the same as the L^p metric associated with the smoothness class: for example, in the case of the Sobolev spaces H^s, one will consider the behaviour of $\|f - f_j\|_{L^2}$ or $\|g_j\|_{L^2}$. The quantity $\|f - f_j\|_{L^p}$ is a *linear approximation* error, in the sense that f_j is an approximation of f in the linear space V_j, and is usually obtained through a linear operation applied to f (typically a projection onto V_j). These characterization results are essentially based on the combination of two types of estimates: a *direct* (or Jackson type) inequality and an *inverse* (or Bernstein type) inequality. We show how to establish such inequalities in the case of the multiresolution spaces that were built in Chapter 2. We discuss several variants of these results, as well as their adaptation to function spaces defined on bounded domains with prescribed boundary conditions. We finally discuss their application to the multilevel preconditioning of elliptic operators.

In Chapter 4, we focus on *adaptive approximation*: the function f is now approximated by a combination

$$f_N = \sum_{(j,k)\in\Lambda_N} c_{j,k}\psi_{j,k},$$

where the set Λ_N has cardinality N and is allowed to depend on the function f, in order to improve the approximation. In practice, Λ_N typically represents the N largest contributions in the wavelet decomposition of f, in the metric where the error is to be measured. Such data-dependent approximation procedures are intrisically *nonlinear*. As in the case of linear approximation, the decay properties of the error $\|f - f_N\|_{L^p}$ are related to the smoothness properties of f, but the corresponding smoothness classes are now associated with an L^q metric with $q < p$. Similar ideas apply to the wavelet discretization of partial differential and integral operators: sparse approximations can be obtained by simple thresholding procedures applied to the matrix entries. Nonlinear approximation also allows one to draw a first comparison between wavelets and adaptive mesh refinement from the point of view of approximation theory. We next discuss the relation between these results and adaptive algorithms for PDE's. In the context of numerically solving PDE's a specific difficulty is that one does not know in advance which are the largest coefficients the solution. Adaptive wavelet methods aim to track this information within the numerical solution process. We describe the two main lines of research which have been followed in the development of such methods since the 1990's, and we compare them to more classical adaptive strategies. We conclude by pointing out some intrinsic shortcomings and remaining challenges in adaptive wavelet methods.

Among the basic references that have influenced the present book, let us mention the two well-known monographs on wavelets by DAUBECHIES [1992] and MEYER [1990], as well as the survey papers by DAHMEN [1997] and DEVORE [1998].

Beside these important references, it goes without saying that many of the results which are presented here have appeared in publications by numerous mathematicians. Most of the time, however, their formulation has been revisited for consistency purpose. We also have chosen to avoid "minimal" or general assumptions in the statement of some of these results, in the cases where these assumptions make the proof more complex without being really useful to numerical applications. In each chapter, several remarks and the last section aim to present an historical account of these contributions, and to give "pointers" toward related theoretical results and algorithmic techniques. The author is aware of the rapid evolution of the subject and of its multiple connexions with other scientific areas. He has tried to mention the most significant contributions in the context of numerical analysis, together with the constraint of keeping this book self-contained and of reasonable size.

Notations

We list below symbols that appear frequently throughout the book. If not below, their definitions appear at the first place where they are introduced.

Spaces
$C^m(\Omega)$ (integer m): Spaces of m-times continuously differentiable functions
$C^s(\Omega)$ (non-integer $s > 0$): Hölder spaces
$L^p(\Omega)$: Lebesgue spaces
$W^{m,p}(\Omega)$: Sobolev spaces ($H^s := W^{s,2}$)
$B^s_{p,q}(\Omega)$: Besov spaces
Π_n: polynomials of total degree n
V_j and $\tilde{V}_j$: primal and dual multiresolution spaces at scale j.
W_j and $\widetilde{W}_j$: primal and dual detail (or wavelet) spaces at scale j.

Functions
$f_{j,k} := 2^{dj/2} f(2^j \cdot -k)$: L^2-scaling of f defined on $\mathbb{R}^d$.
$\psi_{j,k}$ and $\tilde{\psi}_{j,k}$: primal and dual wavelets of resolution 2^{-j}.
$\varphi_{j,k}$ and $\tilde{\varphi}_{j,k}$: primal and dual scaling functions of resolution 2^{-j}.
φ_λ and $\tilde{\varphi}_\lambda$ for $\lambda \in \Gamma_j$: compact notations for the primal and dual scaling functions of resolution 2^{-j}.
ψ_λ and $\tilde{\psi}_\lambda$ for $\lambda \in \nabla_j$: compact notations for the primal and dual wavelets of resolution 2^{-j}.
$|\lambda|$: resolution level of the index λ ($|\lambda| = j$ if $\lambda \in \Gamma_j$ or $\lambda \in \nabla_j$)
$\nabla := \cup_{j \geq -1} \nabla_j$ with $\nabla_{-1} = \Gamma_0$: set of all wavelet indices.
$\nabla^J := \cup_{-1 \leq j < J} \nabla_j$: set of all wavelet indices up to level $J - 1$.
ψ_λ and $\tilde{\psi}_\lambda$ for $\lambda \in \nabla$: primal and dual wavelets at all scales, including scaling functions at the coarsest level.
$\langle f, g \rangle$: duality product ($\int f\overline{g}$ if $f \in L^p$ and $g \in L^{p'}$, $1/p + 1/p' = 1$).

Operators
P_j, $\tilde{P}_j$, Q_j, $\tilde{Q}_j$: projectors onto V_j, $\tilde{V}_j$, W_j, $\widetilde{W}_j$.

A^*: dual of the operator A.
$\mathcal{F}f(\omega) = \hat{f}(\omega) := \int_{\mathbb{R}^d} f(x)e^{-i\omega\cdot x}dx$: Fourier transform of f.

Sets
$\#E$ (or $|E|$): cardinality of a finite set E.
meas(E) (or $|E|$): Lebesgue's measure of a domain $E \subset \mathbb{R}^d$.
diam(E): diameter of a domain $E \subset \mathbb{R}^d$.
dist$(A, B) = \inf_{x\in A, y\in B} |x - y|$: distance between the sets A and B.
Supp(f): support of the function f.

Equivalences
If $A(u)$ and $B(u)$ are positive functions of a set u of parameters, we shall often use the notation

$$A(u) \lesssim B(u),$$

to express that there exists a constant $C > 0$ such that $A(u) \leq CB(u)$ independently of the parameters. For example $\|P_j f\| \lesssim \|f\|$ means that the operator P_j is bounded independently of j. We also use the notation

$$A(u) \sim B(u),$$

to express that $A(u) \lesssim B(u)$ and $B(u) \lesssim A(u)$.

Chapter 1

Basic examples

1.1 Introduction

Before entering the general theory of multiscale decompositions, we shall study two basic examples and show their potential range of applications in numerical analysis as well as their inherent limitations. Our goal is to identify some important features that will be studied with more details, in the general theory developed in the next chapters. This chapter is descriptive in essence and some of its results are quoted without a detailed proof when they are particular cases of theorems that are proved in the next chapters.

In the two examples that we want to study, the approximation f_j of a univariate function f at the scale 2^{-j} will respectively be piecewise constant and piecewise affine on dyadic intervals

$$I_{j,k} := [k2^{-j}, (k+1)2^{-j}[, \quad k \in \mathbb{Z}. \tag{1.1.1}$$

In §1.2 and §1.3, we introduce these approximations and we show how they are related to decompositions in two elementary wavelet bases: the Haar system and the Schauder hierarchical basis. We describe the decomposition and reconstruction algorithms that can be used to compute the coefficients of a function in these bases, and we show in §1.4 how these decompositions can be generalized in a natural way to functions of several variables.

Applications of these elementary tools to numerical analysis are discussed in the next sections: in §1.5, we discuss the application of the Haar and Schauder bases to adaptive approximation of functions. The main idea is that a simple thresholding procedure on the coefficients of a function f in these bases amounts to building an adaptive grid for the discretization of this function. A similar thresholding procedure is applied to sparsify certain integral operators, which usually yield fully populated matrices in their standard finite element discretization. In §1.6, we recall some basic facts on the finite element discretization of a simple model second order elliptic problem, which is known to yield an ill-conditioned system, and we show how multiscale decompositions can be used to design simple preconditioners and optimal iterative solvers.

We conclude by reviewing the most important features of these basic examples that suggest a more general theory. We also point out their inherent limitations: low approximation order and severe restrictions on the class of differential and integral operators that they can handle.

1.2 The Haar system

Let f be a function in $L^2(\mathbb{R})$. We can define piecewise constant approximations f_j of f at scale 2^{-j} by

$$f_j(x) = 2^j \int_{I_{j,k}} f(t)dt, \quad \text{for all } x \in I_{j,k}, \ k \in \mathbb{Z}, \tag{1.2.1}$$

i.e. f is approximated by its mean value on each interval $I_{j,k}$, $k \in \mathbb{Z}$. Let us make three simple comments on this particular choice for f_j:

1. We first remark that the choice of the mean value makes f_j the L^2-orthogonal projection of f onto the space

$$V_j = \{f \in L^2 \ ; \ f \text{ is constant on } I_{j,k}, \ k \in \mathbb{Z}\}. \tag{1.2.2}$$

Indeed, an orthogonal basis for V_j is given by the family

$$\varphi_{j,k} := 2^{j/2}\chi_{I_{j,k}} = 2^{j/2}\varphi(2^j \cdot -k), \ k \in \mathbb{Z}, \tag{1.2.3}$$

where $\varphi := \chi_{[0,1]}$, and clearly f_j can be written

$$f_j = \sum_{k\in\mathbb{Z}} \langle f, \varphi_{j,k}\rangle \varphi_{j,k}, \tag{1.2.4}$$

with the usual notation $\langle f, g\rangle = \int f(t)\overline{g(t)}dt$. We will thus denote f_j by $P_j f$ where P_j is the orthogonal projector onto V_j. We shall also use the notation

$$c_{j,k} = c_{j,k}(f) := \langle f, \varphi_{j,k}\rangle = 2^{j/2} \int_{I_{j,k}} f(t)dt, \tag{1.2.5}$$

for the normalized mean values which are the coordinates of $P_j f$ in the basis $(\varphi_{j,k})_{k\in\mathbb{Z}}$.

2. We also note that this approximation process is *local*: the value of $P_j f$ on $I_{j,k}$ is only influenced by the value of f on the same interval. In particular, we can still use (1.2.1) to define $P_j f$ when f is only locally integrable, or when f is only defined on a bounded interval such as $[0,1]$ (in that case, $P_j f$ makes sense only for $j \geq 0$). As an example, we display in Figure 1.2.1 the function $f(x) = \sin(2\pi x)$ and its approximation $P_4 f$ on $[0,1]$.

3. Finally, since $V_j \subset V_{j+1}$, it is clear that $P_{j+1} f$ contains "more information" on f than the coarser approximation $P_j f$: the mean values on the intervals of size $2^{-j-1}k$ entirely determine those on the coarser intervals of double size 2^{-j} by taking their averages. More precisely, we have

$$P_j f_{|I_{j,k}} = (P_{j+1} f_{|I_{j+1,2k}} + P_{j+1} f_{|I_{j+1,2k+1}})/2. \tag{1.2.6}$$

We can also define the orthogonal projection $Q_j f := P_{j+1} f - P_j f$ onto W_j, the orthogonal complement of V_j into V_{j+1}. From (1.2.6), it is clear that $Q_j f$ "oscillates" in the sense that

$$Q_j f_{|I_{j+1,2k}} = -Q_j f_{|I_{j+1,2k+1}}. \tag{1.2.7}$$

As an example, Figure 1.2.2 shows the components $P_3 f$ and $Q_3 f$ on $[0,1]$, for $f(x) = \sin(2\pi x)$: we have thus decomposed $P_4 f$ of Figure 1.2.1 into the sum of the coarser approximation $P_3 f$ and the oscillating component $Q_3 f$.

The oscillation property (1.2.7) allows us to expand $Q_j f$ into

$$Q_j f = \sum_{k\in\mathbb{Z}} d_{j,k}\psi_{j,k}, \tag{1.2.8}$$

where $\psi_{j,k} := 2^{j/2}\psi(2^j \cdot -k)$ and

$$\psi(x) = \chi_{[0,1/2[} - \chi_{[1/2,1[}. \tag{1.2.9}$$

Since the $\psi_{j,k}$, $k \in \mathbb{Z}$ are also an orthonormal system, they constitute an orthonormal basis for W_j and we thus have

$$d_{j,k} = d_{j,k}(f) = \langle f, \psi_{j,k}\rangle. \tag{1.2.10}$$

We thus have re-expressed the "two-level" decomposition of $P_{j+1} f$ into the coarser approximation $P_j f$ and the additional fluctuations $Q_j f$, according to

$$\sum_k c_{j+1,k}\varphi_{j+1,k} = \sum_k c_{j,k}\varphi_{j,k} + \sum_k d_{j,k}\psi_{j,k}. \tag{1.2.11}$$

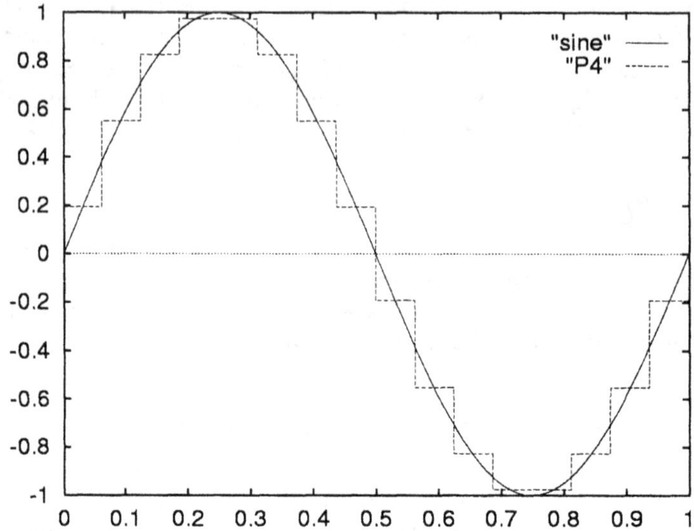

Figure 1.2.1: The function $f(x) = \sin(2\pi x)$ and its approximation $P_4 f$

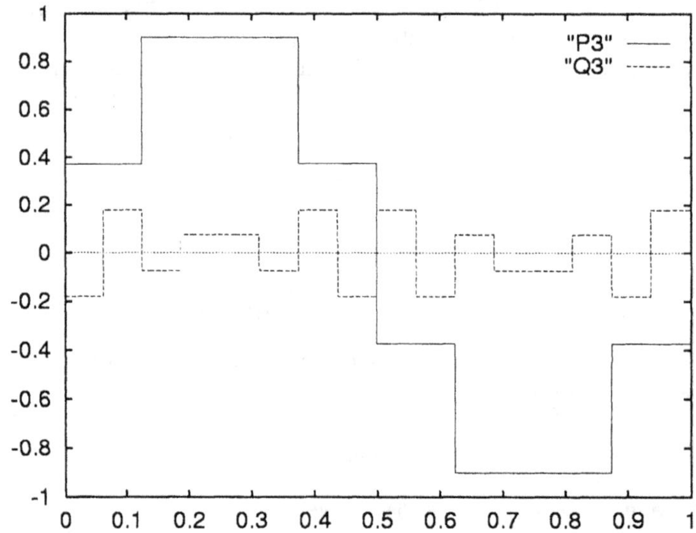

Figure 1.2.2: The approximation $P_3 f$ and the fluctuation $Q_3 f$

This decomposition can be iterated on an arbitrary number of levels: if $j_0 < j_1$, we can rewrite the orthogonal decomposition

$$P_{j_1} f = P_{j_0} f + \sum_{j_0 \le j < j_1} Q_j f, \tag{1.2.12}$$

according to

$$\sum_k c_{j_1,k}\varphi_{j_1,k} = \sum_k c_{j_0,k}\varphi_{j_0,k} + \sum_{j_0 \le j < j_1} \sum_k d_{j,k}\psi_{j,k}. \tag{1.2.13}$$

Both (1.2.12) and (1.2.13) express an additive *multiscale decomposition* of $P_{j_1} f$ into a coarser approximation and a succession of fluctuations at intermediate scales. Formula (1.2.13), however, gives a local description of each contribution and should be viewed as an orthonormal change of basis in V_{j_1}: both $\{\varphi_{j_1,k}\}_{k\in\mathbb{Z}}$ and $\{\varphi_{j_0,k}\}_{k\in\mathbb{Z}} \cup \{\psi_{j,k}\}_{j_0 \le j < j_1, k\in\mathbb{Z}}$ are orthonormal bases for V_{j_1}, and any function in V_{j_1} has thus a unique decomposition in each of these bases.

Note the different role played by the functions φ and ψ: the former is used to characterize the approximation of a function at different scales, while the latter is needed to represent the fluctuation between successive levels of approximation. In particular, we have $\int \psi = 0$, reflecting the oscillatory nature of these fluctuations. In the more general multiresolution context that will be developed in the next chapter, φ will be called *scaling function* and ψ the *mother wavelet*, in the sense that all the wavelets $\psi_{j,k}$ are generated from translations and dilations of ψ. We display in Figure 1.2.3 the function ψ and two of its translated-dilated versions.

Clearly the union of the approximation spaces V_j is dense in $L^2(\mathbb{R})$, i.e.

$$\lim_{j\to+\infty} \|f - P_j f\|_{L^2} = 0, \tag{1.2.14}$$

for all $f \in L^2(\mathbb{R})$. Combining (1.2.14) with (1.2.13), we obtain that the orthonormal family $\{\varphi_{j_0,k}\}_{k\in\mathbb{Z}} \cup \{\psi_{j,k}\}_{j\ge j_0, k\in\mathbb{Z}}$ is a complete orthonormal system of $L^2(\mathbb{R})$. Any function $f \in L^2(\mathbb{R})$ can thus be decomposed into

$$f = \sum_k c_{j_0,k}(f)\varphi_{j_0,k} + \sum_{j\ge j_0}\sum_k d_{j,k}(f)\psi_{j,k}, \tag{1.2.15}$$

where the series converges in L^2.

Remark 1.2.1 *Since we have an orthonormal basis, the L^2-convergence in (1.2.15) is unconditional: one can permutate the terms or change their signs without affecting the convergence of the series. This might not be true when*

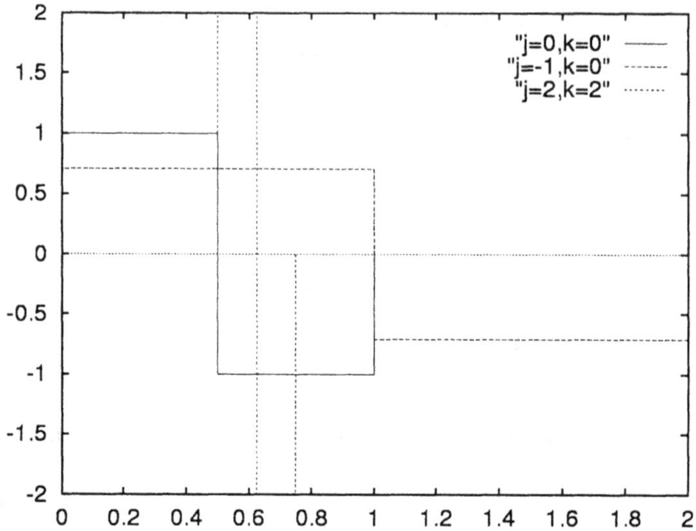

Figure 1.2.3: The wavelets $\psi = \psi_{0,0}$, $\psi_{-1,0}$ and $\psi_{2,2}$

we measure convergence in another norm: if f is uniformly continuous, clearly $P_j f$ converges uniformly to f as j goes to $+\infty$, so that we can define a summation process by letting j_1 go to $+\infty$ in (1.2.13), yet a permutation in the terms might lead to a divergent series. The same remark applies for the L^1 convergence of the series of an L^1 function.

Remark 1.2.2 *We can of course apply the same method to decompose a function that is defined on a bounded interval. In particular, φ and the $\psi_{j,k}$'s for $j \geq 0$ and $0 \leq k < 2^j$ constitute an orthonormal basis for $L^2([0,1])$. This basis was introduced in 1909 by A. Haar, and is therefore refered to as the Haar system.*

If the function f is defined on the whole of $\mathbb{R}$, we can also let the coarsest level j_0 go to $-\infty$ in (1.2.15). It turns out that the coarse scale approximation vanishes in the L^2 sense, when passing to this limit: for any f in $L^2(\mathbb{R})$, we have

$$\lim_{j \to -\infty} \|P_j f\|_{L^2} = 0. \tag{1.2.16}$$

In order to prove (1.2.16), we first suppose that $f = \chi_{[a,b]}$, for some $a < b$. Then, when j is negative and large enough so that $2^{-j} \geq \max\{|a|, |b|\}$, we have $c_{j,k} = \langle f, \varphi_{j,k} \rangle = 0$ if k is not 0 or -1 (the functions f and $\varphi_{j,k}$ have disjoint supports). We thus have, for this range of j,

$$\|P_j f\|_{L^2}^2 = |c_{j,0}|^2 + |c_{j,-1}|^2 \le 2[|a-b|2^{j/2}]^2,$$

which implies (1.2.16) for $f = \chi_{[a,b]}$ and thus for any finite linear combination of such characteristic functions.

For an arbitrary $f \in L^2(\mathbb{R})$ and $\varepsilon > 0$, there exists a function g which is of the type

$$g = \sum_{i=0}^{N} c_i \chi_{[a_i,b_i]}, \tag{1.2.17}$$

such that

$$\|f - g\|_{L^2} \le \varepsilon, \tag{1.2.18}$$

(finite linear combinations of characteristic functions of finite intervals are dense in $L^2(\mathbb{R})$). Since $\|P_j g\|_{L^2}$ tends to zero as j goes to $-\infty$, we have $\|P_j g\|_{L^2} \le \varepsilon$ for j large enough negatively, and therefore

$$\|P_j f\|_{L^2} \le \|P_j g\|_{L^2} + \|P_j(f-g)\|_{L^2} \le \|P_j g\|_{L^2} + \|f-g\|_{L^2} \le 2\varepsilon.$$

Since ε is arbitary, this proves (1.2.16) for a general function $f \in L^2(\mathbb{R})$.

An immediate consequence of (1.2.16) is the following: *the set* $\{\psi_{j,k}\}_{j,k\in\mathbb{Z}}$ *constitutes an orthonormal basis for* $L^2(\mathbb{R})$. This fact might appear as a paradox: a function f can be expanded in terms of basis functions that all have vanishing integral, even if $\int f \neq 0$. This is possible, simply because the convergence of $P_j f$ to zero as j goes to $-\infty$ holds in L^2 *but not in* L^1, for a general function f in $L^1 \cap L^2$. Indeed, according to the definition of P_j by (1.2.1), we clearly have $\int P_j f = \int f$, so that $P_j f$ cannot go to zero in L^1 as soon as f has a non-zero integral.

In the context of applications of multiscale decompositions to numerical simulation, one does not make so much use of the full wavelet basis $\{\psi_{j,k}\}_{j,k\in\mathbb{Z}}$ and of the fact that $P_j f$ goes to 0 in L^2 as j tends to $-\infty$. Indeed, most problems are confined to a bounded domain that limits the coarseness of the analysis scale. We shall thus mainly consider multiscale decompositions of the type (1.2.15) for a fixed minimal scale level j_0, which will often be chosen to be $j_0 = 0$. Of more importance to the numerician is the behaviour of $P_j f$ as j grows, in particular approximation results dealing with the size of the error $f - P_j f$. This question will be addressed in a more general context in chapters 2 and 3.

From a computational point of view, if we want to implement the decomposition of a function in the Haar system, we also need to limit the resolution level from above: we shall start our analysis from an approximation $f_{j_1} = \sum_k c_{j_1,j}\varphi_{j,k}$ of a function f at the finest resolution level j_1. We then have to deal with the following problem: how can we compute the coefficients $c_{j_0,k}$ and $d_{j,k}$, $j_0 \le j < j_1$ from the coefficients $c_{j_1,k}$ at the finest scale ?

Note that this problem makes sense even if f_{j_1} is an approximation of f in V_{j_1} that differs from its projection $P_{j_1}f$: we are then interested in decomposing this particular approximation into multiscale components. This point is crucial, since in numerical analysis applications, the approximate solution of a problem is rarely the L^2-orthogonal projection of the true solution.

To answer this question, we start from the "two level" decomposition (1.2.11). From (1.2.6) and the L^2 normalization of the functions $\varphi_{j,k}$ and $\psi_{j,k}$, we derive the simple relations

$$c_{j,k} = (c_{j+1,2k} + c_{j+1,2k+1})/\sqrt{2}, \tag{1.2.19}$$

and

$$d_{j,k} = (c_{j+1,2k} - c_{j+1,2k+1})/\sqrt{2}. \tag{1.2.20}$$

Another way to derive relations (1.2.19) and (1.2.20) is to inject the identities $\varphi_{j,k} = (\varphi_{j+1,2k} + \varphi_{j+1,2k+1})/\sqrt{2}$ and $\psi_{j,k} = (\varphi_{j+1,2k} - \varphi_{j+1,2k+1})/\sqrt{2}$ inside $c_{j,k} = \langle f, \varphi_{j,k}\rangle$ and $d_{j,k} = \langle f, \psi_{j,k}\rangle$. These relations show the local aspect of the change of basis in (1.2.11): the vectors $(c_{j,k}, d_{j,k})$ and $(c_{j+1,2k}, c_{j+1,2k+1})$ are related by a simple orthogonal transformation.

The inverse transform is given by

$$c_{j+1,2k} = (c_{j,k} + d_{j,k})/\sqrt{2}, \tag{1.2.21}$$

and

$$c_{j+1,2k+1} = (c_{j,k} - d_{j,k})/\sqrt{2}. \tag{1.2.22}$$

From these four relations, we can derive simple decomposition and reconstruction algorithms.

Decomposition algorithm:

1. Start from finest scale approximation coefficients $c_{j_1,k}$.
2. Compute the sequences $c_{j_1-1,k}$ and $d_{j_1-1,k}$ using (1.2.19) and (1.2.20).
3. Store the details $d_{j_1-1,k}$ and iterate the decomposition on the approximations $c_{j_1-1,k}$.

4. Iterate the decomposition from fine to coarse scales.

5. Stop the decomposition when the coarsest level coefficients $c_{j_0,k}$ and $d_{j_0,k}$ are computed.

Reconstruction algorithm:

1. Start from the coarsest level coefficients $c_{j_0,k}$ and $d_{j_0,k}$.

2. Compute the sequence $c_{j_0+1,k}$ using (1.2.21) and (1.2.22).

3. Iterate the reconstruction using the details $d_{j_0+1,k}$.

4. Iterate the reconstruction from coarse to fine scales.

5. Stop the reconstruction when the finest approximation coefficients $c_{j_1,k}$ are computed.

In practice, the function to be decomposed is defined (and approximated at the finest level) on a bounded domain. In turn, the sequence $c_{j,k}$ has finite length, allowing a computer to perform these algorithms. For example, if the function f is defined on the interval $[0,1]$, these algorithms operate on 2^{j_1} coefficients. We remark that both decomposition and reconstruction algorithms perform 2^{j+1} operations between the level j and $j+1$. This is illustrated in Figure 1.2.4 in the case where $j_1 = 3$ and $j_0 = 0$, by the "pyramidal" representation of these algorithms: each "branch" indicates the influence of a coefficient at a given scale j on another one at the level $j+1$ or $j-1$ (the dashed branches correspond to the computations involving the detail coefficients).

The total number of operations for both algorithms is therefore given by $\sum_{j=j_0}^{j_1} 2^{j+1} = \mathcal{O}(N)$ where $N = 2^{j_1}$ is the length of the finest approximation sequence. This complexity is "optimal", in the sense that the number of required operations is of the same order as the number of computed values. A general linear transform would require $\mathcal{O}(N^2)$ operations. From a linear algebra point of view, the Haar decomposition and reconstruction algorithm can thus be viewed as a smart factorisation of a linear transform. A similar situation is encountered when implementing the discrete Fourier transform by the Fast Fourier Transform algorithm which has complexity $\mathcal{O}(N \log N)$.

This first example of a multiscale scheme is rarely used in numerical applications, due to the poor approximation properties of piecewise constant functions, as well as their lack of smoothness. However, its particular simplicity allows us to identify several important features of multiscale decompositions that we shall require in a more general setting: approximations take place in a *nested sequence* of spaces V_j allowing us to compute $P_j f$ from

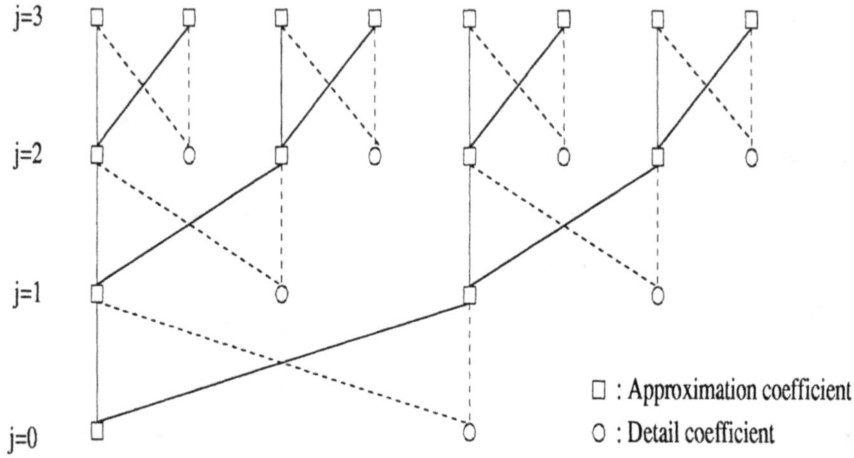

Figure 1.2.4: Pyramidal representation of the Haar transform

$P_{j+1}f$, both approximations and fluctuations are characterized at each scale by *local bases*, and multiscale decompositions and reconstructions in these bases are performed by *fast algorithms* with $\mathcal{O}(N)$ complexity.

1.3 The Schauder hierarchical basis

We now turn to the approximation of a function f by a continuous, piecewise affine function on the intervals $I_{j,k}$, $k \in \mathbb{Z}$. Such an approximation can be defined in a unique way by its values at the points $2^{-j}k$. Assuming that f is a continuous function, we are allowed to impose

$$f_j(2^{-j}k) = f(2^{-j}k), \quad k \in \mathbb{Z}, \tag{1.3.1}$$

i.e. choose f_j to be the *linear interpolation* of f at the scale 2^{-j}. If we denote by P_j the interpolation operator that maps f onto f_j, we remark that P_j is a "projector" onto the space

$$V_j = \{f \in C^0(\mathbb{R}) \ ; \ f \text{ is affine on } I_{j,k}, \ \ k \in \mathbb{Z}\}, \tag{1.3.2}$$

in the sense that $P_jf = f$ whenever $f \in V_j$. However, it is by no means an orthogonal projector, and it is not even bounded in L^p for $p < \infty$ (for example, if $f(x) = \max\{0, 1 - |x|\}$, the sequence $f_n = f(n\cdot)$, $n > 0$, goes to zero in L^p but is uniformly mapped onto f by P_0).

As in the case of piecewise constant approximation, we start by some simple comments.

A natural basis for V_j is given by the functions $\varphi_{j,k}$, $k \in \mathbb{Z}$, where $\varphi(x) = \max\{0, 1 - |x|\}$ and $\varphi_{j,k}(x) = 2^{j/2}\varphi(2^j \cdot -k)$. It allows us to expand $P_j f$ in a unique way according to

$$P_j f = \sum_{k\in\mathbb{Z}} c_{j,k}\varphi_{j,k}, \tag{1.3.3}$$

with

$$c_{j,k} = c_{j,k}(f) := 2^{-j/2} f(2^{-j}k). \tag{1.3.4}$$

This basis is sometimes called a *nodal basis*, since each function $\varphi_{j,k}$ is associated with a "node" of the mesh $\Gamma_j := 2^{-j}\mathbb{Z}$, and the value of a function $f_j \in V_j$ at this node determines its coordinate. Clearly, this is not an orthonormal basis (due to the overlapping supports of $\varphi_{j,k}$ and $\varphi_{j,k+1}$), and the convergence of (1.3.3) is meant in a pointwise sense. However, it is easy to check that $\varphi_{j,k}$ is a *Riesz basis* for V_j: if $f_j \in V_j \cap L^2$, then $\|f_j\|_{L^2}$ and $(\sum_k |c_{j,k}(f_j)|^2)^{1/2}$ are equivalent norms and the series in (1.3.3) converges in L^2. This property will be further analyzed in the more general setting of the next chapter, and we thus postpone any further considerations on non-orthogonal bases in infinite dimensional spaces.

This approximation process is also local, and (1.3.1) allows us to define $P_j f$ when f is defined on a bounded interval such as $[0,1]$, if $j \geq 0$. Note that an homogeneous Dirichlet boundary condition $f(0) = f(1) = 0$ is then preserved by the interpolation process, which amounts to using the approximation space V_j^D spanned by those $\varphi_{j,k}$ which have their support contained in $[0,1]$. As an example, Figure 1.3.1 displays the graphs of the function $f(x) = \sin(2\pi x)$ and its approximation $P_3 f$ on $[0,1]$.

The approximation $P_j f$ can be derived by applying P_j on the finer approximation $P_{j+1} f$, since we have

$$2^{j/2} c_{j,k} = f(2^{-j}k) = 2^{(j+1)/2} c_{j+1,2k}. \tag{1.3.5}$$

Therefore $P_j P_{j+1} = P_j$, a property which is not guaranteed by general non-orthogonal projectors (in contrast $P_{j+1}P_j = P_j$ is a direct consequence of the nestedness of the spaces V_j).

We can again define the "details" $Q_j f := P_{j+1} f - P_j f$. From (1.3.5) we see that $Q_j f(2^{-j}k) = 0$ and

$$Q_j f(2^{-j}(k + \frac{1}{2})) = f(2^{-j}(k + \frac{1}{2})) - \frac{f(2^{-j}k) + f(2^{-j}(k+1))}{2}. \tag{1.3.6}$$

In other words, $Q_j f$ measures the distance between the exact value of f at the fine grid point $2^{-j}(k + 1/2) \in \Gamma_{j+1}$ and an approximate value which is the average between its two closest neighbours on the coarser grid Γ_j. Note that this difference is exactly the second order finite difference between these three points. As an example, Figure 1.3.2 shows the components $P_2 f$ and $Q_2 f$ on $[0,1]$, for $f(x) = \sin(2\pi x)$: we have decomposed $P_3 f$ of Figure 1.3.1 into the sum of the coarser approximation $P_2 f$ and the additional details $Q_2 f$. Note that, in contrast to the case of piecewise constant approximations, we do not have $\int Q_j f = 0$.

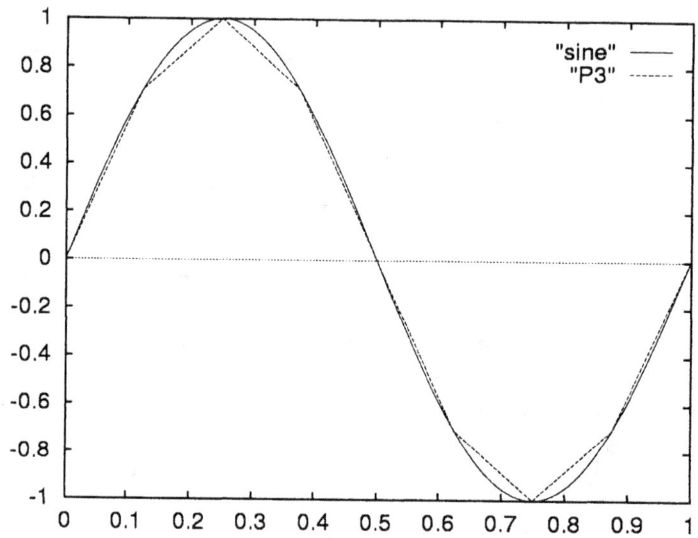

Figure 1.3.1: The function $f(x) = \sin(2\pi x)$ and its approximation $P_3 f$

From our last remark, we see that we can expand Q_j into local contributions according to

$$Q_j f = \sum_{k\in\mathbb{Z}} d_{j,k}\psi_{j,k}, \tag{1.3.7}$$

with

$$\psi = \varphi(2x-1). \tag{1.3.8}$$

In this case, the "wavelets" $\psi_{j,k}$, $k \in \mathbb{Z}$ are thus simply the nodal basis functions at level $j+1$ that are associated with the grid points in $\Gamma_{j+1} \setminus \Gamma_j$.

For $j_1 > j_0$, we can thus express the multiscale decomposition

$$P_{j_1} f = P_{j_0} f + \sum_{j_0 \le j < j_1} Q_j f, \tag{1.3.9}$$

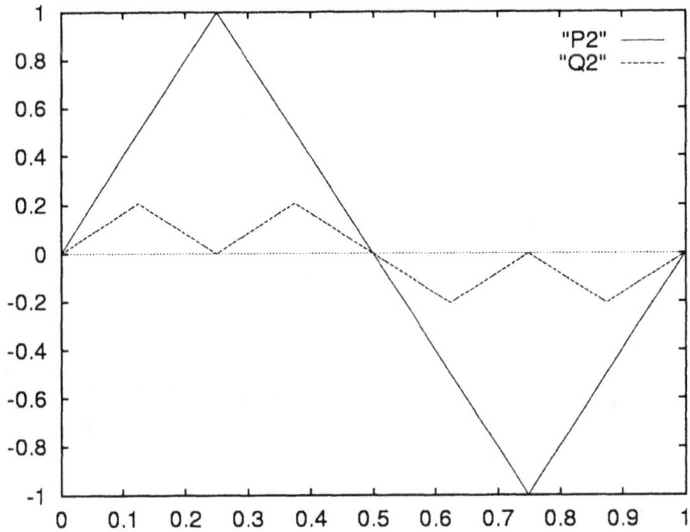

Figure 1.3.2: The approximation P_2f and the fluctuation Q_2f

in terms of local contributions according to

$$\sum_k c_{j_1,k}\varphi_{j_1,k} = \sum_k c_{j_0,k}\varphi_{j_0,k} + \sum_{j_0\le j<j_1}\sum_k d_{j,k}\psi_{j,k}. \tag{1.3.10}$$

As in the case of the Haar system, we have described a change of basis, between the nodal basis $\{\varphi_{j_1,k}\}_{k\in\mathbb{Z}}$ and the *multiscale basis* defined $\{\varphi_{j_0,k}\}_{k\in\mathbb{Z}} \cup \{\psi_{j,k}\}_{j_0\le j<j_1,k\in\mathbb{Z}}$.

This non-orthogonal change of basis can easily be adapted to the treatment of continuous functions defined on a bounded interval, say $[0,1]$. In this case, we define the spaces V_j only for $j \ge 0$, and the nodal and hierarchical basis functions are simply defined as restrictions to $[0,1]$ of those nodal and hierarchical basis functions which do not vanish on this interval. We can also decide to incorporate the homogeneous Dirichlet boundary conditions $f(0) = f(1) = 0$ in the definition of V_j, which corresponds to keeping only the basis functions that are fully supported in $[0,1]$. As an example, we display the nodal and hierachical bases in Figures 1.3.3 and 1.3.4, in the case where $j = 3$.

If we let j_1 go to $+\infty$, the interpolation $P_{j_1}f$ converges pointwise to f, provided that f is a continuous function. It also converges uniformly to f on any compact set. In the case of a continuous function f defined on a

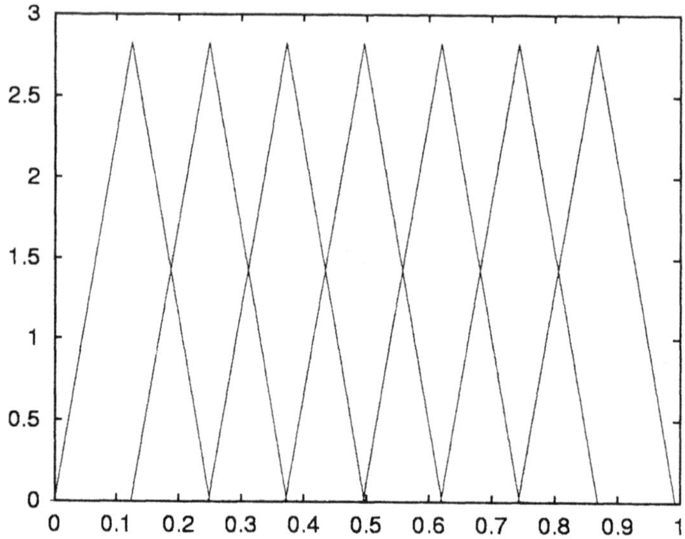

Figure 1.3.3: The nodal basis of V_3

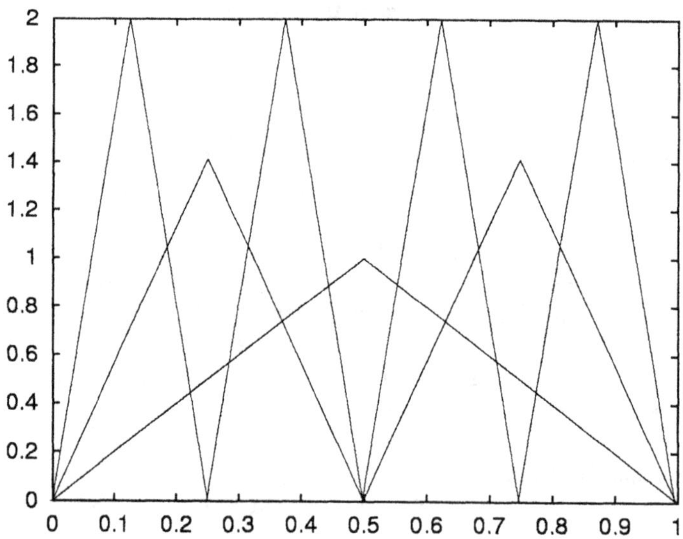

Figure 1.3.4: The hierarchical basis of V_3

bounded interval, the summation in k is finite at each resolution level, and this allows us to express f as the uniform limit of the infinite series

$$f = \sum_k c_{j_0,k}\varphi_{j_0,k} + \sum_{j \geq j_0} \sum_k d_{j,k}\psi_{j,k}. \tag{1.3.11}$$

Note that the convergence in (1.3.11) corresponds to the particular summation process of "piling-up" the details as j_1 go to $+\infty$: a permutation of the terms in the series might ruin it, in contrast to the expansion of an L^2 function f in the orthonormal Haar system which converges to f in L^2 whatever the order of summation.

From a computational point of view, the change of basis in (1.3.10) can be implemented by the same strategy that was described for the Haar transform, i.e. a "fine to coarse" multiscale decomposition, and a "coarse to fine" reconstruction of the nodal representation.

As in the case of the Haar system, we might operate the multiscale decomposition on any fine scale approximation $f_{j_1} = \sum_k c_{j_1,k}\varphi_{j_1,k}$ of f in V_{j_1} that differs from the interpolation $P_{j_1}f$: for example, f_{j_1} could be a Galerkin-type approximation to f, in the context of solving a partial differential equation.

These multiscale algorithms are based on the following simple recursions that are derived from (1.3.6), taking into account the L^2-normalization of $\varphi_{j,k}$ and $\psi_{j,k}$:

$$c_{j,k} = \sqrt{2}c_{j+1,2k}, \quad d_{j,k} = \sqrt{2}[c_{j+1,2k+1} - (c_{j+1,2k} + c_{j+1,2k+2})/2], \tag{1.3.12}$$

for the decomposition, and

$$c_{j+1,2k} = c_{j,k}/\sqrt{2}, \quad c_{j+1,2k+1} = [d_{j,k} + (c_{j,k} + c_{j,k+1})/2]/\sqrt{2}, \tag{1.3.13}$$

for the reconstuction. We display in Figure 1.3.5 the influence diagram for the decomposition and reconstruction algorithms, between two level $j+1$ and j. In contrast to Figure 1.2.4, note the slight difference between the decomposition and reconstruction diagrams, which reflects the non-orthogonality of the discrete transform.

Clearly, these algorithms are of complexity $\mathcal{O}(N)$, similar to the Haar transform. In the theoretical construction of the Schauder basis, as well as in its practical implementation, we have thus encountered the three main features that we identified in the Haar system in the end of the previous section.

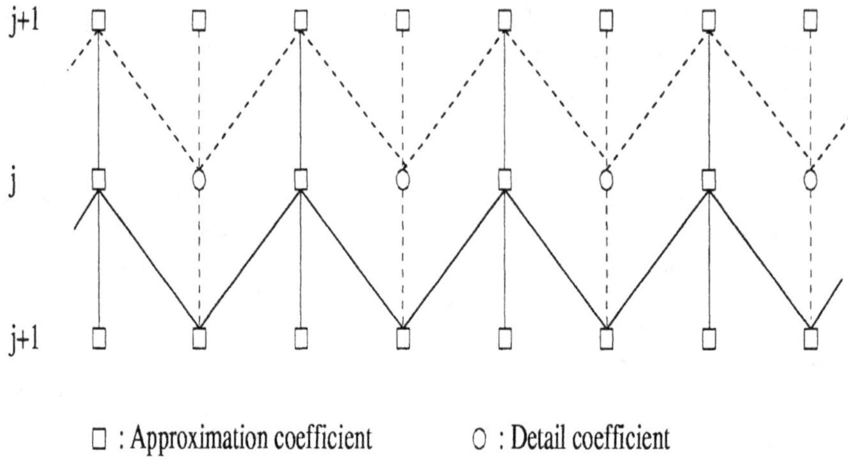

Figure 1.3.5: Decomposition $j+1 \mapsto j$ and reconstruction $j \mapsto j+1$ for the hierarchical Schauder basis

1.4 Multivariate constructions

If we want to generalize the previous multiscale schemes to functions of several variables, we can imagine many different possibilities. We shall focus on the two most frequently used approaches which are respectively based on tensor product and nested triangulations. For the sake of notational simplicity, we shall describe these generalizations in the case of functions of two variables. The extension to more than two variables does not present any additional difficulty and will be sketched for each approach.

In the case of piecewise constant approximation, a natural generalization of (1.2.1) for $f \in L^2(\mathbb{R}^2)$ is to define $P_j f$ on each dyadic square

$$S_{j,k} := I_{j,m} \times I_{j,n} \quad k = (m,n) \in \mathbb{Z}^2, \tag{1.4.1}$$

as the mean value of f on this square.

As in the univariate case, P_j is an orthogonal projection onto a space of piecewise constant functions, now generated by the separable orthonormal basis

$$\phi_{j,k}(x,y) = \varphi_{j,m}(x)\varphi_{j,n}(y) = 2^j\phi(2^j(x,y)-k), \; k = (m,n) \in \mathbb{Z}^2, \tag{1.4.2}$$

where $\varphi = \chi_{[0,1]}$ and $\phi = \chi_{[0,1]^2}$. In other words, the approximation space V_j is the *tensor product space* $V_j^x \otimes V_j^y$, where V_j^x and V_j^y represent the univariate spaces defined in §1.2, in the variables x and y.

If f_{j+1} is a function in V_{j+1}, i.e.

$$f_{j+1} = \sum_{k \in \mathbf{Z}^2} c_{j+1,k} \phi_{j+1,k}, \tag{1.4.3}$$

we can derive its approximation at the coarser scale

$$P_j f_{j+1} = \sum_{k \in \mathbf{Z}^2} c_{j,k} \phi_{j,k}, \tag{1.4.4}$$

using

$$c_{j,(m,n)} = \frac{1}{2}(c_{j+1,(2m,2n)} + c_{j+1,(2m+1,2n)} + c_{j+1,(2m,2n+1)} + c_{j+1,(2m+1,2n+1)}).$$

We now would like to characterize, as in the univariate case, the detail component $Q_j f = P_{j+1} f - P_j f$ by means of wavelets, that are supported on $S_{j,k}$. However, since $S_{j,k}$ supports one basis function in V_j and four basis functions in V_{j+1}, it is clear that we need three wavelets instead of one to span the complement space. These wavelets can be constructed as follows:

$$\psi^a(x,y) = \psi(x)\varphi(y), \quad \psi^b(x,y) = \varphi(x)\psi(y), \quad \psi^c(x,y) = \psi(x)\psi(y). \tag{1.4.5}$$

Indeed, one easily checks (using Fubini's theorem), the orthogonality relations

$$\langle \psi^\alpha_{j,k}, \phi_{j,k} \rangle = \langle \psi^\alpha_{j,k}, \psi^\beta_{j,k} \rangle = 0, \quad \alpha, \beta = a, b, c \text{ and } \alpha \neq \beta. \tag{1.4.6}$$

Thus, $\{\psi^a_{j,k}, \psi^b_{j,k}, \psi^c_{j,k}\}_{k \in \mathbf{Z}^2}$ constitutes an orthonormal basis for the orthogonal complement W_j of V_j into V_{j+1}.

We have in fact used a "distributive" property of tensor product with respect to the orthogonal sum of spaces, in the sense that

$$\begin{aligned} V_{j+1} &= V^x_{j+1} \otimes V^y_{j+1} \\ &= (V^x_j \oplus W^x_j) \otimes (V^y_j \oplus W^y_j) \\ &= (V^x_j \otimes V^y_j) \oplus (W^x_j \otimes V^y_j) \oplus (V^x_j \otimes W^y_j) \oplus (W^x_j \otimes W^y_j) \\ &= V_j \oplus W^a_j \oplus W^b_j \oplus W^c_j = V_j \oplus W_j, \end{aligned}$$

each detail subspace W^α_j being generated by the shifts and dilates of the corresponding wavelet ψ^α, $\alpha = a, b, c$.

This particular representation of the details, allows us to implement the decomposition algorithm using a two step recursion to go from level $j+1$ to level j:

1. Compute the averages and details along the x-direction to obtain the intermediate values:

$$\begin{aligned} a_{j,(m,n)} &= [c_{j+1,(2m,n)} + c_{j+1,(2m+1,n)}]/\sqrt{2}, \\ b_{j,(m,n)} &= [c_{j+1,(2m,n)} - c_{j+1,(2m+1,n)}]/\sqrt{2}. \end{aligned} \tag{1.4.7}$$

2. Compute the averages and details along the y-direction to obtain the wavelets coefficients:

$$\begin{aligned} c_{j,(m,n)} &= [a_{j,(m,2n)} + a_{j,(m,2n+1)}]/\sqrt{2}, \\ d^a_{j,(m,n)} &= [b_{j,(m,2n)} + b_{j,(m,2n+1)}]/\sqrt{2}, \\ d^b_{j,(m,n)} &= [a_{j,(m,2n)} - a_{j,(m,2n+1)}]/\sqrt{2}, \\ d^c_{j,(m,n)} &= [b_{j,(m,2n)} - b_{j,(m,2n+1)}]/\sqrt{2}. \end{aligned} \tag{1.4.8}$$

Clearly, we can apply the same idea to implement the reconstruction algorithm from level j to $j+1$: recompose successively along the y and x directions. We could also have interchanged the order of apperance of the x and y directions.

Remark 1.4.1 *Note that the wavelet coefficients have a directional interpretation: the coefficients $d^a_{j,k}$ (resp. $d^b_{j,k}$) indicate the fluctuations in the x (resp. y) direction, i.e. they are particularily sensitive to a vertical (resp. horizontal) edge on the graph of $f(x,y)$. This fact is directly vizualized, if we apply this type of multiscale decomposition on images, as it will be explained in the next section.*

Remark 1.4.2 *As in the univariate case, we have obtained that for $j_0 \le j_1$ the functions*

$$\{\phi_{j_0,k}\}_{k\in\mathbb{Z}^2} \cup \{\psi^\alpha_{j,k}\}_{j_0\le j<j_1, k\in\mathbb{Z}^2, \alpha=a,b,c}, \tag{1.4.9}$$

constitute an orthonormal multiscale basis for V_{j_1}.

Remark 1.4.3 *Note that another basis could be obtained by taking all possible tensor products of two functions in $\{\varphi_{j_0,k}\}_{k\in\mathbb{Z}} \cup \{\psi_{j,k}\}_{j_0\le j<j_1,k\in\mathbb{Z}}$. This leads to a "full tensor product" basis which incorporates functions of the type $\psi_{j,k}(x)\psi_{j',k'}(y)$, which oscillate at different scales in the directions x and y. The corresponding decomposition (resp. reconstruction) algorithm operates in one direction from the finest to coarsest (resp. coarsest to finest) resolution level, and then performs the same operations in the other direction.*

Remark 1.4.4 *As in the univariate case, we can let j_1 go to $+\infty$ (and j_0 to $-\infty$), and derive a multiscale basis for $L^2(\mathbb{R}^2)$.*

Remark 1.4.5 *The generalization of this construction to more than 2 variables is straightforward: the spaces V_j are generated by the orthogonal basis*

$$\phi_{j,k}(x) = \varphi_{j,k_1}(x_1) \cdots \varphi_{j,k_d}(x_d), \tag{1.4.10}$$

and the wavelets are given by

$$\psi^{\varepsilon}(x) = \psi^{\varepsilon_1}(x_1) \cdots \psi^{\varepsilon_d}(x_d), \tag{1.4.11}$$

with $\varepsilon = (\varepsilon_1, \cdots, \varepsilon_d) \in \{0,1\}^d - (0, \cdots, 0)$ *and where we have set* $\psi^0 = \varphi$ *and* $\psi^1 = \psi$. *We can thus characterize the details by* $2^d - 1$ *wavelets.*

We can also apply the above described tensor product technique on piecewise affine univariate approximation, i.e. define the space V_j as spanned by the nodal basis $\phi_{j,k}(x,y) = \varphi_{j,m}(x)\varphi_{j,n}(y)$, $k = (m,n) \in \mathbb{Z}^2$, where $\varphi = \max\{1 - |x|, 0\}$. The resulting bivariate approximation spaces are the well known rectangular Lagrange $\mathbb{Q}_1$ finite elements: on each square $S_{j,k}$, the approximation f_j has the form $a_k + b_k x + c_k y + d_k xy$, and since it is continuous, f_j is completely determined by its values on $2^{-j}\mathbb{Z}^2$: we have $f_j = \sum_{k\in\mathbb{Z}^2} c_{j,k}\phi_{j,k}$, with $c_{j,k} = 2^{-j} f_j(2^{-j}k)$. We can also define in a unique way the interpolation operator P_j that maps a continuous function f onto the space V_j with $P_j f(2^{-j}k) = f(2^{-j}k)$, $k \in \mathbb{Z}^2$.

The same reasoning as in the case of the Haar system shows that formula (1.4.5) applied to the hat function φ and the Schauder wavelet $\psi = \varphi(2\cdot -1)$ yields a suitable wavelet basis to describe the difference $I_{j+1}f - I_j f$. We obtain a multiscale basis of the form (1.4.9) and we can also use the same separable technique to implement the decomposition and reconstruction algorithm.

In summary, the tensor product construction inherits the essential features of the corresponding univariate multiresolution approximation and wavelet basis: nested approximation spaces, local bases, fast algorithms.

Another natural multivariate counterpart to the piecewise affine construction consists in preserving the affine nature of the approximation, by using $\mathbb{P}_1$ triangular elements, i.e. approximation by continuous piecewise affine functions on triangles. This approach is well adapted to the approximation of functions defined on domains such as polygons that can easily be triangulated. We shall thus consider functions that are defined on a domain Ω which is either $\mathbb{R}^2$ or a bounded polygonal domain.

In order to build approximation spaces V_j that have the nestedness property $V_j \subset V_{j+1}$, we can start from an initial planar triangulation $\mathcal{T}_0$ of Ω, with two basic properties.

1. **Conformity:** the edge of a triangle either coincides with the edge of another triangle, or it is contained in the boundary of Ω.

2. **Quasi-uniformity:** there exist strictly positive constants c and C such that

$$c \leq \inf_{T \in \mathcal{T}_0} r(T) \leq \sup_{T \in \mathcal{T}_0} R(T) \leq C, \tag{1.4.12}$$

where $r(T)$ (resp. $R(T)$) is the radius of the largest (resp. smallest) circle contained in (resp. containing) the triangle T.

Note that in the case where the domain is a bounded polygon, the second property simply means that we start with $\mathcal{T}_0$ of finite cardinality. We then build a family of *nested triangulation* by a classical refinement procedure: $\mathcal{T}_{j+1}$ is defined from $\mathcal{T}_j$ by adding a mesh point in the middle of each edge and subdividing each triangle into four sub-triangles, as shown in Figure 1.4.1 (see also Figure 2.13.5 of Chapter 2).

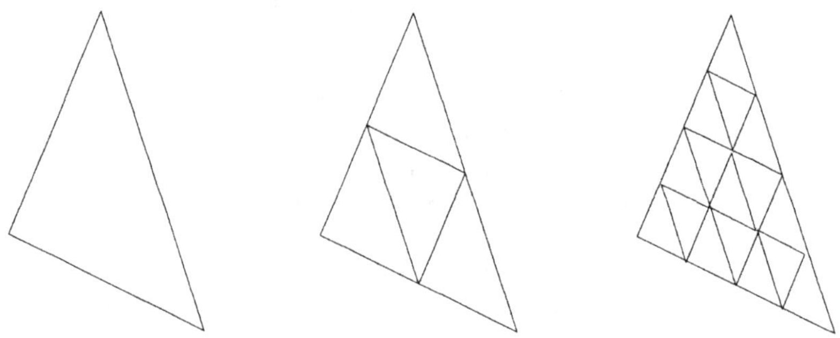

Figure 1.4.1: Subdivision of a triangle of $\mathcal{T}_0$

This refinement procedure generates conformal quasi-uniform triangulations $\mathcal{T}_j$, $j \geq 0$, with the property

$$c2^{-j} \leq \inf_{T \in \mathcal{T}_j} r(T) \leq \sup_{T \in \mathcal{T}_j} .R(T) \leq C2^{-j} \tag{1.4.13}$$

The spaces of $\mathbb{P}_1$ finite elements

$$V_j = \{f \in C^0(\Omega) \ ; \ f_{|T}(x,y) = a_T + b_T x + c_T y, \ T \in \mathcal{T}_j\}, \tag{1.4.14}$$

associated with these triangulations obviously satisfy the nestedness property $V_j \subset V_{j+1}$.

If we denote by Γ_j the set of all vertices of the triangulation $\mathcal{T}_j$, we can again define a simple nodal basis: for $\gamma \in \Gamma_j$, we define $\phi_{j,\gamma}$ to be the unique function in V_j such that

$$\phi_{j,\gamma}(\gamma') = 2^j \delta_{\gamma,\gamma'}, \quad \gamma' \in \Gamma_j. \tag{1.4.15}$$

As in the univariate case, we have chosen to normalize these functions in such a way that their L^2 norm is bounded independently of j. Any $f_j \in V_j$ can thus be decomposed in a unique way according to

$$f_j = \sum_{\gamma \in \Gamma_j} c_{j,\gamma} \phi_{j,\gamma}, \tag{1.4.16}$$

with

$$c_{j,\gamma} = 2^{-j} f_j(\gamma). \tag{1.4.17}$$

We can also define in a unique way the interpolation operator P_j that maps a continuous function f onto the space V_j with the constraint $P_j f(\gamma) = f(\gamma)$, $\gamma \in \Gamma_j$.

We remark that, since $[P_{j+1}f - P_j f](\gamma) = 0$, for all $\gamma \in \Gamma_j$, we can use the functions $\phi_{j+1,\gamma}$, $\gamma \in \Gamma_{j+1} \setminus \Gamma_j$, i.e. the fine scale basis functions corresponding to the vertices that have been added in the refinement process from level j to $j+1$, to characterize this difference. Each of these functions is used to recover the value at the corresponding point of $\Gamma_{j+1} \setminus \Gamma_j$. Clearly, this representation is not redundant, since these functions are independent.

Iterating this decomposition on several levels, say between j_1 and j_0 with $j_0 < j_1$, we thus obtain a multiscale basis for V_{j_1} with the following form:

$$\{\phi_{j_0,\gamma}\}_{\gamma \in \Gamma_{j_0}} \cup \{\phi_{j+1,\gamma}\}_{j_0 \le j < j_1, \gamma \in \Gamma_{j+1} \setminus \Gamma_j}. \tag{1.4.18}$$

As in all previous constructions, we can use simple relations between two successive levels of resolution, in order to derive fine to coarse (decomposition) and coarse to fine (reconstruction) fast algorithms. We denote by $d_{j+1,\gamma}$, $\gamma \in \Gamma_{j+1} \setminus \Gamma_j$ the coefficient of f in the corresponding hierarchical basis function, and by $a(\gamma)$ and $b(\gamma)$, the two points of Γ_j such that γ was built as the mid-point of the edge joining $a(\gamma)$ and $b(\gamma)$. Taking into account the L^2 normalization of the basis functions, we obtain

$$c_{j,\gamma} = 2c_{j+1,\gamma}, \quad \gamma \in \Gamma_j, \tag{1.4.19}$$

and

$$d_{j+1,\gamma} = c_{j+1,\gamma} - [c_{j+1,a(\gamma)} + c_{j+1,b(\gamma)}]/2, \quad \gamma \in \Gamma_{j+1} \setminus \Gamma_j, \tag{1.4.20}$$

for decomposition, and

$$c_{j+1,\gamma} = c_{j,\gamma}/2, \quad \gamma \in \Gamma_j, \tag{1.4.21}$$

and

$$c_{j+1,\gamma} = d_{j+1,\gamma} + [c_{j,a(\gamma)} + c_{j,b(\gamma)}]/4, \quad \gamma \in \Gamma_{j+1} \setminus \Gamma_j, \tag{1.4.22}$$

for reconstruction.

Remark 1.4.6 *As in the case of the univariate Schauder basis, $P_j f$ converges uniformly to f if f is uniformly continuous: any such function can thus be expanded as a uniformly converging series in the hierarchical basis.*

Remark 1.4.7 *In order to generalize this construction to more than two variables, a natural idea is to consider piecewise affine functions on conforming simplicial partitions. However, in contrast to triangles, one cannot refine a d-simplex into 2^d simplices of similar shape when $d > 2$. In turn there is no straightforward analog to the hierarchical basis in this general case.*

Remark 1.4.8 *As in the univariate case, it is also possible to incorporate homogeneous Dirichlet boundary conditions in V_j, without affecting the construction of the multiscale basis: both nodal and hierachical basis functions are then associated only to the nodes that are in the interior of Ω.*

1.5 Adaptive approximation

One of the main interests of multiscale decompositions can be stated by the following heuristic: a function f that is smooth, except at some isolated singularities, should have a *sparse* representation in a multiscale basis, i.e. only a small number of numerically significant coefficients should carry most of the information on f. Indeed, the wavelet coefficient $d_{j,k}$ measures the local fluctuation of f at scale 2^{-j} near $2^{-j}k$, or equivalently in the support of $\psi_{j,k}$. As j grows, i.e. resolution gets finer, these fluctuations should decay quickly to zero in the smooth regions, and should be larger only when the support of $\psi_{j,k}$ contains a singularity of f, i.e. for a small number of indices k. Such a behaviour also suggests that one could approximate f by keeping only a small number of wavelet coefficients, and reduce significantly the complexity of the description of f without affecting its accuracy.

This *compression property* will be analyzed in detail in chapters III and IV. The goal of this section is simply to illustrate it with some simple examples.

As a first example, we consider the function $f(x) = \sqrt{|\cos(2\pi x)|}$ on $[0,1]$. A first sampling of f on 8192 points is used to plot its graph on Figure 1.5.1, which displays its linear interpolation $P_j f$, as defined in §1.3, at resolution level $j = 13$. From this fine scale approximation, we have applied the decomposition algorithm in the Schauder basis. In Figure 1.5.2, we display the points $(2^{-j}k, j)$ corresponding to the wavelet coefficients $|d_{j,k}| \geq 5 \times 10^{-3}$. As expected, the contributions from the high scales tend to concentrate near the singularities at $x = 1/4$ and $3/4$.

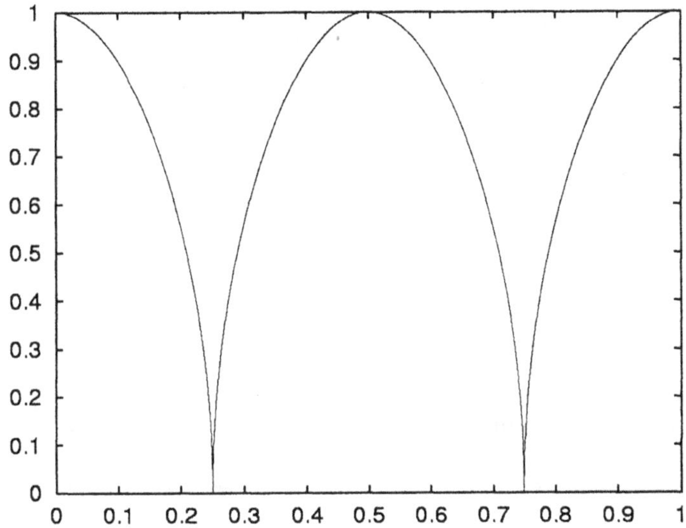

Figure 1.5.1: The function $f(x) = \sqrt{|\cos(2\pi x)|}$ at resolution 2^{-13}

In Figure 1.5.3, we have reconstructed the function f from the coefficients that are marked in Figure 1.5.2. The number of preserved coefficients is 50, i.e. a reduction of the initial complexity by a factor above 150. The result of a similar thresholding operation with the Haar system is displayed on Figure 1.5.4 (the threshold is kept the same, and the number of preserved coefficients is then 116). It is no surprise that the quality of the approximation is visually (and numerically) better with piecewise affine functions.

In both cases, we observe that the approximation is refined near the singularities: an adaptive grid, that takes into account the local fluctuations of the function, is automatically generated by the thresholding of wavelet coefficients.

Remark 1.5.1 *We can provide a simple analysis of the stronger decay (as $j \to +\infty$) of the wavelet coefficients in the regions where f is smooth. In*

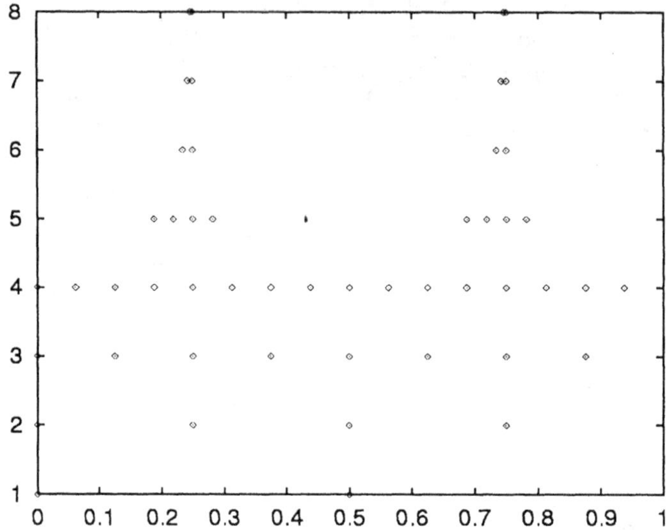

Figure 1.5.2: Schauder basis coefficients above threshold 5×10^{-3}

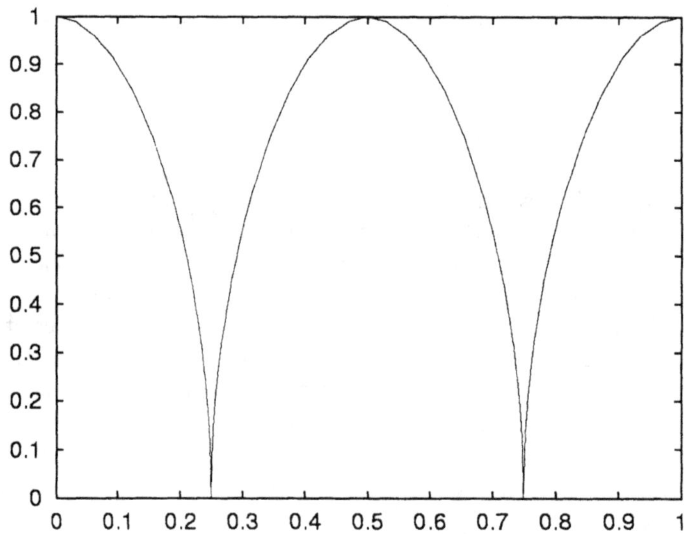

Figure 1.5.3: Reconstruction from the 50 Schauder basis coefficients above 5×10^{-3}

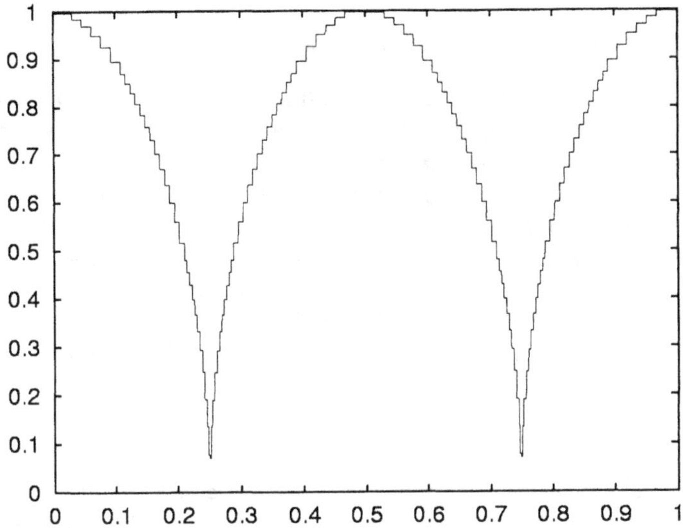

Figure 1.5.4: Reconstruction from the 116 Haar basis coefficients above 5×10^{-3}

the case of the Haar system, when f is C^1 on the support $I_{j,k}$ of $\psi_{j,k}$, we can use the oscillation property of ψ to derive the estimate

$$\begin{aligned} |d_{j,k}| &= |\int_{I_{j,k}} f(x)\psi_{j,k}(x)dx| \\ &= |\int_{I_{j,k}} [f(x) - f(2^{-j}k)]\psi_{j,k}(x)dx| \\ &\leq [\sup_{I_{j,k}} |f'|]2^{-j} \int |\psi_{j,k}(x)|dx \\ &\leq [\sup_{I_{j,k}} |f'|]2^{-3j/2}. \end{aligned}$$

We have thus combined the smoothness of f with the oscillatory nature of ψ to derive a better estimate than the simple $|d_{j,k}| \leq 2^{-j/2} \sup_{I_{j,k}} |f|$. Note that in the case where the function f is only C^α on $I_{j,k}$ with $0 < \alpha < 1$ (i.e. $|f(x) - f(y)| \leq C|x-y|^\alpha$), then the same computation will lead to an estimate of $|d_{j,k}|$ in $\mathcal{O}(2^{-(\alpha+1/2)j})$. This explains why more coefficients are preserved at the finest scales near the singularities of our example which have the Hölder exponent $\alpha = 1/2$. The results of Chapter 3 will actually confirm the intuition that the degree of local smoothness can be measured through the decay of the wavelet coefficients. Note that in the case of the Schauder basis, we can obtain a better estimate if $f \in C^2$: using that $d_{j,k}$ is a normalized second order finite difference, one easily derives $|d_{j,k}| \leq 2^{-5j/2}[\sup_{I_{j,k}} |\frac{d^2 f}{dx^2}|]$.

This better estimate explains why more coefficients have been discarded in the thresholding process with the Schauder basis.

Remark 1.5.2 *Note that the thresholding procedure is a nonlinear operation: the indices (j,k) of the preserved coefficients depend on the function to be approximated. In particular, this means that one needs to store both the values of the preserved coefficients and their indices, in order to describe such an adaptive approximation. Another natural way of defining such nonlinear approximations is by prescribing the number of preserved coefficients rather than a threshold, i.e. define f_N to be the approximation of the above f by retaining its N largest wavelet coefficients. The sparsity of the multiscale representation can then be measured through the decay of $\|f_N - f\|_{L^2}$ as N goes to $+\infty$, i.e. the supremum of all s such that*

$$\|f_N - f\|_{L^2} \le CN^{-s}. \tag{1.5.1}$$

In the particular case of the above example, one can easily derive from the estimates on $|d_{j,k}|$ that this error decays like N^{-1} or N^{-2} when using respectively the Haar system or the Schauder basis. General results on nonlinear approximations will be presented in Chapter 4. In particular these results will imply that (1.5.1) holds with s arbitrarily large for the above example, provided that one uses a sufficiently high order accurate wavelet basis. This is in contrast with linear approximation which defines f_N by retaining the N first coefficients of f, i.e. $f_N := P_j f$ when $N = 2^j$. In this case, one essentially cannot improve on the rate N^{-1} in the above example, even when using high order wavelets, due to the presence of the singularities (for the Schauder basis, a "super-convergence" phenomenon still occurs in our example since the singularities are situated at coarse mesh points, which results in the artificially better rate $N^{-2}log(N)$). This is also in contrast with nonlinear approximation in other bases such as Fourier series: the Fourier coefficients $c_N(f)$ in the above example are not sparse in the sense that they behave like $\mathcal{O}(|N|^{-3/2})$ for all N. In turn, both linear and nonlinear approximation L^2-errors behave like N^{-1}.

We shall now illustrate the compression properties of multiscale decompositions in the case of bivariate functions associated with the mathematical representation of *images*. A digital black and white image is a bidimensional array $I(m,n)$ measuring the grey level intensity at each point (or "pixel": picture element) (m,n). As an example, Figure 1.5.5 displays a 512×512 image, each pixel being quantized on 8 bits, i.e. 256 possible grey levels (0 for black, 255 for white). The tensor product multiscale decomposition which generalizes the Haar system for functions of two variables is particularily adapted to the representation of such images: we can identify the digital image in Figure 1.5.5 with a function in V_9, and proceed to the multiscale decomposition using the separable algorithm that we described in §1.4.

Figure 1.5.5: Digitized picture: 512×512 pixels and 256 grey levels

We display the organization of a decomposition on four levels in Figure 1.5.6: the coefficients of the coarsest approximation (in V_5) appear in the upper left corner, while the rest of the array contains the wavelet coefficients at intermediate resolutions.

In the framework of image processing, one sometimes normalizes the basis functions in L^1 instead of L^2: $\phi_{j,k}(x,y) = 2^{2j}\phi(2^j x - k_x, 2^j y - k_y)$, for $k = (k_x, k_y)$, and similarily for ψ. This normalization allows us to vizualize the multiscale decomposition of our image as another image: the approximation coefficients are exactly the averages of the image on squares of pixels, and thus also range between 0 and 255, as well as the absolute values of wavelet coefficients. We display this decomposition image in Figure 1.5.7: the coefficients of the coarsest approximation appear as a simplified version of the picture. The rest of the array contains the absolute values of wavelet coefficients: as expected, it is mostly sparse, except near the edges. As it was remarked about tensor product decomposition (Remark 1.4.1), vertical and horizontal edges are matched by a specific wavelet.

On Figure 1.5.8, we have reconstructed the image with the 2000 largest coefficients (after renormalization in L^2), i.e. a parameter reduction above

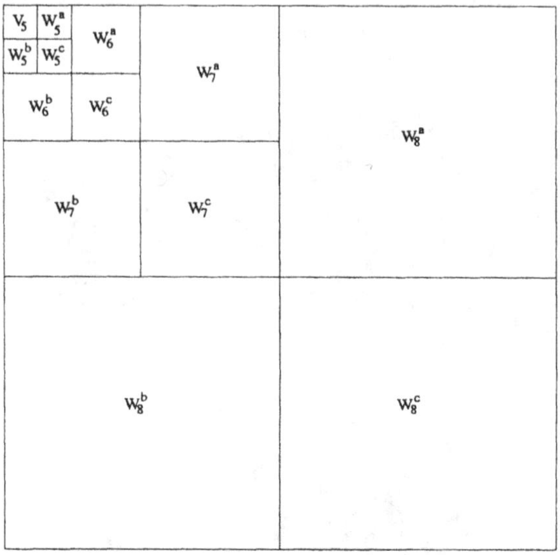

Figure 1.5.6: Tensor product multiscale decomposition

Figure 1.5.7: Multiscale decomposition of an image

100.

Figure 1.5.8: Reconstruction from 2000 largest coefficients

Clearly, the Haar system is not very well adapted for the task of representing images with a few coefficients: visual artefacts appear, reflecting the square-shaped discontinuities of the generating functions. However, we again observe that the thresholding strategy generates an adaptive approximation of the image, in the sense that the resolution level is increased near the edges.

Finally, we want to show that multiscale decomposition can also be applied to "compress" operators in integral equations. Such equations arise in numerous contexts, either as a direct modelling of a physical process, or as alternative formulations of partial differential equations. They involve the application or the inversion of an integral operator T defined by a formula of the type

$$Tf(x) = \int K(x,y)f(y)dy, \tag{1.5.2}$$

where the kernel $K(x,y)$ is a function which is supported over all ranges of x and y. The distributed nature of $K(x,y)$ has the following immediate consequence: the usual discretizations of T - based on finite element methods or straightforward sampling of $K(x,y)$ - result in fully populated matrices that are heavy to store, to apply or to invert.

In order to understand how a multiscale decomposition can "sparsify" the representation of T, let us consider a simple case where x and y range in $I = [0,1]$. We denote by V_J, $J \geq 0$, the space of piecewise constant functions defined in §1.2 and adapted to I, and $\varphi_{J,k}$, $k = 0, \cdots, 2^J - 1$, its orthonormal basis. We then define a discretization of T on V_J as the matrix

$$T_J = (\langle T\varphi_{J,m}, \varphi_{J,n}\rangle)_{m,n=0,\cdots,2^J-1}. \quad (1.5.3)$$

This matrix appears naturally in two different situations.

1. **Approximate the action of T on a function:** given f, find an approximation g_J in V_J of $g = Tf$. For this, we can start by approximating f by $f_J \in V_J$ and then define $g_J = P_J T f_J$, where P_J is the orthogonal projection. The computation of the coordinates vector G_J of g_J (in the basis $\varphi_{J,k}$) is then done by applying T_J on the coordinate vector F_J of f_J.

2. **Approximate the action of T^{-1} on a function:** given f, find an approximation $g_J \in V_J$ of g solution of $Tg = f$. The Galerkin method consists in searching for $g_J \in V_J$ such that $\langle Tg_J, u_J\rangle = \langle f, u_J\rangle$ for all $u_J \in V_J$, i.e. solving $T_J G_J = F_J$, where F_J is the coordinate vector of $f_J = P_J f$, and G_J the coordinate vector of the unknown g_J.

The well-posedness of the system in the second problem, as well as the error estimates that could be obtained in some prescribed norm for both problems, are of course dependent on the specific nature of the operator T and the data f.

Since the matrix elements of T_J are given by

$$T_J(m,n) = \langle T\varphi_{J,m}, \varphi_{J,n}\rangle = \int K(x,y)\varphi_{J,m}(x)\varphi_{J,n}(y)dx\,dy, \quad (1.5.4)$$

it is clear that the distributed nature of $K(x,y)$ will result in a full matrix. For example, if we have a uniform bound $|K(x,y)| \leq C$, we are ensured that (1.5.2) defines a bounded operator in $L^2(I)$, and that the operators T_J are bounded independently of J. We also obtain from (1.5.4) the estimate $|T_J(m,n)| \lesssim 2^{-J}$, but this estimate does not a-priori allow us to approximate T_J in operator norm by a sparse matrix.

If we now use the multiscale basis, i.e. $\varphi_{j_0,k}$, $k = 0, \cdots 2^{j_0} - 1$, and $\psi_{j,k}$, $j_0 \leq j < J$, $k = 0, \cdots, 2^j - 1$, to reformulate both problems, we obtain a new matrix S_J, whose elements are typically given by

$$S_J(j,l,m,n) = \langle T\psi_{j,m}, \psi_{l,n}\rangle = \int K(x,y)\psi_{j,m}(x)\psi_{l,n}(y)dx\,dy, \quad (1.5.5)$$

for $j_0 \le j, l \le J-1$, $m = 0, \cdots, 2^j - 1$, $n = 0, \cdots, 2^l - 1$ (with analogous expressions for those elements involving the basis functions $\varphi_{j_0,k}$). We see from (1.5.5) that S_J is simply obtained by applying on T_J the "full tensor product" wavelet decomposition that was described in Remark 1.4.3: the discretized kernel is "processed" like a digital picture. The structure of the resulting matrix is displayed in Figure 1.5.9, in the case where $J = 4$ and $j_0 = 1$.

We can thus hope to gain some sparsity when the kernel has some smoothness properties. In particular, if $K(x, y)$ is C^1 on the support of $\psi_{j,m}(x)\psi_{l,n}(y)$, we can use the same method as in Remark 1.5.1, to derive the estimate

$$|S_J(j, l, m, n)| \le [\sup_{I_{j,m} \times I_{l,n}} |\nabla K|] 2^{-2\max\{j,l\}}, \tag{1.5.6}$$

which, in contrast to the crude estimate that we had for T_J might allow us to discard many coefficients while preserving a good approximation to S_J.

As an example, we consider a single layer logarithmic potential operator that relates the density of electric charge on an infinite cylinder of unit radius $\{(z, e^{i2\pi x}),\ z \in \mathbb{R},\ x \in [0,1]\}$ to the induced potential on the same cylinder, when both functions are independent of the variable z. The associated kernel

$$K(x, y) = \log|e^{i2\pi x} - e^{i2\pi y}|. \tag{1.5.7}$$

is singular on the diagonal $\{x = y\}$, but integrable.

Starting from the discretization T_9 in V_9, we compute the multiscale matrix S_9 and we define a sparse matrix $\tilde{S}_9$ by setting to zero all entries of S_9 with modulus less than $10^{-2} \times 2^{-9}$. We display in Figure 1.5.10 the location of the preserved coefficients: on each subblock of S_9, that corresponds to a couple of scale (j, l), it is no surprise that we find important coefficients near the diagonal, since it corresponds to the singular part of the kernel.

This approximation of S_9 contains approximately 30000 non-zero entries, i.e. a compression factor near 10. To evaluate its accuracy, we can use the following simple version of the Schur lemma for matrices (see Chapter 4, Lemma 4.6.1, for the general version)

$$||A||^2 \le [\sup_j \sum_i |A_{i,j}|][\sup_i \sum_j |A_{i,j}|], \tag{1.5.8}$$

which yields the error estimate in operator norm

$$||S_9 - \tilde{S}_9|| \le 10^{-2}. \tag{1.5.9}$$

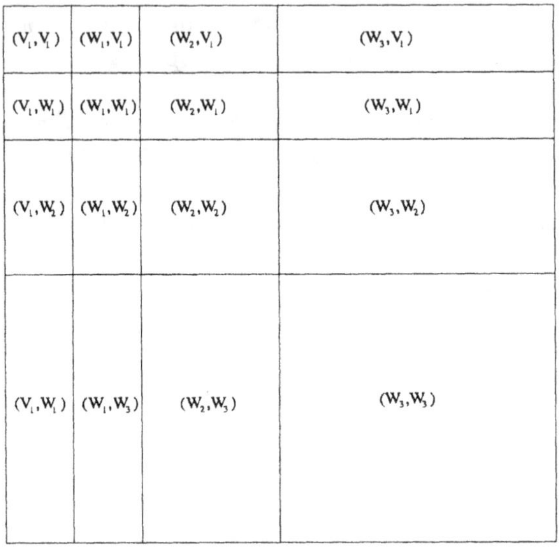

Figure 1.5.9: Multiscale discretization of a kernel

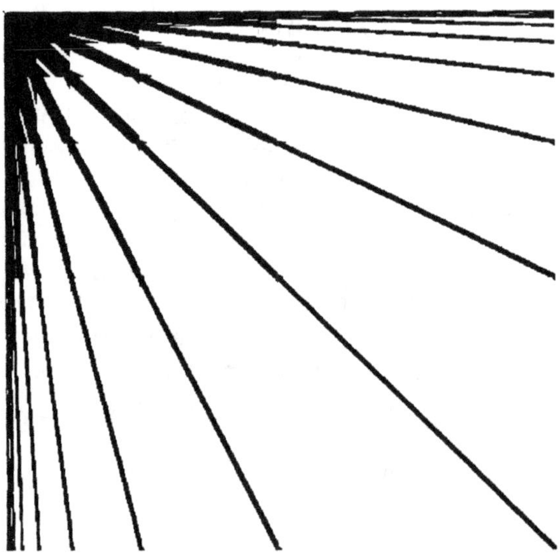

Figure 1.5.10: Sparsification of a logarithmic potential kernel

Remark 1.5.3 *The single layer potential is known to be an $H^{1/2}$-elliptic operator, which therefore maps $H^{-1/2}$ into its dual $H^{1/2}$. In Chapter 4, we shall see that it is possible to take into account this smoothing property in the compression process, in order to obtain more appropriate error bounds on $\|S_J - \widetilde{S}_J\|_{H^{-1/2}\to H^{1/2}}$.*

Remark 1.5.4 *The singular kernel in (1.5.7) satisfies $K(x,y) = k(x-y)$ where $k(\cdot)$ is a 1-periodic function, and in turn its discretization T_9 is also a Toepliz cyclic matrix, i.e. $T_9(m,n) = t((m-n)[\mathrm{mod}\ 2^9])$. A more appropriate change of basis, in that case, would be provided by the discrete Fourier transform, since it would lead to a diagonal matrix. However, it is important to note that the sparsification property of the multiscale decomposition does not rely on a structural property of the operator such as translation invariance, but on the analytical smoothness properties of the kernel. As a consequence, we can expect more robustness, in comparison to Fourier decomposition: a perturbation of the circular shape of the cylinder in the above example should not affect the sparsity of S_9, while it might generate a fully populated matrix when using the Fourier basis. We shall see in §4.6 of Chapter 4 that fairly general classes of operators can be sparsified by wavelet techniques.*

1.6 Multilevel preconditioning

One of the interests of multiscale methods is that they yield fast solvers for linear elliptic problems. We shall illustrate this point with a simple example involving the piecewise affine multiresolution spaces of §1.3 and §1.4 in order to motivate a more detailed study that will be the object of Chapter 3.

Our model problem is the elliptic boundary value problem,

$$-\Delta u = f \ \text{ in } \Omega, \ \ u = 0 \ \text{ on } \partial\Omega, \tag{1.6.1}$$

where $\Omega \subset \mathbb{R}^2$ is a polygonal domain and $f \in L^2(\Omega)$.

We first recall some theoretical facts about this problem. Under the assumptions made on f and Ω, it is well known (see for example GRISVARD [1983]), that (1.6.1) has a unique solution u in the Sobolev space $H^2(\Omega)$. We recall that $H^m(\Omega)$ consists of those functions $u \in L^2(\Omega)$ whose distributional partial derivatives $\partial^\alpha u$ also belong to $L^2(\Omega)$ for $|\alpha| \le m$, and constitutes a Hilbert space when equipped with the norm

$$\|u\|_{H^m} := \Big(\sum_{|\alpha|\le m} \|\partial^\alpha u\|_{L^2}^2\Big)^{1/2}. \tag{1.6.2}$$

The solution of (1.6.1) satisfies the a-priori estimate

$$\|u\|_{H^2} \le C\|f\|_{L^2}, \tag{1.6.3}$$

where C depends only on Ω. We also recall that u is the unique solution of the *variational formulation* of problem (1.6.1):

$$a(u,v) = \langle f,v\rangle, \text{ for all } v \in H_0^1(\Omega), \tag{1.6.4}$$

where $\langle f,v\rangle = \int_\Omega f(x)v(x)dx$ and $a(u,v)$ is the bilinear form

$$a(u,v) = \int_\Omega \nabla u(x)\nabla v(x)dx. \tag{1.6.5}$$

Recall that $H_0^1(\Omega)$ denotes the closed subspace of $H^1(\Omega)$ of those functions with vanishing trace on the boundary $\partial\Omega$, or equivalently the closure in $H^1(\Omega)$ of the space $\mathcal{D}(\Omega)$ of smooth functions compactly supported in Ω. The existence and uniqueness of the variational solution $u \in H_0^1$ follows from the Lax-Milgram lemma, remarking that the bilinear form a is continuous on $H_0^1 \times H_0^1$ and satisfies the ellipticity property

$$a(u,u) \ge C\|f\|_{H^1}^2, \text{ for all } f \in H_0^1(\Omega), \tag{1.6.6}$$

which directly stems from Poincaré's inequality. Note that continuity and coercivity imply that the energy norm $\|u\|_a := \sqrt{a(u,u)}$ and the Sobolev norm $\|u\|_{H^1}$ are equivalent norms on H_0^1.

We next recall some basic facts concerning the discretization of this problem by the finite element method. We consider here the $\mathbb{P}_1$ finite element spaces V_J, $J \ge 0$, that were introduced at the end of §1.4: a triangulation $\mathcal{T}_J$ being generated from an initial coarse triangulation $\mathcal{T}_0$ of Ω by J steps of uniform refinements, V_J consists of those continuous functions which are piecewise affine on $\mathcal{T}_J$ and vanish on $\partial\Omega$. The continuity and ellipticity of a on V_J ensure the existence and uniqueness of the Galerkin approximation $u_J \in V_J$ to the solution u, that solves

$$a(u_J,v_J) = \langle f,v_J\rangle, \text{ for all } v_J \in V_J. \tag{1.6.7}$$

Since we also have $a(u,v_J) = \langle f,v_J\rangle$ for all $v_J \in V_J$, u_J should be viewed as the projection of u onto V_J with respect to the energy norm $\|v\|_a$. Using the equivalence of $\|\cdot\|_a$ and $\|\cdot\|_{H^1}$ and classical finite element approximation results (see, for example, CIARLET [1978]), we can thus derive the error estimate

$$\|u-u_J\|_{H^1} \lesssim \inf_{v_J\in V_J} \|u-v_J\|_{H^1} \lesssim 2^{-J}\|u\|_{H^2}. \tag{1.6.8}$$

Such an estimate relates the a-priori accuracy $\varepsilon = \varepsilon(J) = \|u - u_J\|_{H^1}$ of the *Galerkin approximation* to the solution of problem (1.6.1) with the number of parameters $N_J := \dim(V_J) \sim 2^{2J}$ that are used to describe this approximation: we need (at most) $N(\varepsilon) = \mathcal{O}(\varepsilon^{-2})$ parameters (with a constant proportional to $\|u\|_{H^2}^2$) to reach the accuracy ε.

In the practical implementation of the finite element method, it is also important to evaluate the *computational cost*, i.e. the total number $\mathcal{N}(\varepsilon)$ of arithmetic operations that is needed to compute u_J, possibly up to the prescribed accuracy ε. Clearly $\mathcal{N}(\varepsilon)$ should be at least larger than $N(\varepsilon)$, since there are at least $N(\varepsilon)$ unknowns to compute, but we can expect of a good resolution method to make $\mathcal{N}(\varepsilon)$ not substantially larger than $N(\varepsilon)$. In order to compute u_J, we recall the nodal basis $\{\phi_{J,\gamma}; \gamma \in \widetilde{\Gamma}_J\}$, indexed by the set of interior vertices $\widetilde{\Gamma}_J = \Gamma_J \setminus \partial\Omega$. We choose an L^2 normalization for this basis, i.e.

$$\phi_{J,\gamma}(\gamma') = 2^J \delta_{\gamma,\gamma'}, \quad \gamma' \in \Gamma_J. \tag{1.6.9}$$

Defining U_J as the coordinate vector of u_J in this basis, the system resulting from (1.6.7) has the form

$$A_J U_J = F_J, \tag{1.6.10}$$

where $F_J = (\langle f, \phi_{j,\gamma}\rangle)_{\gamma\in\widetilde{\Gamma}_J}$, and A_J is the positive definite stiffness matrix defined by

$$A_J(\gamma,\gamma') = a(\phi_{j,\gamma}, \phi_{j,\gamma'}), \quad \gamma,\gamma' \in \widetilde{\Gamma}_J. \tag{1.6.11}$$

For large values of J, direct methods for the computation of U_J become unpractical and a more reasonable strategy is to rely on an iterative algorithm. The simplest algorithm that can be applied to solve (1.6.10) is the *Richardson iteration*: starting from an inital guess U_J^0 of U_J (for example $U_J^0 = 0$), one defines at each step n,

$$U_J^{n+1} = U_J^n + \tau(F_J - A_J U_J^n). \tag{1.6.12}$$

The parameter $\tau > 0$ is used to stabilize the algorithm: Since U_J is clearly a fixed point of the Richardson iteration, the error $E_J^n = U_J^n - U_J$ satisfies

$$E_J^n = (I - \tau A_J)E_J^n = \cdots = (I - \tau A_J)^n E_J^0, \tag{1.6.13}$$

and will thus decay to 0 if we tune the parameter τ so that the contraction factor given by the spectral radius $r_J = \rho(I - \tau A_J)$ is strictly less than 1. A natural choice is the value $\tau = 2/(\lambda_{J,\max} + \lambda_{J,\min})$ which minimizes r_J, where $\lambda_{J,\max}$ and $\lambda_{J,\min}$ are respectively the largest and smallest eigenvalues of A_J. The contraction factor is thus governed by the condition number $\mathcal{K}(A_J) = \mathcal{K}_J = \lambda_{J,\max}/\lambda_{J,\min}$ since we then have

$$r_J = \frac{\mathcal{K}_J - 1}{\mathcal{K}_J + 1}. \tag{1.6.14}$$

In our case of interest, it is well known that the condition number $\mathcal{K}_J$ behaves in $\mathcal{O}(2^{2J})$ as J grows. Defining u_J^n to be the element in V_J with coordinates U_J^n, i.e. the approximate solution computed after n steps, we can accept an error $\|u_J - u_J^n\|_{H^1}$ of the same order of magnitude as the approximation error $\varepsilon(J) \sim 2^{-J}\|u\|_{H^2}$. This allows us to estimate the number of required iterations, in the case we take 0 as an initial guess: we have then

$$\|u_J - u_J^n\|_{H^1} \le r_J^n \|u_J\|_{H^1} \le C r_J^n \|u\|_{H^1}, \tag{1.6.15}$$

since the Galerkin projection is stable in H^1. We thus need to choose the number of iterations $n(J)$ so that r_J^n is of the order $\varepsilon(J)$, i.e.

$$(1 - C2^{-2J})^{n(J)} \sim 2^{-J}. \tag{1.6.16}$$

Taking the logarithm of this estimate, we obtain that the number of required iterations is of the order $J2^{2J}$. The sparsity of A_J allows us to perform each iteration in $\mathcal{O}(2^{2J})$ operation. We thus conclude that the accuracy $\varepsilon \sim 2^{-J}$ can be achieved with a *computational cost*

$$\mathcal{N}(\varepsilon) \sim J2^{4J} \sim \log(\varepsilon)\varepsilon^{-4}. \tag{1.6.17}$$

In summary, the bad conditioning of the stiffness matrix results in a computational cost that is much higher than the number of parameters that are used to describe the solution with a prescribed accuracy. In that sense, the method that we have described is far from being optimal.

A natural generalization of the Richardson iteration consists in replacing the scalar factor τ in (1.6.12) by the action of an invertible operator, i.e.

$$U_J^{n+1} = U_J^n + B_J(F - A_J U_J^n). \tag{1.6.18}$$

With such a generalization, we can hope to improve the reduction factor which is now given by the spectral radius $r_J = \rho(I - B_J A_J)$. We call B_J a *preconditioner* for A_J. Note that the simple choice $B_J = A_J^{-1}$ would make the iteration converge in one step, but is meaningless since it amounts to a direct inversion of A_J which is precisely the task that we want to avoid here. A *good* preconditioner B_J should thus have two basic properties.

1. **Simplicity:** it should be easy to apply. In particular, an iteration of (1.6.18) should have the same optimal order $\mathcal{O}(N_J)$ of computational cost as the simpler (1.6.12).

2. **Efficiency:** it should keep the reduction factor away from 1, if possible independently of J. This property means that the condition

number $\widetilde{\mathcal{K}}_J = \mathcal{K}(B_J A_J)$ is better controlled than $\mathcal{K}(A_J)$ as J grows, if possible bounded independently of J, since up to renormalizing B_J by a proper scalar multiplicative constant we obtain $r_J = \frac{\widetilde{\mathcal{K}}_J - 1}{\widetilde{\mathcal{K}}_J + 1}$ as the new reduction factor.

In the case where B_J is symmetric, note the second property can also be expressed by $\langle A_J U, U\rangle_d \sim \langle B_J^{-1} U, U\rangle_d$, where $\langle\cdot,\cdot\rangle_d$ stands for the discrete inner product between coordinates vectors, since this is equivalent to $\langle B_J^{1/2} A_J B_J^{1/2} U, U\rangle_d \sim \langle U, U\rangle_d$, i.e. $\mathcal{K}(B_J A_J) = \mathcal{K}(B_J^{1/2} A_J B_J^{1/2}) = \mathcal{O}(1)$. For this reason A_J and B_J^{-1} are said to be *spectrally equivalent*. Note that since $\mathcal{K}(B_J^{-1} A_J^{-1}) = 1/\mathcal{K}(B_J A_J)$, the operators A_J^{-1} and B_J are also spectrally equivalent.

Multiscale decomposition methods provide such preconditioners: we shall use them to build operators B_J that can be applied in $\mathcal{O}(N_J)$ operations, and such that $\mathcal{K}(B_J A_J)$ is independent of J, i.e. behaves in $\mathcal{O}(1)$. The analysis of such preconditioners makes an important use of the ability to characterize the functions of H_0^1 from the numerical properties of their multicale decompositions.

Let us first illustrate this point on the simpler one dimensional version of (1.6.1), namely

$$-u'' = f \text{ in }]0,1[\text{ and } u(0) = u(1) = 0. \tag{1.6.19}$$

In this case we apply the Galerkin method on the space V_J introduced in §1.3. In order to define the system associated with the discrete problem, we have the choice between representing the solution in the nodal basis and in the Schauder hierarchical basis.

At this stage, a key observation is that the hierarchical basis wavelets $\psi_{j,k}$ defined by (1.3.8) satisfy

$$\psi'_{j,k} = 2^{j+1} h_{j,k}, \tag{1.6.20}$$

where $h = \chi_{[0,1/2[} - \chi_{[1/2,1[}$ is the Haar wavelet. In turn, the Schauder hierarchical basis is *orthogonal* with respect to the energy inner product $a(u,v) = \int_0^1 u'v'$ associated with problem (1.6.19). If u is a function in $H_0^1(]0,1[)$ and if we consider its expansion in the Schauder basis

$$u = \sum_{j\geq 0} \sum_{k=0}^{2^j-1} d_{j,k}\psi_{j,k}, \tag{1.6.21}$$

(which makes sense since such a function is always continuous), we thus have the identity

$$\begin{aligned}\int |u'|^2 &= \int |\sum_{j,k} 2^{j+1} d_{j,k} h_{j,k}|^2 \\ &= \sum_{j\geq 0} 2^{2j+2} \sum_k |d_{j,k}|^2 \\ &= \sum_{j\geq 0} 2^{2j+2} \|P_{j+1}u - P_j u\|^2_{L^2([0,1])},\end{aligned}$$

where P_j is the interpolation operator introduced in §1.3. A consequence of this fact it that if we use the Schauder basis in order to discretize (1.6.19) in V_J, the resulting stiffness matrix $\widetilde{A}_J$ is *diagonal* (still with condition number behaving like $\mathcal{O}(2^{2j})$). In this particular case the inversion of the system becomes trivial, but we can also view this as a way to precondition the stiffness matrix A_J if we want to solve the problem in the nodal basis: we have $\widetilde{A}_J = R_J^* A_J R_J$ where R_J represent the reconstruction operator that maps the coefficients of a function in the Schauder basis to those of the same function in the nodal basis, so that $A_J^{-1} = R_J \widetilde{A}_J^{-1} R_J^*$. The operator $B_J := R_J \widetilde{A}_J^{-1} R_J^*$ qualifies as a good preconditioner according to our previous prescriptions: it clearly result on the best possible condition number $\widetilde{\mathcal{K}} = 1$ since it is the inverse of A_J, and the application of each of his three components has optimal complexity $\mathcal{O}(N_J)$ if we use the fast algorithm described in §1.3 to implement R_J and its transpose.

If we now return to the two dimensional case, a rapid check reveals that the hierarchical basis constructed in §1.4 is *not* orthogonal in the sense of the energy inner product. We could still hope more modestly for an equivalence of the type

$$\|u\|^2_{H^1(\Omega)} \sim \|P_0 u\|^2_{L^2(\Omega)} + \sum_{j\geq 0} 2^{2j} \|Q_j u\|^2_{L^2(\Omega)}, \tag{1.6.22}$$

where $Q_j = P_{j+1} - P_j$ and P_j is the corresponding interpolation operator. However, this cannot hold either, since the functions of H_0^1 might be discontinuous and unbounded in more than one dimension, so that $P_j f$ might be meaningless.

The results of Chapter 3 will nevertheless establish this type of equivalence for more general function spaces and multiscale decompositions in arbitrary dimensions. For the particular case of H_0^1, it will be shown that (1.6.22) holds in higher dimensions, provided that for the definition of P_j one replaces interpolation by some appropriate projection chosen among a fairly general class of projectors onto the finite element space V_j. In particular, we shall see that the L^2-orthogonal projector onto V_j satisfies this property. If in addition we can expand at each level the details according to a wavelet basis $Q_j u = \sum_{\lambda\in\nabla_j} d_\lambda \psi_\lambda$ with the uniform stability property

$$\|Q_j u\|^2_{L^2} \sim \sum_{\lambda\in\nabla_j} |d_\lambda|^2, \tag{1.6.23}$$

we shall thus have

$$\|u\|_{H^1(\Omega)}^2 \sim \sum_{j\geq -1} 2^{2j} \sum_{\lambda\in\nabla_j} |d_\lambda|^2, \tag{1.6.24}$$

as for the Schauder basis in one dimension. Here by convention, we incorporate the nodal basis of V_0 in the multiscale basis for the indices $\lambda \in \nabla_{-1}$, in order to include the term $\|P_0u\|_{L^2(\Omega)}^2$ in the above sum.

The equivalences (1.6.22) and (1.6.24) should be at least understood intuitively at this point, by thinking of P_ju as the part of u that essentially contains frequencies below 2^j. Let us for instance define P_j in the one dimensional case as the orthogonal projection onto band limited functions such that $\hat{u}(\omega) = 0$ if $|\omega| > 2^j$, i.e. the filtering operator $P_ju = u_j$ with

$$\hat{u}_j(\omega) = \hat{u}(\omega)\chi_{[-2^j,2^j]}, \tag{1.6.25}$$

where $\hat{u}(\omega) = \int u(t)e^{-i\omega t}dt$ is the Fourier transform of u. Using Parseval's identity, we can decompose the H^1 norm of u according to

$$\begin{aligned}\|u\|_{H^1}^2 &\sim \int(1+|\omega|^2)|\hat{u}(\omega)|^2d\omega \\ &= \int_{-1}^1(1+|\omega|^2)|\hat{u}(\omega)|^2d\omega + \sum_{j\geq 0}\int_{2^j<|\omega|<2^{j+1}}(1+|\omega|^2)|\hat{u}(\omega)|^2d\omega \\ &\sim \int_{-1}^1|\hat{u}_0(\omega)|^2d\omega + \sum_{j\geq 0}2^{2j}\int_{2^j<|\omega|<2^{j+1}}|(\hat{u}_{j+1}-\hat{u}_j)(\omega)|^2d\omega \\ &\sim \|P_0f\|_{L^2(\Omega)}^2 + \sum_{j\geq 0}2^{2j}\|P_{j+1}u - P_ju\|_{L^2(\Omega)}^2.\end{aligned}$$

and we have thus proved an equivalence of the same type as (1.6.22). The results of Chapter 3 will make clear that such equivalences can also be obtained with approximation operators onto finite element spaces, which do not correspond exactly to a "cut-off" in frequency, but are better fitted to numerical simulation problems.

Remark 1.6.1 *In the case where P_j is an orthogonal projector, using that $\|u - P_ju\|_{L^2}^2 = \sum_{l\geq j}\|Q_lu\|_{L^2}^2$, we obtain*

$$\begin{aligned}\sum_{j\geq 0}2^{2j}\|Q_ju\|_{L^2}^2 &\leq \sum_{j\geq 0}2^{2j}\|u-P_ju\|_{L^2}^2 \\ &= \sum_{j\geq 0}2^{2j}\sum_{l\geq j}\|Q_lu\|_{L^2}^2 \\ &= \sum_{l\geq 0}[\sum_{j=0}^l 2^{2j}]\|Q_lu\|_{L^2}^2 \\ &\lesssim \sum_{l\geq 0}2^{2l}\|Q_lu\|_{L^2}^2\end{aligned}$$

A reformulation of (1.6.22) is therefore the norm equivalence

$$\|u\|_{H^1(\Omega)}^2 \sim \|P_0u\|_{L^2}^2 + \sum_{j\geq 0}2^{2j}\|u-P_ju\|_{L^2}^2. \tag{1.6.26}$$

which shows that $u \in H_0^1$ implies a bit more than the classical error estimate $\|u - P_j u\|_{L^2} \lesssim 2^{-j}\|u\|_{H^1}$. It also expresses the equivalence between a smoothness property and a decay property of the multiscale approximation error. This point of view will be developed in full generality in Chapter 3.

Let us now explain how (1.6.24) can be used to build a simple preconditioner for A_J. We again denote by $\widetilde{A}_J = R_J^* A_J R_J$ the stiffness matrix in the multiscale basis

$$\Psi_J := \{\psi_\lambda,\ \lambda \in \nabla_j\ ;\ j = -1, \cdots, J\}. \tag{1.6.27}$$

If u is a function in V_J and if U is the coordinate vector corresponding to the expansion $u = \sum_{j<J} \sum_{\lambda \in \nabla_j} d_\lambda \psi_\lambda$ in the multiscale basis, we thus have by (1.6.24)

$$\langle \widetilde{A}_J U, U \rangle_d = a(u,u) \sim \|U\|_{H^1}^2 \sim \sum_{j<J} 2^{2j} \sum_{\lambda \in \nabla_j} |d_\lambda|^2 = \langle D_J U, U \rangle_d. \tag{1.6.28}$$

where D_J is a simple diagonal operator corresponding to a scaling of d_λ by 2^{2j} if $\lambda \in \nabla_j$. The operators A_J and D_J are thus spectrally equivalent, i.e. $\mathcal{K}(D_J^{-1}\widetilde{A}_J) = \mathcal{K}(R_J D_J^{-1} R_J^* A_J)$ behaves in $\mathcal{O}(1)$.

We thus see that $B_J := R_J D_J^{-1} R_J^*$ is a good candidate to be a preconditioner for A_J, provided that R_J and its transpose can be applied in $\mathcal{O}(N_J)$ operation, as in the case of the hierarachical basis. Such an optimal complexity for R_J can essentially be obtained if the multiscale basis is local.

One of the main objectives of the next chapter is to address the effective construction of such local multiscale bases that will yield norm equivalences of the type (1.6.22)

1.7 Conclusions

Several important features of multiscale decomposition schemes have been identified and exploited in the applications that were presented in this chapter:

1. **Nested approximations:** a specific multiresolution approximation process onto a *hierachy of spaces* $V_j \subset V_{j+1} \subset \cdots$ is defined through projectors P_j.

2. **Bases:** the approximations and the fluctuations can be further decomposed into *local* bases.

3. **Stability:** Certain stability properties are satisfied by these bases, either within one resolution level such as in (1.6.23), or across the different levels, such as in (1.6.24).

4. **Local approximation properties:** we can estimate the accuracy of the multiscale approximations of a function in the regions where it is smooth, and improve this accuracy by local *refinements* near its singularities which are operated by simple *thresholding* procedures.

5. **Smoothness:** The approximation spaces might also have certain smoothness properties, which are usually required for the discretization of partial differential equations.

6. **Algorithms:** Multiscale decompositions and reconstructions are performed by fast $\mathcal{O}(N)$ algorithms.

On the other hand these schemes suffer from numerous drawbacks. The Haar system, based on piecewise constant functions, has no C^0-smoothness and poor approximation properties. In turn, it cannot be used to discretize PDE's. The smoothness and approximation properties of piecewise affine functions are also one step higher, yet the Schauder hierarchical basis suffers from a lack of stability across levels, in the sense that the L^2 norm of a function cannot be estimated from the ℓ^2 norm of its coefficients. As was indicated in §1.6, it is also inadequate to characterize H^1 smoothness in the multivariate setting. On the other hand, we remarked that an orthogonal decomposition based on piecewise affine functions involves non local computations.

These remarks motivate the development of more general multiscale approximation and decomposition schemes. These generalizations should preserve the important features that we identified, while allowing a higher degree of smoothness, higher order approximations and better stability properties.

1.8 Historical notes

In 1873, Dubois-Reymond proved the existence of a continuous periodic function f such that its Fourier series diverges at some point x. This striking result shows how an orthonormal basis that only consists of smooth functions might yet be unadequate for the analysis and the approximation of function spaces different from L^2.

In contrast, the Haar system, introduced in the beginning of the century (in the PhD. dissertation of Haar, 1909), has the property that the series of a continuous function on $[0, 1]$ converges uniformly. However, the Haar basis does not constitute a proper basis for the space $C^0([0, 1])$, since its elements are not contained in this space.

The Schauder hierarchical basis, which is simply obtained by taking the primitives of the Haar basis functions, was introduced and studied by Faber and Schauder between 1910 and 1920. In contrast to the Haar system, it does constitute a proper basis for the space $C^0([0, 1])$, but cannot be used to decompose a general L^2 function.

Generally speaking, a *Schauder basis* in a Banach space X is a family e_n, $n \geq 0$ such that any $x \in X$ has a unique expansion $\sum_{n \geq 0} x_n e_n$ that converges in X. The basis is said to be *unconditional* if and only if, the convergence of the series is maintained if x_n is replaced by y_n such that $|y_n| \leq |x_n|$. In the setting of function spaces, this means that one can characterize X from the size properties of the coefficients in the basis e_n. Several classical separable Banach spaces, in particular $L^1(I)$ and $C^0(I)$, are known to possess no unconditional basis and there even exist some more pathological Banach spaces which do not possess any Schauder basis. A review of these different concepts of basis in a Banach space can be found in YOUNG [1980].

As it will be shown in Chapter 3, the Schauder basis is well adapted to the characterization of the Hölder space C^α, $0 < \alpha < 1$, for which it constitutes an unconditional basis. This remarkable property played an important role in the analysis of Brownian motion by Levy in the 1930's, who used the representation in the Schauder basis to estimate (in an almost sure sense) the regularity of its trajectories.

The use of the hierarchical basis in the context of numerical analysis, for the preconditioning of systems arising from elliptic boundary value problems, is more recent: the idea of replacing the nodal basis by the hierarchical basis was proposed in BABUSHKA and AL. [1982] in the univariate case, and studied in the multivariate case by YSERANTANT [1986]. These contributions, although they should be viewed as part of the important development of multigrid methods, emphasize the point of view of a "multilevel decomposition of finite element spaces", which is not central in the mainstream development of multigrid techniques.

Chapter 2

Multiresolution approximation

2.1 Introduction

In this chapter, we shall study in detail the construction and the properties of general multiresolution approximation and decomposition schemes. These schemes are based on hierarchies of nested spaces $V_j \subset V_{j+1}$, that reproduce the main features that we have identified in the simple examples of Chapter 1, while allowing additional properties of approximation, smoothness and stability.

A first idea in generalizing the constructions of the previous chapter is to use finite element spaces V_j, built on partitions - such as triangulations or quadrangulations in the two dimensional case - with mesh size of the order 2^{-j}, and try to impose the nestedness property, which leads to certain restrictions on the type of partitions and finite element spaces to be used. This approach has the advantage of inheriting the structural simplicity of finite element spaces. We recall below four characteristic features of these spaces.

1. **Local characterization:** On each element of the partition, the approximating functions have the simple explicit expression of a *polynomial* that depends on a fixed number of nodal values.

2. **Bases:** Local *nodal* bases for these spaces can be constructed in a simple way, and the stability properties of these bases is easy to analyze, using affine tranformations onto a reference element.

3. **Smoothness:** The smoothness of the functions in the finite element spaces is well understood. It is controlled by the matching conditions that are imposed at the nodes located on the interfaces between different elements.

4. **Accuracy:** The approximation properties of these spaces are also well understood. They are controlled by the amount of polynomial exactness, i.e. the maximal d such that $\Pi_d \subset V_j$ where Π_d is the space of polynomials of total degree less than d.

However, the construction of multiscale bases in the finite element context is by no means a straightforward operation. Such bases should both be related to nodal bases with simple and fast transformation algorithms and be well adapted to characterize smoothness classes in terms of norm equivalences that would generalize (1.6.24). In many instances of finite element spaces, the construction of such bases remains an open question.

An alternative approach is based on ideas that were introduced and developed more recently: the approximation spaces V_j will be defined through a basis, which is generated (in the one dimensional case) by the translations and dilations $\varphi(2^j \cdot -k)$, $k \in \mathbb{Z}$ of a single *scaling function* φ. In contrast to the finite element case, the function φ is not necessarly piecewise polynomial, but it needs to satisfy an equation of the type

$$\varphi(x) = \sum_{k\in\mathbb{Z}} h_k \varphi(2x-k), \tag{2.1.1}$$

which expresses the nestedness of the spaces V_j. Solutions of such equations are called *refinable functions*, and the associated hierachy of approximation spaces is called a *multiresolution analysis*. We shall require in addition the existence of a dual scaling function $\tilde{\varphi}(x) = \sum_{k\in\mathbb{Z}} \tilde{h}_k \varphi(2x-k)$ that will allow us to build local projectors of the form

$$P_j f = \sum_{k\in\mathbb{Z}} \langle f, \tilde{\varphi}_{j,k}\rangle \varphi_{j,k}. \tag{2.1.2}$$

This second strategy yields in a natural way the construction of biorthogonal wavelet bases and fast transformation algorithms. It is easily generalizable to the multivariate setting by means of tensor product strategies, similar to those used in §1.4. On the other hand, it is implicitly tied to the setting of uniform tensor product discretization, and it requires a good understanding of the construction, properties and practical manipulation of refinable functions. It also requires particular adaptations, if we want to approximate or decompose functions that are defined on bounded domains, with prescribed boundary conditions.

Since finite elements are classical material, that has been documented in numerous textbooks (e.g. CIARLET [1978], STRANG and FIX [1973], BRENNER and SCOTT [1983]), we shall devote a substantial part of this chapter to explain the construction of multiscale decompositions from the second approach.

In the last section, we shall return to the first approach: multiscale decompositions will be directly derived from nested sequences of finite element spaces, or also from nested discretizations by point values or cell averages equipped with certain inter-scale operators. While these last tools are probably the most intuitive and relevant to the numerician, their interpretation in terms of the biorthogonal wavelets framework developed throughout this chapter is very beneficial for the rigorous analysis of their performance in many applications.

This chapter is organized as follows: In §2.2, we introduce the general concept of multiresolution analysis. In §2.3, we focus more precisely on the equation (2.1.1). We show how any solution of (2.1.1) is a potential candidate for generating a multiresolution approximation. In §2.4 and §2.5, we show respectively how to compute this solution and how it can be used for numerical approximation, even when it is not known explicitly. We show in §2.6 that these concepts lead in a natural way to the construction of wavelet bases, and fast decomposition and reconstruction algorithms.

The next sections §2.7, §2.8 and §2.9 are concerned with the problem of analyzing some important properties of refinable functions - smoothness and polynomial reproduction, stability, orthonormality, biorthogonality - from the data of the coefficients h_n in (2.1.1). We then use these results in §2.10 to build important examples that satisfy interpolation or orthonormality properties, and we discuss examples involving spline functions in §2.11.

For the sake of notational simplicity we have chosen to focus on multiresolution approximation techniques for functions of one variable that are defined on the whole of $\mathbb{R}$. We thus postpone to §2.12 the main techniques that are used to adapt these constructions to multivariate functions and to bounded domains with specific boundary conditions. Finally §2.13 deals with the construction of multiscale decompositions from the point of view of classical discretizations: point values sampling, cell averages and finite elements.

2.2 Multiresolution analysis

The following definition was first introduced in MALLAT [1989].

Definition 2.2.1 *A multiresolution analysis is a sequence of closed subspaces of $L^2(\mathbb{R})$, such that the following properties are satisfied:*

1. *The sequence is nested, i.e. for all j,*

$$V_j \subset V_{j+1}. \tag{2.2.1}$$

2. *The spaces are related to each other by dyadic scaling, i.e.*

$$f \in V_j \Leftrightarrow f(2\cdot) \in V_{j+1} \Leftrightarrow f(2^{-j}\cdot) \in V_0. \tag{2.2.2}$$

3. *The union of the spaces is dense, i.e. for all f in $L^2(\mathbb{R})$*

$$\lim_{j\to+\infty} \|f - P_j^o f\|_{L^2} = 0, \tag{2.2.3}$$

   where P_j^o is the orthonormal projection onto V_j.

4. *The intersection of the spaces is reduced to the null function, i.e.*

$$\lim_{j\to-\infty} \|P_j^o f\|_{L^2} = 0. \tag{2.2.4}$$

5. *There exists a function $\varphi \in V_0$ such that the family*

$$\varphi(\cdot - k), \quad k \in \mathbb{Z}, \tag{2.2.5}$$

   is a Riesz basis of V_0.

By definition, a family $\{e_k\}_{k\in\mathbb{Z}}$ is a *Riesz basis* of a Hilbert space H, if and only if it spans H, i.e. the finite linear combinations of the e_k are dense in H, and if there exist $0 < C_1 \leq C_2$ such that for all finitely supported sequences (x_k), we have

$$C_1 \sum_k |x_k|^2 \leq \|\sum_k x_k e_k\|_H^2 \leq C_2 \sum_k |x_k|^2. \tag{2.2.6}$$

This property expresses the "stability" of the expansion in this basis with respect to the coordinates. It also means that the application

$$T: \ (c_k)_{k\in\mathbb{Z}} \mapsto \sum_{k\in\mathbb{Z}} c_k e_k, \tag{2.2.7}$$

defines an isomorphism from $\ell^2(\mathbb{Z})$ to H.

The following three facts concerning Riesz bases are easy to check: firstly, the series $\sum_{k\in\mathbb{Z}} x_k e_k$ converges *unconditionally* in L^2 (i.e. its terms can be permutated without affecting the convergence) if and only if $\sum_{k\in\mathbb{Z}}|x_k|^2$ is finite. Secondly, any $x \in H$ can be decomposed in a unique way according to $x = \sum_{k\in\mathbb{Z}} x_k e_k$ with $(x_k)_{k\in\mathbb{Z}}$ in $\ell^2(\mathbb{Z})$ and the equivalence in (2.2.6) also holds for such infinite linear combinations. Finally, there exists a unique *biorthogonal* Riesz basis $\{\tilde{e}_k\}_{k\in\mathbb{Z}}$, (defined by $\tilde{e}_k = (TT^*)^{-1}e_k$) such that $\langle e_k, \tilde{e}_l\rangle = \delta_{k,l}$, and the coordinates of x in the basis are given by $x_k = \langle x, \tilde{e}_k\rangle$. We therefore have for all $x \in H$

$$x = \sum_{k\in\mathbb{Z}} \langle x, \tilde{e}_k\rangle e_k = \sum_{k\in\mathbb{Z}} \langle x, e_k\rangle \tilde{e}_k. \tag{2.2.8}$$

In this sense, Riesz bases are very close to orthonormal bases, which appear as a particular case for which $e_k = \tilde{e}_k$ and $C_1 = C_2 = 1$.

Remark 2.2.1 *A consequence of the Riesz basis property is*

$$C_1\|x\|_H^2 \le \sum_{k\in\mathbb{Z}} |\langle x, e_k\rangle|^2 \le C_2\|x\|_H^2, \tag{2.2.9}$$

since the adjoint mapping

$$T^* : x \mapsto (\langle x, e_k\rangle_H)_{k\in\mathbb{Z}}, \tag{2.2.10}$$

is also bounded and invertible from H to $\ell^2(\mathbb{Z})$. However (2.2.9) is weaker than (2.2.6) since it only reflects the left invertibility of T^*. Sequences $(e_k)_{k\in\mathbb{Z}}$ that satisfy (2.2.9) are called frames. The property (2.2.9) ensures that a vector x can be reconstructed from its inner products $\langle x, e_k\rangle$. However, in contrast to Riesz bases, this reconstruction is not unique, and frames can be redundant: take the example of $(0,1)$, $(1,0)$, $(-1,0)$ and $(0,-1)$ in $\mathbb{R}^2$ that satisfies (2.2.9) with $C_1 = C_2 = 2$. Frames were introduced in the 1950's in DUFFIN *and* SCHAEFFER *[1952] in the context of non-harmonic Fourier series and irregular sampling. An introduction to frames and their basic properties can also be found in* YOUNG *[1980] and in* DAUBECHIES *[1992].*

In our case of interest, we shall sometimes call L^2-*stable* a function φ such that $\varphi(\cdot - k)$, $k \in \mathbb{Z}$ forms a Riesz basis of its L^2-span. We then derive from property (2.2.2) that the family

$$\varphi_{j,k} = 2^{j/2}\varphi(2^j \cdot -k), \quad k \in \mathbb{Z}, \tag{2.2.11}$$

is a Riesz basis for the space V_j: any $f_j \in V_j$ can be written $f_j = 2^{j/2}f_0(2^j\cdot)$ with $f_0 = \sum_{k\in\mathbb{Z}} c_k\varphi(\cdot - k) \in V_0$, and thus $f_j = \sum_{k\in\mathbb{Z}} c_k\varphi_{j,k}$. Moreover, since $\|f_j\|_{L^2} = \|f_0\|_{L^2}$, we see that the constants C_1 and C_2 in the equivalence (2.2.6) are independent of j. The function φ is called a *scaling function*, since we need to use its scaled version to generate the space V_j.

Remark 2.2.2 *From the definition of a multiresolution analysis, it is clear that $f \in V_j$ implies that $f(\cdot - kh) \in V_j$ for all $k \in \mathbb{Z}$, where we have set $h = 2^{-j}$. A space V that satisfies such a property for some $h > 0$ is called shift invariant, with respect to the shift h. A shift-invariant space that is furthermore generated by the translations $\varphi(\cdot - nh)$, $n \in \mathbb{Z}$ (such as our V_j spaces) of a single function is called principal shift invariant. The study of shift-invariant spaces and their property can be done in a more general context than multiresolution approximation (see e.g.* STRANG *and* FIX *[1969],* DE BOOR, DEVORE *and* RON *[1994]). In most of these works, as in several results in this chapter, Fourier analysis plays an important role, due to the following observation: if $f = \sum_{n\in\mathbb{Z}} c_n\varphi(\cdot - nh) \in V$, we have*

$$\hat{f}(\omega) = [\sum_{n\in\mathbb{Z}} c_n e^{-inh\omega}]\hat{\varphi}(\omega). \tag{2.2.12}$$

The Fourier transform of V can thus be viewed as the simple multiplication of $\hat{\varphi}$ with a space of 2π-periodic functions (which is exactly $L^2([0, 2\pi])$ in the case where φ is L^2-stable since the c_n can be any ℓ^2 sequence), and several properties of V can be "read" on the Fourier transform of φ.

Since $V_0 \subset V_1$, the scaling function φ can be expanded in terms of the basis of V_1, i.e.

$$\varphi(x) = \sum_{n\in\mathbb{Z}} h_n\varphi(2x - n), \tag{2.2.13}$$

where the sequence $(h_n)_{n\in\mathbb{Z}}$ is in $\ell^2(\mathbb{Z})$ by the stability property. The equation (2.2.13) known as *refinement equation*, or *two-scale difference equation*, will play a crucial role in this chapter. A function that satisfies such an equation is called a *refinable function*, and the coefficients h_n in (2.2.13) are called the *refinement coefficients*. The terms "refinement" and "refinable" will be justified in §2.4 (Remark 2.4.1), where it is shown that such functions are related to iterative refinement schemes that are typical tools of computer-aided geometric design.

Our first examples of scaling functions have an explicit expression, while we shall further encounter more elaborate examples which are defined implicitely as solutions of a two-scale difference equation.

Example 1 The "box function" $\varphi = \mathcal{X}_{[0,1]}$ has been considered in §1.2 of Chapter 1. It generates a multiresolution analysis that consists of the spaces V_j defined in (1.2.2). It satisfies the two-scale difference equation

$$\varphi(x) = \varphi(2x) + \varphi(2x - 1), \tag{2.2.14}$$

i.e. $h_0 = h_1 = 1$, else $h_k = 0$. A particular feature of this scaling function is that $\varphi_{j,k}$ is an orthonormal basis for V_j. We shall use the following terminology to summarize this property:

Definition 2.2.2 *A scaling function $\varphi \in L^2$ is said to be orthonormal, when it satisfies*

$$\langle \varphi(\cdot - k), \varphi(\cdot - l)\rangle = \delta_{k,l}, \quad k,l \in \mathbb{Z}. \tag{2.2.15}$$

Example 2 The "hat function" $\varphi(x) = \max\{1 - |x|, 0\}$ was used in §1.3 of Chapter 1. It generates a multiresolution analysis that consists of the spaces

$$V_j = \{f \in L^2 \cap C^0;\ \ f_{|I_{j,k}} \in \Pi_1\}. \tag{2.2.16}$$

The proof that its translations form a Riesz basis is a particular case of Corollary 2.2.2 below that proves this property for more general spline functions. The associated refinement equation is

$$\varphi(x) = \varphi(2x) + [\varphi(2x-1) + \varphi(2x+1)]/2, \tag{2.2.17}$$

i.e. $h_0 = 1$, $h_1 = h_{-1} = 1/2$, else $h_k = 0$. A particular feature of this scaling function is that $\varphi(k) = \delta_{0,k}$, $k \in \mathbb{Z}$, so that a function $f_j \in V_j$ satisfies

$$f_j = \sum_{k\in\mathbb{Z}} f_j(2^{-j}k)\varphi(2^j\cdot - k) = \sum_{k\in\mathbb{Z}} 2^{-j/2} f_j(2^{-j}k)\varphi_{j,k}. \tag{2.2.18}$$

We use the following terminology to summarize this property:

Definition 2.2.3 *A scaling function $\varphi \in L^2 \cap C^0$ is said to be interpolatory, when it satisfies*

$$\varphi(k) = \delta_{0,k}, \quad k \in \mathbb{Z}. \tag{2.2.19}$$

Example 3 The spaces of *band-limited* functions

$$V_j = \{f \in L^2\ ;\ \mathrm{Supp}(\mathcal{F}f) \subset [-2^j\pi, 2^j\pi]\}, \tag{2.2.20}$$

(with $\mathcal{F}f(\omega) = \hat{f}(\omega)$) constitute a multiresolution analysis: the first four properties in definition 2.2.1 are immediate to check (using that the projector P_j is defined by $\mathcal{F}P_j f = \chi_{[-2^j\pi, 2^j\pi]}\mathcal{F}f$). The Fourier transform of V_0 has a natural orthonormal basis $e_n(x) = (2\pi)^{-1/2}e^{inx}\chi_{[-\pi,\pi]}$, $n \in \mathbb{Z}$. Applying the inverse Fourier transform and invoking Parseval identity, we obtain an orthonormal basis for V_0 of the form $\varphi(\cdot - k)$, $k \in \mathbb{Z}$, with

$$\varphi(x) = \frac{\sin(\pi x)}{\pi x}. \tag{2.2.21}$$

We notice furthermore that the function φ is also interpolatory. This allows us to derive the associated scaling equation

$$\varphi(x) = \sum_{n\in\mathbb{Z}} \frac{2\sin(\pi n/2)}{\pi n}\varphi(2x-n). \tag{2.2.22}$$

This last scaling function strongly differs from the two first in the sense that is has infinite support and slow decay. For these reasons it is not of much use in numerical simulation that often requires local properties. These examples also suggest that the supports of φ and of the sequence h_n are the same. This property will be proved in §2.4.

Our first result is a general characterization of the properties of stability, orthonormality and interpolation for the system $\{\varphi(\cdot-k)\}_{k\in\mathbb{Z}}$ based on the Fourier transform.

Theorem 2.2.1 *If $\varphi \in L^2$, the series*

$$\sum_{n\in\mathbb{Z}} |\hat{\varphi}(\omega+2n\pi)|^2, \tag{2.2.23}$$

converges in $L^1(I)$ for any compact set I, to an L^1_{loc}, 2π-periodic function. The function φ is L^2-stable if and only if there exist two constants $C_1, C_2 > 0$, such that

$$C_1 \le \sum_{n\in\mathbb{Z}} |\hat{\varphi}(\omega+2n\pi)|^2 \le C_2, \tag{2.2.24}$$

almost everywhere. Moreover, φ is an orthonormal scaling function if and only if

$$\sum_{n\in\mathbb{Z}} |\hat{\varphi}(\omega+2n\pi)|^2 = 1, \tag{2.2.25}$$

almost everywhere. If we have, for some $\varepsilon > 0$,

$$|\hat{\varphi}(\omega)| \lesssim (1+|\omega|)^{-1-\varepsilon}, \tag{2.2.26}$$

then φ is interpolatory if and only if

$$\sum_{n\in\mathbb{Z}} \hat{\varphi}(\omega+2n\pi) = 1, \tag{2.2.27}$$

almost everywhere.

Proof Since $\hat{\varphi}$ is in L^2, the series

$$S_\varphi(\omega) = \sum_{n \in \mathbb{Z}} |\hat{\varphi}(\omega + 2n\pi)|^2, \tag{2.2.28}$$

converges in $L^1([-k\pi, (k+1)\pi])$ for all $k \in \mathbb{Z}$, and thus in $L^1(I)$ for any bounded interval I. The limit S_φ is clearly 2π-periodic.

If $f = \sum_{|k|<N} c_k \varphi(\cdot - k)$ is a finite linear combination of the integer shifts of φ, then we can express its L^2 norm as follows:

$$\begin{aligned} \|f\|_{L^2}^2 &= (2\pi)^{-1}\|\hat{f}(\omega)\|^2 \\ &= (2\pi)^{-1}\int_{-\infty}^{+\infty} |[\sum_{|k|<N} c_k e^{-ik\omega}]\hat{\varphi}(\omega)|^2 d\omega \\ &= (2\pi)^{-1}\int_{-\pi}^{\pi} |\sum_{|k|<N} c_k e^{-ik\omega}|^2 S_\varphi(\omega) d\omega \\ &= (2\pi)^{-1}\int_{-\pi}^{\pi} |m(\omega)|^2 S_\varphi(\omega) d\omega, \end{aligned}$$

where $m(\omega) = \sum_{|k|<N} c_k e^{-ik\omega}$. On the other hand, we have that

$$\sum_{|k|<N} |c_k|^2 = (2\pi)^{-1}\int_{-\pi}^{\pi} |m(\omega)|^2 d\omega. \tag{2.2.29}$$

Now $\{\varphi(\cdot - k)\}_{k \in \mathbb{Z}}$ is a Riesz basis for its span if and only if the two above quantities are equivalent for any trigonometric polynomial $m(\omega)$ of arbitrary degree N. By density, this also means that these quantities are equivalent for any periodic function $m(\omega)$ which is square integrable on $[0, 2\pi]$, which clearly holds if and only if $S_\varphi(\omega)$ is almost everywhere bounded below and above by the constants C_1 and C_2 of (2.2.6).

We also remark that the Fourier coefficients of $S_\varphi(\omega)$ are given by

$$\begin{aligned} S_k &= (2\pi)^{-1}\int_{-\pi}^{\pi} S_\varphi(\omega) e^{-ik\omega} d\omega \\ &= (2\pi)^{-1}\int_{-\infty}^{+\infty} |\hat{\varphi}(\omega)|^2 e^{-ik\omega} d\omega \\ &= \langle \varphi(\cdot - k), \varphi \rangle. \end{aligned}$$

It follows that $S_\varphi(\omega) = 1$ almost everywhere if and only if φ is an orthonormal scaling function.

It is clear that the decay condition (2.2.26) implies that φ coincides almost everywhere with a continuous and bounded function that we can identify with φ, and that the series

$$R_\varphi(\omega) = \sum_{n \in \mathbb{Z}} \hat{\varphi}(\omega + 2n\pi), \tag{2.2.30}$$

converges in L^1 on every compact set. The Fourier coefficients of $R_\varphi(\omega)$ are given by

$$R_k = \frac{1}{2\pi}\int_{-\pi}^{\pi} R_\varphi(\omega) e^{-ik\omega} d\omega = \frac{1}{2\pi}\int_{-\infty}^{+\infty} \hat{\varphi}(\omega) e^{-ik\omega} d\omega = \varphi(-k). \tag{2.2.31}$$

It follows that $R_\varphi(\omega) = 1$ almost everywhere if and only if φ is an interpolatory scaling function. ◇

Remark 2.2.3 *It should be noted that under the assumption of the convergence of (2.2.23), the expansion of the Fourier series*

$$S_\varphi(\omega) = \sum_{k\in\mathbb{Z}} \langle \varphi(\cdot - k), \varphi\rangle e^{ik\omega}, \quad (2.2.32)$$

does not converge better than in $L^2([0, 2\pi])$, unless we make some additional assumptions on the localization properties of φ, in order to derive decay properties on the Fourier coefficients in (2.2.32). Similar observations can be made for

$$R_\varphi(\omega) = \sum_{k\in\mathbb{Z}} \varphi(k)e^{-ik\omega}, \quad (2.2.33)$$

which is not sure to converge even in $L^1([0, 2\pi])$ under the sole assumption (2.2.26).

Remark 2.2.4 *In the proof of Theorem 2.2.1, we see that if we only have the right part of (2.2.24) with some $C_2 > 0$, we still obtain the right part of (2.2.6), i.e. the mapping T defined by (2.2.7) is bounded. By duality, we also obtain that $\varphi(\cdot - k)$, $k \in \mathbb{Z}$ is a Bessel sequence of L^2, i.e. satifies*

$$\sum_{k\in\mathbb{Z}} |\langle f, \varphi(\cdot - k)\rangle|^2 \lesssim \|f\|_2^2, \quad (2.2.34)$$

for all $f \in L^2$.

Remark 2.2.5 *One can easily check that, under the assumption (2.2.26), $R_\varphi(\omega) \geq C_R > 0$ implies $S_\varphi(\omega) \geq C_S > 0$. In particular, an interpolatory scaling function is always L^2-stable.*

A simple corollary to Theorem 2.2.1, deals with the effective construction of orthonormal and interpolatory scaling functions. We recall that $S_\varphi(\omega)$ and $R_\varphi(\omega)$ are the 2π-periodic functions defined by (2.2.28) and (2.2.30).

Corollary 2.2.1 *Let $\varphi \in L^2$ be a scaling function that satisfies (2.2.24). Then the function φ^o defined by*

$$\hat{\varphi}^o(\omega) = [S_\varphi(\omega)]^{-1/2}\hat{\varphi}(\omega), \quad (2.2.35)$$

is in V_0 and orthonormal. The function φ^d defined by

$$\hat{\varphi}^d(\omega) = [S_\varphi(\omega)]^{-1}\hat{\varphi}(\omega), \quad (2.2.36)$$

is in V_0 and generates a dual Riesz basis, i.e. $\langle\varphi(\cdot-k),\varphi^d(\cdot-l)\rangle=\delta_{k,l}$.

If $\varphi\in L^2$ is such that (2.2.26) holds and $R(\omega)\geq C>0$, then the function φ^i defined by

$$\hat{\varphi}^i(\omega)=[R_\varphi(\omega)]^{-1}\hat{\varphi}(\omega), \tag{2.2.37}$$

is in V_0 and interpolatory.

When φ has compact support, one can define S_φ and R_φ directly through (2.2.32) and (2.2.33), and if φ is also bounded, the resulting functions φ^o, φ^d and φ^i have exponential decay at infinity.

Proof It is clear that the functions $[S_\varphi(\omega)]^{-1/2}$, $[S_\varphi(\omega)]^{-1}$ and $[R_\varphi(\omega)]^{-1}$ are in $L^2([0,2\pi])$, when R_φ and S_φ are bounded from below by some constant $C>0$. Based on the characterization of $\mathcal{F}V_0$ (see Remark 2.2.2), we conclude that the functions φ^o, φ^d and φ^i are in V_0.

By construction it is clear that φ^o and φ^i satisfy respectively the criteria of Theorem 2.2.1 for orthonormality and interpolation. Next, we remark that the series

$$U_{\varphi,\varphi^d}(\omega)=\sum_{n\in\mathbb{Z}}\overline{\hat{\varphi}}\hat{\varphi}^d(\omega+2n\pi), \tag{2.2.38}$$

converges in L^1 to 1 on every compact set, and that its Fourier coefficients are given by

$$\begin{aligned}U_k&=(2\pi)^{-1}\int_{-\pi}^{\pi}U_{\varphi,\varphi^d}(\omega)e^{-ik\omega}d\omega\\&=(2\pi)^{-1}\int_{-\infty}^{+\infty}\overline{\hat{\varphi}(\omega)}\hat{\varphi}^d(\omega)e^{-ik\omega}d\omega\\&=\langle\varphi^d(\cdot-k),\varphi\rangle.\end{aligned}$$

Since $U_k=\delta_{0,k}$, we conclude that φ^d generates the dual Riesz basis.

Finally, we remark that if φ has compact support, then both R_φ and S_φ (now directly defined by (2.2.32) and (2.2.33)) are trigonometric polynomials. We consider the corresponding meromorphic functions $r_\varphi(z)$ and $s_\varphi(z)$ such that $R_\varphi(\omega)=r_\varphi(e^{i\omega})$ and $S_\varphi(\omega)=s_\varphi(e^{i\omega})$. Since these functions do not vanish on the unit circle, the functions $[s_\varphi(e^{i\omega})]^{-1/2}$, $[s_\varphi(e^{i\omega})]^{-1}$ and $[s_\varphi(e^{i\omega})]^{-1}$ have analytic extensions on some ring $\{a^{-1}\leq|z|\leq a\}$, $a>1$. It follows that the Fourier coefficients of these functions have exponential decay (majorized by Ca^{-n}) at infinity.

Thus, φ^o, φ^d and φ^i are combinations of $\varphi(\cdot-k)$ with coefficients that have exponential decay at infinity. If φ is bounded and compactly supported, it follows that the functions φ^o, φ^d and φ^i have exponential decay at infinity. In this particular case, one can directly verify the property $\langle\varphi(\cdot-k),\varphi^d\rangle=\langle\varphi^o(\cdot-k),\varphi^o\rangle=\varphi^i(k)=\delta_{0,k}$, by introducing the Grammian matrix $G=(\langle\varphi(\cdot-m),\varphi(\cdot-n)\rangle)_{m,n\in\mathbb{Z}}$ and the collocation matrix

$H = (\varphi(m-n))_{m,n\in\mathbb{Z}}$ which have a banded structure. From their definition, the functions $\varphi^d(\cdot-k)$, $\varphi^o(\cdot-k)$ and $\varphi^i(\cdot-k)$, $k\in\mathbb{Z}$ are constructed by applying the matrices G^{-1}, $G^{-1/2}$ and H^{-1} (which are then well defined) on the functions $\varphi(\cdot-k)$, $k\in\mathbb{Z}$, which immediately implies the duality, orthonormality and interpolation properties. ◇

Let us now apply the Fourier transform on the two-scale difference equation (2.2.13). For this, we define the ***symbol*** of the refinable function φ, as the Fourier series

$$m(\omega) = \frac{1}{2}\sum_{n\in\mathbb{Z}} h_n e^{-in\omega}, \tag{2.2.39}$$

and we obtain the following reformulation of (2.2.13):

$$\hat\varphi(\omega) = m(\omega/2)\hat\varphi(\omega/2). \tag{2.2.40}$$

An immediate consequence of (2.2.40), that we shall use in the next example, is that if φ_1 and φ_2 are two refinable functions in $L^1\cap L^2$ with symbols m_1 and m_2, then the function $\varphi = \varphi_1 * \varphi_2 \in L^1\cap C^0$ is also refinable, with symbol $m = m_1 m_2$. The refinement coefficients are thus given by $h_n = \frac{1}{2}(h^1 * h^2)_n$ where h^1_n and h^2_n are the refinement coefficients of φ_1 and φ_2.

Example 4 We shall apply the previous results to a slightly more elaborate example. We define the B-spline B_N of degree N by $B_0 = \chi_{[0,1]}$ and

$$B_N = B_0 * B_{N-1} = (*)^{N+1}\chi_{[0,1]}. \tag{2.2.41}$$

we note that B_1 is the hat function of example 2, up to an integer shift that does not modify the corresponding multiresolution analysis. The function B_N is supported in $[0, N+1]$, and generates the multiresolution spaces

$$V_j = \{f\in L^2\cap C^{N-1};\ \ f_{|I_{j,k}}\in\Pi_N\}. \tag{2.2.42}$$

Detailed treatments of spline functions and their remarkable properties can be found in DE BOOR [1978] and SCHUMAKER [1981]. According to the previous remark, $\varphi = B_N$ satisfies (2.2.13) with

$$h_n = 2^{-N}\binom{N+1}{n},\ \text{ if } 0\le n\le N+1,\ \ h_n = 0 \text{ otherwise.} \tag{2.2.43}$$

Figure 2.2.1 illustrates the refinement equation in the particular case of the quadratic spline $B_2 = \frac{1}{4}[B_2(2\cdot) + B_2(2\cdot-3)] + \frac{3}{4}[B_2(2\cdot-1) + B_2(2\cdot-2)]$.

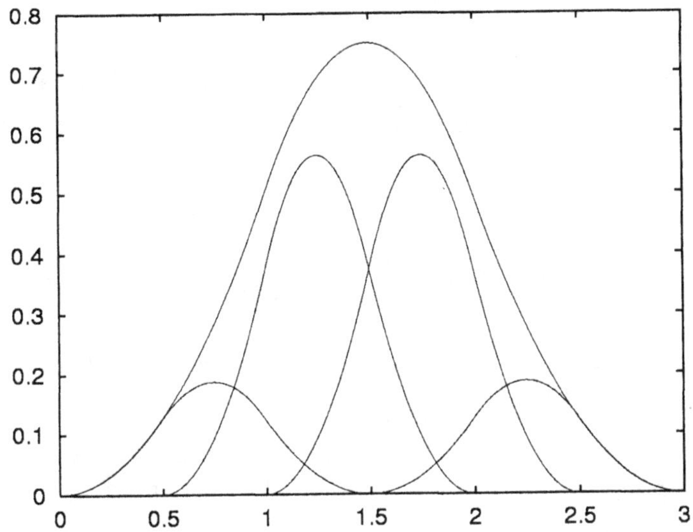

Figure 2.2.1: Decomposition of the quadratic B-spline as a combination of its scaled versions

Corollary 2.2.2 *The function $\varphi = B_N$ is L^2-stable. The associated orthonormal and dual scaling functions φ^o, φ^d are well defined in V_0 and have exponential decay at infinity. If N is odd, the associated interpolatory function φ^i is well defined in V_0 and has exponential decay.*

Proof From (2.2.41), we obtain

$$\hat{\varphi}(\omega) = [\hat{B}_0(\omega)]^{N+1} = \left[\frac{e^{i\omega}-1}{i\omega}\right]^{N+1}, \tag{2.2.44}$$

so that

$$|\hat{\varphi}(\omega)|^2 = \left[\frac{\sin(\omega/2)}{\omega}\right]^{2(N+1)}, \tag{2.2.45}$$

does not vanish on $[-\pi, \pi]$. Thus $S_\varphi(\omega)$ does not vanish and the existence and properties of φ^o and φ^d follow from the first part of Corollary 2.2.1.

For odd $N = 2M - 1$, we have

$$\hat{\varphi}(\omega) = e^{-iM\omega}\left[\frac{\sin(\omega/2)}{\omega}\right]^{2M}. \tag{2.2.46}$$

which shows that $R_\varphi(\omega) = e^{-iM\omega} S_{B_{M-1}}(\omega)$. It follows that $R_\varphi(\omega)$ does not vanish, and the existence and properties of φ^i follow from the second part of Corollary 2.2.1. ◇

We have already defined P_j^o to be the L^2-orthogonal projection on V_j. We can use the functions φ^d and φ^o to express P_j^o in two different ways:

$$P_j^o f = \sum_{k\in\mathbb{Z}} \langle f, \varphi_{j,k}^o\rangle \varphi_{j,k}^o, \tag{2.2.47}$$

or

$$P_j^o f = \sum_{k\in\mathbb{Z}} \langle f, \varphi_{j,k}^d\rangle \varphi_{j,k}. \tag{2.2.48}$$

In the case of B-splines of degree $N \geq 1$, however, both φ^o and φ^d are globally supported on $\mathbb{R}$. As a result, the projector P_j^o has a non-local effect and is difficult to manipulate numerically.

In order to circumvent this type of difficulty, we shall allow the use of non-orthogonal projectors, based on the following definition:

Definition 2.2.4 *A refinable function $\tilde{\varphi} = \sum_n \tilde{h}_n \tilde{\varphi}(2\cdot -n)$ in L^2 is dual to φ if it satisfies*

$$\langle \varphi(\cdot - k), \tilde{\varphi}(\cdot - l)\rangle = \delta_{k,l}, \;\; k,l \in \mathbb{Z}. \tag{2.2.49}$$

From a pair $(\varphi, \tilde{\varphi})$ of dual scaling functions, we can define an oblic projector by

$$P_j f = \sum_{k\in\mathbb{Z}} \langle f, \tilde{\varphi}_{j,k}\rangle \varphi_{j,k}. \tag{2.2.50}$$

Indeed, relation (2.2.49) implies that $\langle \varphi_{j,k}, \tilde{\varphi}_{j,l}\rangle = \delta_{k,l}$ so that $P_j f = f$ for all $f \in V_j$. We postpone the study of the convergence of (2.2.50) and of the L^2-boundedness of P_j to the next section.

In contrast to φ^d, the function $\tilde{\varphi}$ is not enforced to be in V_0. Moreover, *it is not uniquely determined* for a given φ. We can hope to use this new flexibility to build a compactly supported dual $\tilde{\varphi}$, so that the new projector P_j has a local action. This hope turns out to be fufilled: a result proved in LEMARIÉ [1997] states that if φ is a compactly supported L^2-stable refinable function, then there always exists a dual scaling function $\tilde{\varphi}$ in L^2 which is also compactly supported.

Note that the dual $P_j^* f = \sum_{k\in\mathbb{Z}} \langle f, \varphi_{j,k}\rangle \tilde{\varphi}_{j,k}$ of P_j is also an oblic projector whose range is the "dual space" $\tilde{V}_j$ spanned by the Riesz basis $\tilde{\varphi}_{j,k}$, $k \in \mathbb{Z}$. Because we have assumed $\tilde{\varphi}$ to be refinable, the spaces $\tilde{V}_j$ are also nested. As a consequence, if $j < l$, we have $P_l^* P_j^* = P_j^*$ and thus $P_j P_l = P_j$. This last property is crucial for numerical computations, since it means that the approximations can be computed recursively from fine to coarse scales.

Remark 2.2.6 *Orthonormal scaling functions can be viewed as a particular case of dual scaling functions for which $\varphi = \tilde{\varphi}$. Interpolatory scaling functions can also be placed in this general setting as a "degenerate case": if φ is interpolatory, we define $\tilde{\varphi}$ to be the Dirac distribution δ centered at the origin, so that (2.2.49) holds in the sense of the duality product between continuous functions and Radon measures. The Dirac distribution satisfies (in a weak sense) $\delta(x) = 2\delta(2x)$ and should thus be considered as refinable. However the resulting "projector" P_j is not L^2-bounded.*

2.3 Refinable functions

The main message of this section is that *any solution φ of a refinement equation (2.2.13) is a potential candidate to generate a multiresolution analysis,* provided that it satisfies some basic properties that we shall study here. By "generate", we mean that we shall directly define V_j as the L^2-span of $\varphi_{j,k}$, $k \in \mathbb{Z}$.

The price to pay in this approach is that, in contrast to finite element bases which are well defined piecewise polynomial functions, the scaling function φ will not in general be given by an explicit formula, but rather as a solution of (2.2.13), *where only the coefficients h_n will be explicitly given.* In this situation, we need to address the following questions: what are the basic properties of φ which ensure that the spaces V_j satisfy the multiresolution analysis axioms of definition 2.2.1 ? How can we compute φ numerically at some point x ? How can we compute accurately inner products such as $\langle f, \varphi_{j,k}\rangle$ or $\langle \varphi'_{j,m}, \varphi'_{j,n}\rangle$, that are needed in Galerkin approximation methods ? How can we predict the properties of φ (such as compact support, L^2-stability or smoothness) from the coefficients h_n ?

In this section, we shall answer the first question. The answer to the second and third questions are the object of the two next sections, while the last question is the object of sections §2.7 to §2.9. We shall then use the results of these sections to design the coefficients h_n in (2.2.13), in such a way that specific properties of φ are fullfilled.

In the rest of this section, as well as in §2.4 to §2.6, φ and $\tilde{\varphi}$ will denote a pair of dual refinable functions that are compactly supported and real-valued. Although most of the results that we shall prove are generalizable to non-compactly supported refinable functions, we prefer to keep this assumption since it will be essential for the practical implementation of related algorithms, and because it removes many technicalities in several proofs. Similarly the real-valued assumption is made here to simplify the notations (in particular to avoid complex conjugate quantities appearing in inner products) and because almost all relevant examples are real-valued.

A first result is that the existence of $\tilde{\varphi}$ ensures the L^2-stability of φ and the L^2-boundedness of the oblic projector P_j defined by (2.2.50).

Theorem 2.3.1 *Under the above assumptions, the functions φ and $\tilde{\varphi}$ are L^2-stable. Moreover, the oblic projector P_j defined by (2.2.50) is L^2-bounded independently of j.*

Proof Let $f = \sum_n c_n \varphi(\cdot - n)$ be a finite linear combination of the functions $\varphi(\cdot - n)$, $n \in \mathbb{Z}$. On the one hand, we have the inequality

$$\|f\|_{L^2}^2 = \sum_{m,n} c_m \overline{c_n} \langle \varphi_{0,m}, \varphi_{0,n} \rangle \leq C \sum_n |c_n|^2, \tag{2.3.1}$$

with $C = [2S - 1]\|\varphi\|_{L^2}^2$, $S = |\mathrm{Supp}(\varphi)|$. On the other hand, we have

$$|c_n|^2 = |\langle f, \tilde{\varphi}(\cdot - n) \rangle|^2 \leq \|\tilde{\varphi}\|_{L^2}^2 \int_{\mathrm{Supp}(\tilde{\varphi}_{0,n})} |f|^2, \tag{2.3.2}$$

by Cauchy-Schwarz inequality, and thus

$$\sum_n |c_n|^2 \leq \tilde{C} \|f\|_{L^2}^2, \tag{2.3.3}$$

with $\tilde{C} = [2\tilde{S} - 1]\|\tilde{\varphi}\|_{L^2}^2$, $\tilde{S} = |\mathrm{Supp}(\tilde{\varphi})|$. A symmetric argument shows that $\tilde{\varphi}$ is also L^2-stable.

The same arguments show that $P_0 f$ is well defined by (2.2.50) if $f \in L^2$ since $(\langle f, \tilde{\varphi}(\cdot - k) \rangle)_{k \in \mathbb{Z}}$ is in ℓ^2 and φ is L^2-stable. Moreover, we have

$$\|P_0 f\|_{L^2}^2 \leq C \sum_{n \in \mathbb{Z}} |\langle f, \varphi_{0,n} \rangle|^2 \leq C\tilde{C} \|f\|_{L^2}^2, \tag{2.3.4}$$

and thus $\|P_0\| \leq \sqrt{C\tilde{C}}$. Now we remark that

$$P_j = S_j P_0 S_{-j}, \tag{2.3.5}$$

where S_j is the scaling operator $f \mapsto 2^{j/2} f(2^j \cdot)$ which preserves the L^2 norm, so that we have $\|P_j\| = \|P_0\| \leq \sqrt{C\tilde{C}}$. $\diamond$.

Remark 2.3.1 *The above results stay valid in weaker assumptions than compact support for φ and $\tilde{\varphi}$. In particular, we can assume that both φ and $\tilde{\varphi}$ are in L^2 and satisfy the right part of (2.2.24) with constants C_2 and $\tilde{C}_2$. Under these assumptions, according to Remark 2.2.4, we can use the one sided L^2-stability and Bessel sequence properties of φ and $\tilde{\varphi}$ to reach the same conclusion. The main interest of the technique that we have used for compactly supported functions is that it is not based on Fourier transform, and will be rather easily adapted to bounded domains and to L^p norms with $p \neq 2$ (Theorem 3.3.1 of Chapter 3).*

The properties that remain to be checked in Definition 2.2.1 are thus (2.2.3) and (2.2.4), i.e. the "limit" properties of the V_j spaces as j goes to infinity. For this, we need to make the distinction between P_j and the orthogonal projector P_j^o which is defined by

$$P_j^o f = \sum_{k\in\mathbb{Z}} \langle f, \varphi_{j,k}^o\rangle \varphi_{j,k}^o, \tag{2.3.6}$$

where φ^o is defined through (2.2.35).

Theorem 2.3.2 *If $f \in L^2$, then*

$$\lim_{j\to-\infty} \|P_j^o f\|_{L^2} = \lim_{j\to-\infty} \|P_j f\|_{L^2} = 0. \tag{2.3.7}$$

Proof Since both P_j and P_j^o are uniformly L^2-bounded in j, by the same reasoning which was used to prove (1.2.16) in §1.2 of Chapter 1, it suffices to prove the result for $f = \chi_{[a,b]}$.

We first show that $\|P_j^o f\|_{L^2}^2 = \sum_{k\in\mathbb{Z}} |\int_a^b \varphi_{j,k}^o(t)dt|^2$ goes to zero as $j \to -\infty$. By Cauchy-Schwartz inequality and a change of variable, for $-j$ large enough so that $2^j(b-a) \le 1$, we obtain

$$\|P_j^o f\|_{L^2}^2 \le \int_{-\infty}^{+\infty} \chi_{A_j}(t)|\varphi^o(t)|^2 dt, \tag{2.3.8}$$

where $A_j = \cup_{k\in\mathbb{Z}}[2^j a + k, 2^j b + k]$. Since $\chi_{A_j}(t)$ tends to zero almost everywhere, we obtain the result by Lebesgue's dominated convergence theorem. Since $\|P_j f\|^2 \le C \sum_{k\in\mathbb{Z}} |\int_a^b \tilde{\varphi}_{j,k}(t)dt|^2$, with C independent of j, the same reasoning proves the result for P_j. ◇

We now turn to the behaviour of $P_j f - f$ and $P_j^o f - f$ as $j \to +\infty$. The following observation (originally due to Lebesgue) shows that we only have to treat one of these

Lemma 2.3.1 *If P is a bounded projector from a Banach space X to a closed subspace Y, then for all $f \in X$,*

$$\inf_{g\in Y} \|f-g\|_X \le \|f - Pf\|_X \le (1 + \|P\|) \inf_{g\in Y} \|f-g\|_X. \tag{2.3.9}$$

with $\|P\| = \sup_{\|f\|_X=1} \|Pf\|_X$,

Proof The left inequality is immediate. For the right inequality, we remark that for $g \in Y$,

$$\|f - Pf\|_X \le \|f-g\|_X + \|Pf - Pg\|_X \le (1 + \|P\|)\|f-g\|_X, \tag{2.3.10}$$

and we conclude by taking the infimum over all $g \in Y$. ◇

Since the projectors P_j are uniformly L^2-bounded (by Theorem 2.3.1), we shall now focus our attention on the behaviour of $P_jf - f$ as $j \to +\infty$.

Theorem 2.3.3 *The approximation operator $P_jf = \sum_k \langle f, \tilde{\varphi}_{j,k}\rangle \varphi_{j,k}$ satisfies*

$$\lim_{j \to +\infty} \|P_jf - f\|_{L^2} = 0, \tag{2.3.11}$$

if and only if

$$\left[\int \tilde{\varphi}\right] \sum_{k \in \mathbb{Z}} \varphi(x-k) = 1, \tag{2.3.12}$$

almost everywhere on $\mathbb{R}$.

Proof Using once again the uniform L^2-boundedness of the projectors P_j and a density argument, we see that the property (2.3.11) is satisfied for all $f \in L^2$ if and only if it is satisfied for all $f = \chi_{[a,b]}$ where $[a,b]$ is an arbitrary bounded interval. We thus consider $f = \chi_{[a,b]}$ for some $a < b$.

At first, let us remark that the 1-periodic function defined by

$$A(x) := \left[\int \tilde{\varphi}\right] \sum_{k \in \mathbb{Z}} \varphi(x-k), \tag{2.3.13}$$

is well defined in L^2_{loc}, since φ has compact support.

If s and $\tilde{s}$ are such that $\mathrm{Supp}(\varphi) \subset [-s,s]$ and $\mathrm{Supp}(\tilde{\varphi}) \subset [-\tilde{s},\tilde{s}]$, we remark that we have exactly $P_jf(x) = 0$ for $x < a - 2^{-j}(s+\tilde{s})$ or $x > b + 2^{-j}(s+\tilde{s})$, i.e. the approximation is exact away from the interval at a minimal distance of order 2^{-j}. Moreover, if j is large enough so that $2^{-j}(s+\tilde{s}) \le (b-a)/2$, then for $x \in [a + 2^{-j}(s+\tilde{s}), b - 2^{-j}(s+\tilde{s})]$, we have exactly

$$P_jf(x) = \sum_k [\int \tilde{\varphi}_{j,k}]\varphi_{j,k} = \left[\int \tilde{\varphi}\right] \sum_k \varphi(2^jx - k) = A(2^jx). \tag{2.3.14}$$

Assuming that $A(x) = 1$ almost everywhere, it follows that for j large enough, the approximation P_jf is thus equal to f except on the intervals $I_j^a = [a - 2^{-j}(s+\tilde{s}), a + 2^{-j}(s+\tilde{s})]$ and $I_j^b = [b - 2^{-j}(s+\tilde{s}), b + 2^{-j}(s+\tilde{s})]$.

The convergence of P_jf to f in L^2 follows then from the analysis of the contribution of these intervals: clearly $\int_{I_j^a} |f|^2 \le |I_j^a| \to 0$ as $j \to +\infty$, and

$$\begin{aligned} \textstyle\int_{I_j^a} |P_jf|^2 &\le \textstyle\int_{\mathbb{R}} |\sum_{|k-2^ja| \le 2s+\tilde{s}} \langle f, \tilde{\varphi}_{j,k}\rangle \varphi_{j,k}|^2 \\ &\lesssim \textstyle\sum_{|k-2^ja| \le 2s+\tilde{s}} |\langle f, \tilde{\varphi}_{j,k}\rangle|^2 \\ &\le (4s + 2\tilde{s})2^{-j} \ \to 0 \ \text{ as } \ j \to +\infty. \end{aligned}$$

Similar properties hold for I_j^b and we thus conclude that (2.3.11) is satisfied.

Assuming now that $A(x)$ differs from 1 on a set of non-zero measure, we choose $j_0, k_0 \in \mathbb{Z}$ such that $I_{j_0,k_0} = [2^{-j_0}k_0, 2^{-j_0}(k_0+1)]$ is contained in the interior of $[a,b]$. For $j \geq j_0$ large enough, we have $P_j f(x) = A(2^j x)$ on I_{j_0,k_0}, so that

$$\begin{aligned} \|f - P_j f\|_{L^2}^2 &\geq \int_{2^{-j_0}k_0}^{2^{-j_0}(k_0+1)} |1 - A(2^j x)|^2 dx \\ &= 2^{-j} \int_{2^{j-j_0}k_0}^{2^{j-j_0}(k_0+1)} |1 - A(x)|^2 dx \\ &= 2^{-j_0} \int_0^1 |1 - A(x)|^2 dx > 0. \end{aligned}$$

We have thus proved the necessity of (2.3.12). ◇

Remark 2.3.2 *In the above proof, we have not made use of the duality property (2.2.49): $\tilde{\varphi}$ could be any L^2-function with compact support. The operator P_j would still be local and L^2-bounded, but not a projector. Our analysis shows that a minimal requirement for such a local approximation operator is that it "reproduces" the constant functions in the sense that*

$$P_j 1 := \sum_k \langle \tilde{\varphi}_{j,k}, 1 \rangle \varphi_{j,k} = 1. \tag{2.3.15}$$

This property implies in particular that the constant functions are generated by the integer translates of φ. Such general approximation operators (when $\tilde{\varphi}$ is not necessarily dual to φ) are known as quasi-interpolant. A simple example can be obtained with the choice $\varphi = \tilde{\varphi} = B_1 = \max\{1 - |x|, 0\}$, which clearly satisfies (2.3.15). Quasi-interpolants have been introduced in DE BOOR *and* FIX *[1973b] in order to study the properties of spline approximation. Similar approximation operators have also been proposed for multivariate finite element spaces, see e.g.* CLEMENT *[1975].*

Remark 2.3.3 *We also did not exploit the refinability of φ and $\tilde{\varphi}$ in the proof of Theorem 2.3.3. It will be shown in §2.8 that when φ is both refinable and L^2-stable, the property (2.3.12) is then automatically fulfilled. For the time being, we shall simply assume that our refinable dual functions φ and $\tilde{\varphi}$ satisfy this property, so that all the multiresolution properties of definition 2.2.1 are satisfied.*

Remark 2.3.4 *It is easy to see that if the union of the V_j spaces is dense in L^2, the same property holds for the dual multiresolution approximation spaces $\tilde{V}_j$ generated by $\tilde{\varphi}$: indeed, if this was not the case, there would exist a non trivial function f orthogonal to all $\tilde{\varphi}_{j,k}$, $j,k \in \mathbb{Z}$, so that $P_j f = 0$ for* *all j, which is in contradiction with (2.3.11). This shows in particular that the dual identity*

$$\Big[\int \varphi\Big] \sum_{k\in\mathbb{Z}} \tilde{\varphi}(x-k) = 1, \tag{2.3.16}$$

holds almost everywhere.

Corollary 2.3.1 *Both functions φ and $\tilde{\varphi}$ have non-zero integral, and satisfy*

$$\Big[\int \varphi\Big]\Big[\int \tilde{\varphi}\Big] = 1. \tag{2.3.17}$$

Up to a renormalization, we can assume that $\int \varphi = \int \tilde{\varphi} = 1$. The coefficients h_n and $\tilde{h}_n$ satisfy the sum rules

$$\sum h_n = \sum \tilde{h}_n = 2, \ \sum(-1)^n h_n = \sum(-1)^n \tilde{h}_n = 0. \tag{2.3.18}$$

Proof Integrating (2.3.12) over $[0,1]$, we obtain

$$1 = \Big[\int \tilde{\varphi}\Big] \int_0^1 \sum_{k\in\mathbb{Z}} \varphi(x-k)dx = \Big[\int \varphi\Big]\Big[\int \tilde{\varphi}\Big], \tag{2.3.19}$$

i.e. (2.3.17) which shows that both integrals cannot vanish, and that we can renormalize the functions φ and $\tilde{\varphi}$ such that $\int \varphi = \int \tilde{\varphi} = 1$, without affecting P_j, the duality (2.2.49) and the refinable nature of these functions.

Integrating both side of the refinement equation (2.2.13) over $\mathbb{R}$, we obtain

$$1 = \int \varphi = \sum_n h_n \int \varphi(2\cdot -n) = \frac{1}{2}[\sum_n h_n] \int \varphi = [\sum_n h_n]/2, \tag{2.3.20}$$

and similarly for the coefficients $\tilde{h}_n$, i.e. the first part of the sum rules (2.3.18).

For the second part, we inject the refinement equation in (2.3.12) which gives

$$\begin{aligned} 1 &= \textstyle\sum_{k\in\mathbb{Z}} \varphi(\cdot - k) \\ &= \textstyle\sum_{k\in\mathbb{Z}}[\sum_n h_n]\varphi(2\cdot -n-2k) \\ &= \textstyle\sum_{k\in\mathbb{Z}}[\sum_n h_{2n}]\varphi(2\cdot -2k) + \sum_{k\in\mathbb{Z}}[\sum_n h_{2n+1}]\varphi(2\cdot -2k-1) \end{aligned}$$

Applying on both sides of this identity the inner product with $\tilde{\varphi}(2\cdot)$ and $\tilde{\varphi}(2\cdot -1)$, we obtain $\sum_n h_{2n} = \sum_n h_{2n+1} = 1$, i.e. the second part of the sum rules (2.3.18) for the coefficients h_n. The same holds for the coefficients $\tilde{h}_n$ by (2.3.16). ◇

Remark 2.3.5 *From the compact support of φ and $\tilde{\varphi}$, it is also clear that the real-valued sequence $h_n = 2\langle\varphi, \tilde{\varphi}(2\cdot -n)\rangle$ has finite support. Moreover, we necessarily have $h_n = 0$ when $n \notin [a,b] = \text{Supp}(\varphi)$. Indeed, if we had for example $h_n \neq 0$ for some $n < a$, we reach a contradiction by taking the smallest n with this property and remarking that $\text{Supp}(\varphi(2\cdot -n))$ alone contributes in the refinement equation for $x \in [(n+a)/2, (n+1+a)/2[$, making φ non-zero on this interval, which contradicts its support property.*

2.4 Subdivision schemes

Having identified these diverse properties of "admissible dual refinable functions", let us now turn to the question of how to compute the numerical values of the function φ. The solution to this problem can be obtained by viewing the function φ as a fixed point of the operator

$$f \mapsto Rf = \sum_n h_n f(2\cdot -n), \tag{2.4.1}$$

and try to obtain it through an iterative method, i.e. as $\varphi = \lim_{j\to+\infty} R^j f_0$ for some well-chosen f_0. We note that the property $\sum_n h_n = 2$ implies the invariance by R of the integral of f when it exists, i.e.

$$\int Rf(x)dx = \int f(x)dx, \tag{2.4.2}$$

so that we need at least $\int f_0 = 1$.

With the simple choice $f_0 = \max\{1-|x|, 0\}$, we obtain after j iterations a function $f_j = R^j f_0$ which is piecewise affine on each interval $I_{j,k}$, $k \in \mathbb{Z}$, and continuous. Our hope is that f_j constitutes a proper piecewise affine approximation of φ at resolution 2^{-j}. The values $s_{j,k} = f_j(2^{-j}k)$, $k \in \mathbb{Z}$, that completely determine f_j, can be obtained by two different recursions both applied to the initial data $s_{0,k} = \delta_{0,k}$, $k \in \mathbb{Z}$: on the one hand, if we directly apply $f_{j+1} = Rf_j$, we get $f_j(2^{-j-1}k) = \sum_n h_n f_j(2^{-j}k - n)$, which yields

$$s_{j+1,k} = \sum_n h_n s_{j,k-2^j n}. \tag{2.4.3}$$

On the other hand, remarking that for a more general f we always have $R^j f = \sum_k s_{j,k} f(2^j x - k)$, and applying this to $f = Rf_0$, we obtain

$$f_{j+1} = \sum_k s_{j+1,k} f_0(2^{j+1}x - k) = \sum_k s_{j,k}[\sum_n h_n f_0(2^j x - 2k - n)], \tag{2.4.4}$$

which yields

$$s_{j+1,k} = \sum_n h_{k-2n} s_{j,n}. \tag{2.4.5}$$

The recursion formulae (2.4.3) and (2.4.5) are of different natures. In particular, (2.4.5) reveals a *local* property of the iteration, in the sense that, if $h_n = 0$ for $|n| > N$, the value of f_j at $2^{-j}k$ is only influenced by the values of f_{j-1} at dyadic points contained in $[2^{-j}k - 2^{-j+1}N, 2^{-j}k + 2^{-j+1}N]$. This property is reflected in Figure 2.4.1 that represents the influence of the values from level to level. In contrast, the recursion (2.4.3) is *global.* Although both recursions produce the same result f_j, the local procedure is often preferred for the following reason: if we are interested in the value of φ at a specific point, or in a specific region, we can use (2.4.5) to localize the computation of f_j in this region.

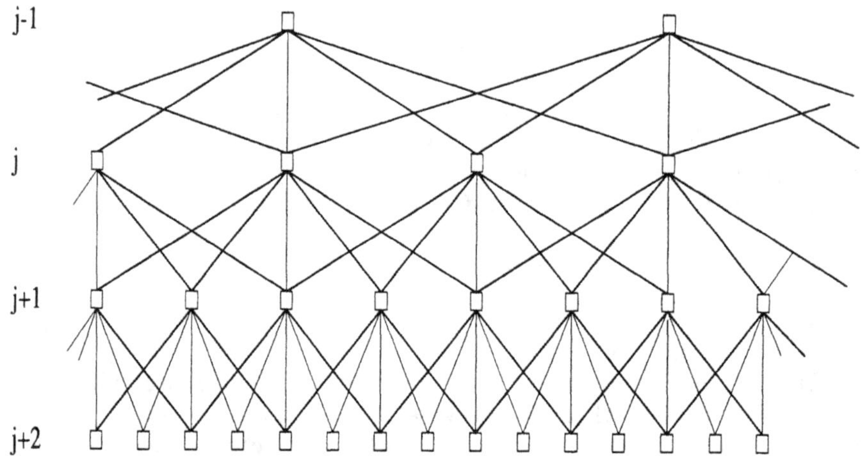

Figure 2.4.1: The subdivision algorithm

The global procedure based on (2.4.3) is known as the *cascade algorithm,* while the local refinement (2.4.5) is an instance of a *subdivision scheme.*

Generally speaking, subdivision schemes are a design tool for the fast generation of smooth curves and surfaces from a set of control points by means of iterative refinements. In the most often considered binary univariate case, one starts from a sequence $s_0(k)$ and obtains at step j a sequence $s_j(2^{-j}k)$, generated from the previous one by linear rules:

$$s_j(2^{-j}k) = 2 \sum_{n \in k+2\mathbb{Z}} h_{j,k}(n) s_{j-1}(2^{-j}(k-n)). \tag{2.4.6}$$

The *masks* $h_{j,k} = \{h_{j,k}(n)\}_{n\in\mathbb{Z}}$ are in general finite sequences, a property that is clearly useful for the practical implementation of (2.4.6).

A natural problem is then to study the convergence of such an algorithm to a limit function. In particular, the scheme is said to be strongly convergent if and only if there exists a continuous function $f(x)$ such that

$$\lim_{j\to+\infty}(\sup_k |s_j(2^{-j}k) - f(2^{-j}k)|) = 0. \tag{2.4.7}$$

A subdivision scheme is said to be stationary and uniform when the masks $h_{j,k}(n) = h_n$ are independent of the parameters j and k. In that case, (2.4.6) rewrites as (2.4.5) with the notation $s_{j,k} = s_j(2^{-j}k)$. It is then sufficient to check the convergence of the subdivision on the initial data $s_{0,k} = \delta_{0,k}$: if φ is the limit function with such data, the limit function with more general data is $f = \sum_k s_0(k)\varphi(\cdot - k)$. General treatment of subdivision schemes, with numerous examples, can be found in CAVARETTA, DAHMEN and MICCHELLI [1991] and DYN [1992].

In our case of interest, we precisely start with the fundamental data $s_{0,k} = \delta_{0,k}$. We shall now use the L^2-stability of φ to analyze the uniform convergence of the subdivision scheme.

Theorem 2.4.1 *If the scaling function φ is continuous, then the associated subdivision algorithm converges uniformly, i.e. f_j converges uniformly to φ.*

Proof We simply remark that, φ being a fixed point of R, we have

$$\varphi(x) = \sum_k s_{j,k}\varphi(2^j x - k), \tag{2.4.8}$$

so that

$$s_{j,k} = 2^{j/2}\langle\varphi, \tilde{\varphi}_{j,k}\rangle = 2^j \int \varphi(x)\tilde{\varphi}(2^j x - k)dx, \tag{2.4.9}$$

where $\tilde{\varphi}$ is the compactly supported dual scaling function. Since $\tilde{\varphi} \in L^1$ and satisfies $\int \tilde{\varphi} = 1$, and since φ is uniformly continuous (by continuity and compact support), this implies the strong convergence property (2.4.7). $\diamond$

Remark 2.4.1 *This way of obtaining φ through iterative refinements justifies the term "refinable function". Conversely, if φ is defined as the limit of a converging subdivision scheme with coefficients h_n and initial data $\delta_{0,k}$, one easily checks that it satisfies the refinement equation (2.2.13) (because it is also the limit of the subdivision of the sequence $s_1(n/2) = h_n$). Thus, we can define a continuous refinable function φ either as a solution of (2.2.13), or as the fundamental limit function of a subdivision scheme.*

The following example, which is of common use in computer graphics, is a good illustration of the relation between subdivision schemes and refinable functions: given a polygonal graph with vertices $p_{0,k} = (x_{0,k}, y_{0,k})$, $k \in \mathbb{Z}$, the *corner cutting algorithm*, introduced by CHAITKIN [1974], consists in a recursive construction of refined polygonal lines $p_{j,k} = (x_{j,k}, y_{j,k})$, $k \in \mathbb{Z}$, $j > 0$, by placing the new points according to

$$p_{j,2k} = \frac{3}{4}p_{j,k} + \frac{1}{4}p_{j,k+1} \text{ and } p_{j,2k+1} = \frac{1}{4}p_{j,k} + \frac{3}{4}p_{j,k+1}. \tag{2.4.10}$$

Figure 2.4.2 reveals the *smoothing* effect of this procedure as j goes to $+\infty$.

From (2.4.10), we see that the sequences $x_{j,k}$ and $y_{j,k}$, are obtained by the iteration of a subdivision scheme with coefficients $h_{-1} = h_0 = 3/4$ and $h_{-2} = h_1 = 1/4$. We notice that the associated scaling function is the quadratic B-spline $\varphi = B_2(x+2)$ defined from (2.2.41). From the stability of this scaling function, we conclude that the corner cutting algorithm converges to the limit curve $p(t) = (\sum_k x_{0,k}B_2(t-k+2), \sum_k y_{0,k}B_2(t-k+2))$. We also display in Figure 2.4.3 the subdivision process with fundamental initial data that converges to B_2.

Remark 2.4.2 *If φ has some additional Hölder regularity C^r, one easily obtains the estimate $\sup_k |s_j(2^{-j}k) - f(2^{-j}k)| \lesssim 2^{-rj}$ for the rate of convergence. Clearly, the uniform convergence of the subdivision is equivalent to the uniform convergence of f_j to f with the choice $f_0 = \max\{1-|x|, 0\}$. It is of course possible to study other types of convergence (for examples in L^p and Sobolev spaces), and to use other choice of initial data f_0. As an example, we display in Figure 2.4.4 a subdivision scheme with coefficients $h_0 = h_4 = -1/4$, $h_1 = h_3 = 1/2$, $h_2 = 3/2$, that can be proved to converge in all L^p spaces for $p < +\infty$ to a discontinuous limit which has logarithmic singularities at dyadic points. Results on diverse type of convergence can be found in several references, e.g.* JIA *and* HAN *[1998] for L^p convergence or* COHEN AND RYAN *[1995, chapter 3] for convergence in smoother norms.*

Remark 2.4.3 *The cascade algorithm, introduced in* DAUBECHIES *[1988], although non-local, has the following advantage: if the initial data are chosen to be $s_{0,k} = \varphi(k)$, then the iteration of (2.4.3) produces the exact values of φ at the dyadic points. We show in the end of the next section (see in particular Remark 2.5.2) how to compute directly the initial values $\varphi(k)$.*

Remark 2.4.4 *If $h_n = 0$ for $n < a$ and $n > b$, then the support of f_j is contained by construction in $[a_j, b_j]$ with $a_j = (1/2+1/4+\cdots+2^{-j})a+2^{-j}$ and $b_j = (1/2+1/4+\cdots+2^{-j})b - 2^{-j}$. Thus φ is supported in $[a,b]$. Combining this with Remark 2.3.5, we see that the support of φ is exactly the interval $[a,b]$ with $a = \min\{n\,;\, h_n \neq 0\}$ and $b = \max\{n\,;\, h_n \neq 0\}$.*

Figure 2.4.2: Corner cutting algorithm

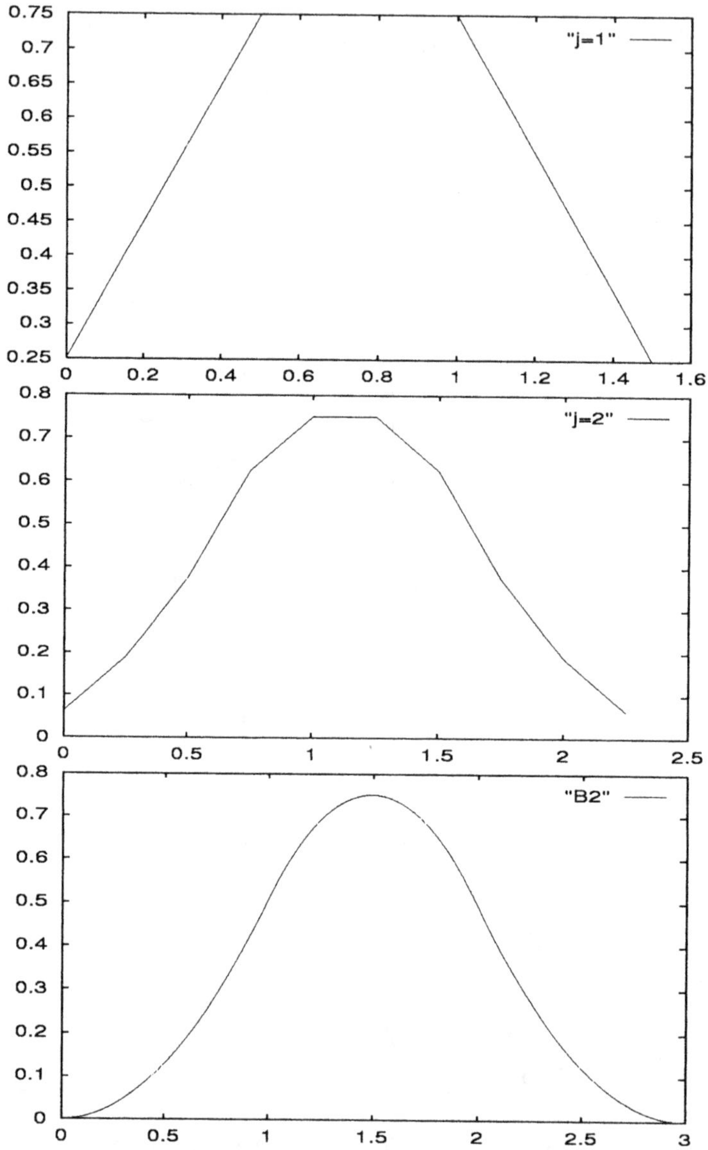

Figure 2.4.3: Convergence to the quadratic B-spline

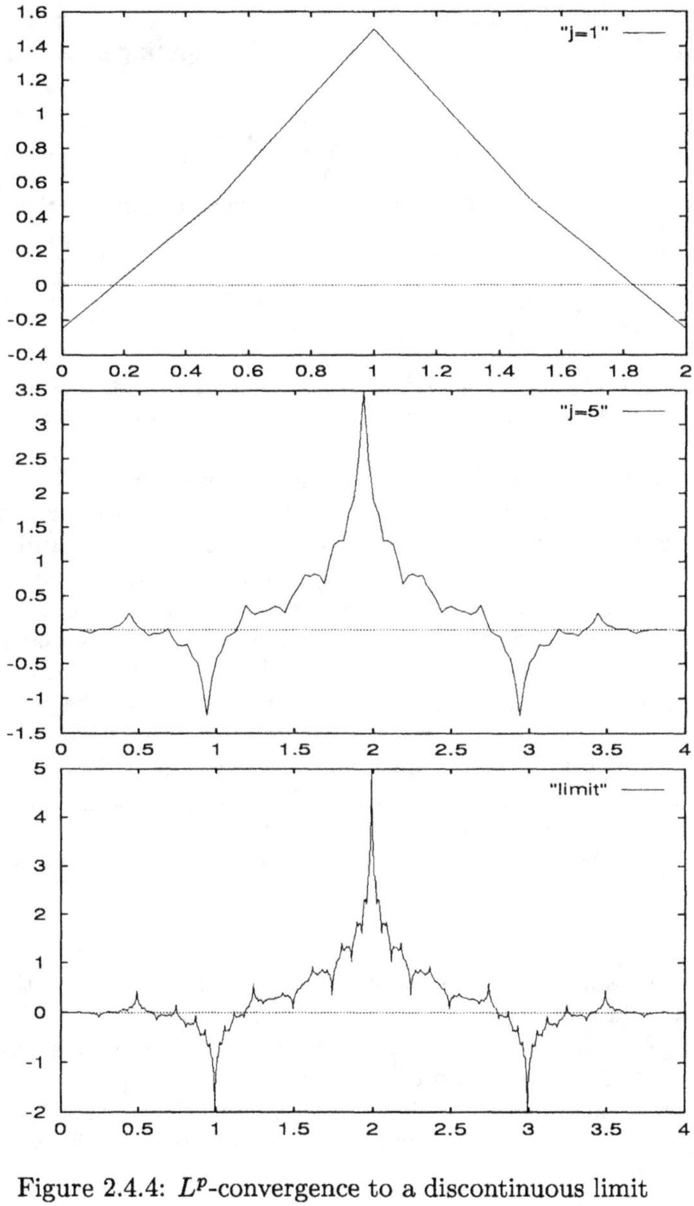

Figure 2.4.4: L^p-convergence to a discontinuous limit

2.5 Computing with refinable functions

In the practice of using the space V_j to discretize partial differential or integral equations, we might face the problem of computing several quantities. Among others we find:

1. The coefficients $\langle f, \varphi_{j,k}\rangle$ (or $\langle f, \tilde{\varphi}_{j,k}\rangle$), $k \in \mathbb{Z}$ for some function f.

2. The inner products $M_{k,l} = \langle \varphi_{j,k}, \varphi_{j,l}\rangle$, that constitute the mass matrix.

3. The elements $R_{k,l} = \langle (\partial/\partial x)^n \varphi_{j,k}, \varphi_{j,l}\rangle$ arising from the discretization of a differential problem.

4. The quantities $T_{k,l} = \int K(x,y)\varphi_{j,k}(x)\varphi_{j,l}(y)dx\,dy$ arising from the discretization of an integral operator with kernel $K(x,y)$.

In all these different situations, we may think of using the subdivision approximation to φ that we have discussed, in order to evaluate these quantities numerically: one could first compute $\varphi_{j,k}$ on a fine grid and then design quadrature rules on this grid to evaluate the above inner products. However, this approach is unpractical when φ has poor smoothness, in particular when it is not continuous: the subdivision converges too slowly to φ and too many points are then required to obtain a good approximation.

A smarter approach consists in injecting the refinement equation in the computation of these quantities. In this section, we shall explain how this strategy works in the context of the above examples.

We first consider the computation of $c_{j,k} = \int f(t)\varphi_{j,k}(t)dt$, assuming that f has C^r-smoothness for some $r > 0$. Since $\int \varphi = 1$ and $\varphi_{j,k}$ is localized near the point $2^{-j}k$, we see that a first crude estimate of $c_{j,k}$ is given by

$$c_{j,k} \simeq 2^{-j/2} f(2^{-j}k). \tag{2.5.1}$$

This estimate amounts to replacing f by a constant function on the support $S_{j,k}$ of $\varphi_{j,k}$. In order to improve on (2.5.1) we can think of replacing f on $S_{j,k}$ by a polynomial P_l of degree $l < r$. We can choose for example P_l to be the Lagrange interpolation of f at points $2^{-j}(k+n)$, $n = 0, \cdots, l$. The coefficients of this polynomial depend linearly on the values $f(2^{-j}(k+n))$, $n = 0, \cdots, l$, and the resulting estimate for $c_{j,k}$ is thus of the type

$$c_{j,k} \simeq \tilde{c}_{j,k} := 2^{-j/2}[\sum_{n=0}^{l} w_n f(2^{-j}(k+n))]. \tag{2.5.2}$$

The weights w_n are simply given by

$$w_n = \int L_n(x)\varphi(x)dx, \tag{2.5.3}$$

where L_n are the Lagrange polynomials of degree l such that $L_n(m) = \delta_{m,n}$, $m, n = 0, \cdots, l$. From the classical results on polynomial interpolation, we can then estimate the quadrature error as follows

$$\begin{aligned} |c_{j,k} - \tilde{c}_{j,k}| &\leq \|\varphi_{j,k}\|_{L^1} \sup_{x\in S_{j,k}} |f(x) - P_l(x)| \\ &\lesssim 2^{-j(r+1/2)} \|f\|_{C^r(S_{j,k})}, \end{aligned}$$

where C is independent of f and j. The computation of the weights w_n amounts to determining the moments

$$M_k = \int x^k \varphi(x)dx. \tag{2.5.4}$$

At this point, we inject the refinement equation, which leads to

$$\begin{aligned} M_k &= \sum_n h_n \int x^k \varphi(2x-n)dx \\ &= 2^{-k-1} \sum_n h_n \int (x+n)^k \varphi(x)dx \\ &= 2^{-k-1} \sum_{l\leq k} [\sum_n h_n \binom{k}{l} n^{k-l}] M_l, \end{aligned}$$

and thus

$$M_k = (1-2^{-k})^{-1} 2^{-k-1} \sum_{l<k} [\sum_n h_n \binom{k}{l} n^{k-l}] M_l. \tag{2.5.5}$$

Since we know the first value $M_0 = 1$, we can use (2.5.5) to compute the exact values of the moments by induction *without the use of an approximation of φ by a subdivision scheme.* This simple idea was used in SWELDENS and PIESSENS [1994] for the design of economical and accurate quadrature formulae.

Note that the computation of $T_{k,l} = \int K(x,y)\varphi_{j,k}(x)\varphi_{j,l}(y)dx\,dy$ can be treated in the same way when $K(x,y)$ is smooth, using the refinable structure of $\phi(x,y) = \varphi(x)\varphi(y) = \sum_{n,m\in\mathbb{Z}} h_n h_m \varphi(2x-n)\varphi(2y-m)$ to compute the multivariate moments.

Remark 2.5.1 *If one wants to improve the accuracy of the quadrature (2.5.2), several methods can be used: (i) increase the degree l of the polynomial interpolation, (ii) use another type of polynomial interpolation (e.g. at Gauss points) or (iii) apply the subdivision algorithm (2.4.3) to express the function $\varphi_{j,k}$ according to*

$$\varphi_{j,k} = 2^{j/2-J/2} \sum_n s_{J-j,n} \varphi_{J,2^{J-j}k+n}, \tag{2.5.6}$$

for some $J > j$. Injecting (2.5.6) in the inner product defining $c_{j,k}$ and using the quadrature rule (2.5.2) for the functions $\varphi_{J,2^{J-j}k+n}$, we can reduce the error estimate to

$$\begin{aligned}\varepsilon_J &= 2^{j/2-J/2}\sum_n |s_{J-j,n}|\,|c_{J,2^{J-j}k+n} - \tilde{c}_{J,2^{J-j}k+n}| \\ &\lesssim 2^{j/2-J/2}2^{-J(r+1/2)}[\sum_n |s_{J-j,n}|]\|f\|_{C^r(S_{j,k})} \\ &\lesssim 2^{j/2}2^{-Jr}[\int |\varphi(x)|(\sum_n |\tilde{\varphi}_{J-j,n}(x)|)dx]\|f\|_{C^r(S_{j,k})} \\ &\lesssim 2^{-J(r+1/2)}\|f\|_{C^r(S_{j,k})}.\end{aligned}$$

When f or K have poor regularity at some isolated regions, one can also perform these refinements up to a level J that is allowed to vary locally (see COHEN *and* EZZINE *[1996]). A similar idea is proposed in* DAHMEN, KUNOTH *and* SCHNEIDER *[1996] for the discretization of singular integral operators.*

We now turn to the computation of the mass matrix $M_{k,l} = \langle \varphi_{j,k}, \varphi_{j,l}\rangle$. Clearly, we only need to compute the values $I_k = \langle \varphi, \varphi(\cdot - k)\rangle$, $k \in \mathbb{Z}$, which are finite in number due to the compact support property. Here again, we inject the refinement relation (2.2.13), so that we obtain

$$\begin{aligned}I_k &= \sum_{m,n} h_n h_m \langle \varphi(2\cdot -m), \varphi(2\cdot -2k-n)\rangle \\ &= \tfrac{1}{2}\sum_{m,n} h_n h_m I_{2k+n-m} \\ &= \sum_n [\tfrac{1}{2}\sum_m h_{n+m-2k}h_m] I_n.\end{aligned}$$

This shows that the I_k's are the coordinates of an eigenvector of the matrix $H = (H_{k,n})$ with $H_{k,n} = \frac{1}{2}\sum_m h_{n+m-2k}h_m$ associated with the eigenvalue $\lambda = 1$. Since I_k is finite dimensional, we only need to solve this eigenvalue problem in a finite dimensional space.

We can proceed in a similar way for the computation of the inner products $R_{k,l} = \langle (\partial/\partial x)^n \varphi_{j,k}, \varphi_{j,l}\rangle$. In this case, we need to compute the values $J_k = \langle (\partial/\partial x)^n \varphi, \varphi(\cdot - k)\rangle$. Differentiating the refinement equation, we obtain

$$J_k = 2^n \sum_n [\frac{1}{2}\sum_m h_{n+m-2k}h_m] J_n, \tag{2.5.7}$$

i.e. the J_k's are the coordinates of an eigenvector associated with the eigenvalue $\lambda = 2^{-n}$ of the same matrix H as above.

In these two examples, two important questions need to be addressed: (i) is there a unique eigenvector (up to a normalization) of H associated with the eigenvalues $1, \cdots, 2^{-n}$? (ii) how should one normalize the eigenvector to obtain the quantities for which we search ?

These examples turn out to be part of a general theory developed in DAHMEN and MICCHELLI [1993], that deals with the computations of quantities of the type

$$I_{k_1,n_1,\cdots,k_d,n_d} = \int (\partial/\partial x)^{n_1} \varphi_1(x-k_1)\cdots(\partial/\partial x)^{n_d}\varphi_d(x-k_d)dx, \quad (2.5.8)$$

where $\varphi_1, \cdots, \varphi_d$ are compactly supported refinable functions.

We shall not develop this theory here, but simply summarize below its main features.

First, the computation of these quantities always amounts to an eigenvalue problem, by injecting the refinement equation as in the above examples. The existence of an eigenvector is ensured if these quantities are well defined, i.e. when $(\partial/\partial x)^{n_1}\varphi_1(x-k_1)\cdots(\partial/\partial x)^{n_d}\varphi_d(x-k_d)$ is integrable. The uniqueness of the eigenvector is the hard part of the theory: it depends crucially on certain stability properties of the scaling functions, and (in particular for multivariate refinable functions) on a smart choice of a restricted space in which the eigenvector should be searched. Finally, the normalization of the eigenvector is made possible by sum rules of the type (2.3.12), that are satisfied by the refinable functions.

In particular, in the case of the mass matrix, it can be shown that the L^2-stability of φ implies that the eigenvalue $\lambda = 1$ is simple (see e.g. COHEN and DAUBECHIES [1992]). The normalization of the corresponding eigenvector is made by (2.3.12) which implies $\sum I_k = 1$.

In the case of a stiffness matrix $J_k = \langle \varphi'', \varphi(\cdot - k)\rangle$, (2.3.12) is not sufficient since it leads to an homogeneous relation $\sum_k J_k = \int \varphi''(x)dx = 0$. One has to use an additional sum rule of the form

$$\sum_k k^2\varphi(x-k) = x^2 + ax + b, \quad (2.5.9)$$

which, as we shall see in §2.8, holds for sufficiently smooth φ. This leads to the normalization rule

$$\sum_k k^2 J_k = \int [x^2+ax+b]\varphi''(x)dx = 2\int \varphi(x)dx = 2, \quad (2.5.10)$$

using integration by part.

Remark 2.5.2 *The computation of the moments M_k can be viewed as a variation on the above theory, since $x^k = 2^{-k}(x/2)^k$ has a refinable structure but is not compactly supported. A particular case of (2.5.8) is also the computation of the values $\varphi(k)$, $k \in \mathbb{Z}$, since it corresponds to an inner product with Dirac δ functions, which, as we observed in Remark 2.2.6, also have a refinable structure. Using the refinement equation we obtain the eigenvalue problem $\varphi(k) = \sum_n h_n\varphi(2k-n)$. The normalization of the right solution is ensured by the condition $\sum_k \varphi(k) = 1$ which is imposed by (2.3.12).*

2.6 Wavelets and multiscale algorithms

Let φ and $\tilde{\varphi}$ be a pair of compactly supported dual scaling functions that satisfy equation (2.2.13) with refinement coefficients h_n and $\tilde{h}_n$, and P_j the oblic projector defined in (2.2.50).

In this basic setting, we can already perform two simple algorithms that are related to specific tasks:

1. **Projection:** Given $f_j = \sum_k c_{j,k}\varphi_{j,k} \in V_j$, we want to compute the coefficients of the approximation $P_{j-1}f_j = \sum_k c_{j-1,k}\varphi_{j-1,k}$ at the next coarser scale. To do so, we inject the refinement equation satisfied by $\tilde{\varphi}$ in the definition of $c_{j-1,k}$ as follows:

$$\begin{aligned} c_{j-1,k} &= 2^{(j-1)/2}\langle f_j, \tilde{\varphi}(2^{j-1}\cdot -k)\rangle \\ &= 2^{(j-1)/2}\langle f_j, \textstyle\sum_{n\in\mathbb{Z}} \tilde{h}_n \tilde{\varphi}(2^j\cdot -2k-n)\rangle \\ &= \tfrac{1}{\sqrt{2}}\textstyle\sum_{n\in\mathbb{Z}} \tilde{h}_n c_{j,2k+n}. \end{aligned}$$

   We thus obtain a "fine to coarse" projection algorithm expressed by the formula

$$c_{j-1,k} = \frac{1}{\sqrt{2}}\sum_{n\in\mathbb{Z}} \tilde{h}_{n-2k}c_{j,n}. \tag{2.6.1}$$

2. **Prediction:** Given the projection $P_{j-1}f_j = \sum_k c_{j-1,k}\varphi_{j-1,k}$, we want to express it in terms of the basis functions at finer scale, i.e. obtain an approximation of the real coefficients $c_{j,k}$ that is predicted from the data at resolution $j-1$. To do so, we now exploit the refinement equation of φ, which gives us

$$\begin{aligned} P_{j-1}f_j &= 2^{(j-1)/2}\textstyle\sum_{k\in\mathbb{Z}} c_{j-1,k}\varphi(2^{j-1}-k) \\ &= 2^{(j-1)/2}\textstyle\sum_{k\in\mathbb{Z}} c_{j-1,k}\sum_{n\in\mathbb{Z}} h_n\varphi(2^j\cdot -2k-n) \\ &= \tfrac{1}{\sqrt{2}}\textstyle\sum_{k\in\mathbb{Z}}[\sum_{n\in\mathbb{Z}} h_{k-2n}c_{j-1,n}]\varphi_{j,k}. \end{aligned}$$

   We thus obtain a "coarse to fine" prediction algorithm expressed by the subdivision formula that we already encountered in §2.4:

$$\hat{c}_{j,k} = \frac{1}{\sqrt{2}}\sum_{n\in\mathbb{Z}} h_{k-2n}c_{j-1,n}. \tag{2.6.2}$$

   (similar to (2.4.3) up to a renormalization) where $\hat{c}_{j,k}$ is the predicted coefficient associated with the function $\varphi_{j,k}$.

If we now consider the multiscale decomposition of a function $f_{j_1} \in V_{j_1}$ into

$$f_{j_1} = f_{j_0} + \sum_{j=j_0}^{j_1-1} g_j, \tag{2.6.3}$$

with $f_j = P_j f_{j_1}$ and $g_j = f_{j+1} - f_j$, we see that the details g_j can be expanded into

$$g_j = \sum_{k \in \mathbb{Z}} (c_{j+1,k} - \hat{c}_{j+1,k}) \varphi_{j+1,k}. \tag{2.6.4}$$

A first representation of $f_{j_1} = \sum_k c_{j_1,k} \varphi_{j_1,k}$ is thus provided by the coarse approximation coefficients $c_{j_0,k}$ and the coefficients $c_{j+1,k} - \hat{c}_{j+1,k}$, $j = j_0, \cdots, j_1 - 1$ which represent the error of prediction. The above algorithms allow us to compute this multiscale representation from fine to coarse scales and to reconstruct the original coefficients $c_{j_1,k}$ by a coarse to fine procedure.

However, this representation cannot be viewed as a change of basis: it introduces redundancy, since the use of the basis $\varphi_{j+1,k}$, $k \in \mathbb{Z}$, to represent the fluctuation does not exploit the fact that these details are in the kernel of P_j. In order to remove this redundancy, we need to introduce a wavelet basis which can represent these fluctuations in a natural way, as in the case of the Haar basis or the hierarchical Schauder basis that we explored in Chapter 1. We shall see that such wavelets can be obtained in a systematic way, once we are given a pair of dual scaling functions. Thus, the main theoretical difficulties in the construction of wavelet bases mostly reside in the selection of an appropriate pair of dual scaling functions.

Recalling the refinement equations satisfied by φ and $\tilde{\varphi}$

$$\varphi(x) = \sum_n h_n \varphi(2x - n) \text{ and } \tilde{\varphi}(x) = \sum_n \tilde{h}_n \tilde{\varphi}(2x - n), \tag{2.6.5}$$

we define new coefficients

$$g_n = (-1)^n \tilde{h}_{1-n} \text{ and } \tilde{g}_n = (-1)^n h_{1-n}. \tag{2.6.6}$$

We then define a pair of dual wavelets by

$$\psi(x) = \sum_n g_n \varphi(2x - n) \text{ and } \tilde{\psi}(x) = \sum_n \tilde{g}_n \tilde{\varphi}(2x - n). \tag{2.6.7}$$

Clearly $\psi \in V_1$ and $\tilde{\psi} \in \tilde{V}_1$. Before showing how these functions can be used to characterize the fluctuations, we shall first prove some preliminary results on the coefficients h_n, $\tilde{h}_n$, g_n and $\tilde{g}_n$.

Lemma 2.6.1 *The coefficients satisfy the identities*

$$\sum_n h_n \tilde{h}_{n+2k} = \sum_n g_n \tilde{g}_{n+2k} = 2\delta_{0,k}, \tag{2.6.8}$$

and

$$\sum_n h_n \tilde{g}_{n+2k} = \sum_n g_n \tilde{h}_{n+2k} = 0. \tag{2.6.9}$$

Proof Injecting the refinement equations (2.6.5) in the duality relation, we find

$$\begin{aligned}\delta_{0,k} &= \langle\tilde{\varphi},\varphi(\cdot-k)\rangle\\ &= \textstyle\sum_{m,n}\tilde{h}_m h_n\langle\tilde{\varphi}(2\cdot-m),\varphi(2\cdot-2k-n)\rangle\\ &= \tfrac{1}{2}\textstyle\sum_{m,n}\tilde{h}_m h_n\langle\tilde{\varphi}(\cdot),\varphi(\cdot-2k-n+m)\rangle\\ &= \tfrac{1}{2}\textstyle\sum_n h_n\tilde{h}_{n+2k}.\end{aligned}$$

The second identity is immediate, since we have

$$\sum_n g_n\tilde{g}_{n+2k}=\sum_n h_{1-n}\tilde{h}_{1-n-2k}=\sum_m h_m\tilde{h}_{m-2k}=2\delta_{0,k}. \qquad (2.6.10)$$

Finally, the last identities are the consequences of the oscillation factor $(-1)^n$ in the definition of g_n and $\tilde{g}_n$. More precisely, we have

$$\begin{aligned}\textstyle\sum_n h_n\tilde{g}_{n+2k} \quad & \textstyle\sum_n(-1)^n h_n h_{1-n-2k}\\ & \textstyle\sum_n h_{2n}h_{1-2n-2k}-\sum_n h_{2n+1}h_{-2n-2k}\\ & \textstyle\sum_n h_{2n}h_{1-2n-2k}-\sum_n h_{-2m-2k+1}h_{2m}=0,\end{aligned}$$

and similarly for $\sum_n g_n\tilde{h}_{n+2k}$. ◇

Remark 2.6.1 *The above lemma can be expressed in a matrix formulation: defining the bi-infinite banded matrix M and $\tilde{M}$ by*

$$M_{2i,j}=h_{j-2i}\ ,\ M_{2i+1,j}=g_{j-2i}\ ,\ \tilde{M}_{i,2j}=\tilde{h}_{i-2j}\ ,\ \tilde{M}_{i,2j+1}=\tilde{g}_{i-2j},$$

we see that (2.6.8) and (2.6.9) can be re-expressed as $M\tilde{M}=\tilde{M}M=2I$, i.e. $\tilde{M}$ is twice the inverse of M. Note that the matrices M and $\tilde{M}$ have a structure close to Toeplitz, with entries alternating row by row for $\tilde{M}$ and column by column for M. In contrast, it is well known that a purely bi-infinite Toeplitz matrix, i.e. of convolution type, cannot be banded with banded inverse, except when it is a multiple of I.

We are now ready to prove the main results of this section. We are interested in characterizing the details $Q_jf=P_{j+1}f-P_jf$. Remark that the operator Q_j satisfies

$$Q_j^2=P_{j+1}^2-P_{j+1}P_j-P_jP_{j+1}+P_j^2=Q_j, \qquad (2.6.11)$$

(using the property $P_jP_{j+1}=P_j$ that we observed at the end of §2.2). Thus Q_j is also a projector on a space W_j which complements V_j in V_{j+1}. The space W_j can also be defined as the kernel of P_j in V_{j+1}.

Theorem 2.6.1 *The functions ψ and $\tilde{\psi}$ satisfy*

$$\langle \psi, \tilde{\psi}(\cdot - k)\rangle = \delta_{0,k}, \tag{2.6.12}$$

and

$$\langle \psi, \tilde{\varphi}(\cdot - k)\rangle = \langle \tilde{\psi}, \varphi(\cdot - k)\rangle = 0. \tag{2.6.13}$$

The projector Q_j can be expanded into

$$Q_j f = \sum_{k\in\mathbb{Z}} \langle f, \tilde{\psi}_{j,k}\rangle \psi_{j,k}, \tag{2.6.14}$$

and the functions $\psi_{j,k}$, $k \in \mathbb{Z}$ constitute a Riesz basis of the complement space W_j.

Proof We simply use the definition of ψ and $\tilde{\psi}$ to expand the first inner product according to

$$\begin{aligned} \langle \psi, \tilde{\psi}(\cdot - k)\rangle &= \textstyle\sum_{m,n} g_m \tilde{g}_n \langle \varphi(2\cdot - m), \tilde{\varphi}(2\cdot - 2k - n)\rangle \\ &= \tfrac{1}{2}\textstyle\sum_{m,n} g_m \tilde{g}_n \langle \varphi, \tilde{\varphi}(\cdot - 2k - n + m)\rangle \\ &= \tfrac{1}{2}\textstyle\sum_n \tilde{g}_{n+2k} g_n = \delta_{0,k}, \end{aligned}$$

by (2.6.8). In a similar way, we use the refinement equation to obtain

$$\langle \psi, \tilde{\varphi}(\cdot - k)\rangle = \sum_n g_{n+2k} \tilde{h}_n = 0, \tag{2.6.15}$$

and

$$\langle \tilde{\psi}, \varphi(\cdot - k)\rangle = \sum_n \tilde{g}_{n+2k} h_n = 0, \tag{2.6.16}$$

by (2.6.9). It is sufficient to prove (2.6.14) for $j = 0$ since the general case follows by scaling.

We already see from (2.6.12) that ψ is L^2-stable, using Theorem 2.3.1, and that $R_0 f = \sum_{k\in\mathbb{Z}} \langle f, \tilde{\psi}_{0,k}\rangle \psi_{0,k}$, is a bounded projector onto the space $S_0 \subset V_1$ spanned by these functions. From (2.6.13) we also see S_0 is contained in W_0 (since it is the kernel of P_0 in V_1). It remains to prove that S_0 coincides with W_0.

Let $f \in L^2$ and $P_1 f = \sum_{k\in\mathbb{Z}} c_{1,k}\varphi_{1,k} \in V_1$. We can compute the coefficients $d_{0,k}$ of $\psi(\cdot - k)$ in $R_0 f$ from $P_1 f$ by

$$d_{0,k} = \langle f, \tilde{\psi}_{0,k}\rangle = \frac{1}{\sqrt{2}} \sum_n \tilde{g}_n \langle f, \tilde{\varphi}_{1,2k+n}\rangle \frac{1}{\sqrt{2}} \sum_n \tilde{g}_{n-2k} c_{1,n}. \tag{2.6.17}$$

From (2.6.17) and (2.6.1), we see that the bi-infinite vector X with coordinates $x_{2k} = c_{0,k}$ and $x_{2k+1} = d_{0,k+1}$ is obtained from Y with coordinates $y_k = c_{1,k}$ by the application of $\tilde{M}$, i.e.

$$X = \frac{1}{\sqrt{2}}\tilde{M}Y. \tag{2.6.18}$$

We can then develop $R_0f + P_0f$ in the basis $\varphi_{1,k}$, according to

$$\begin{aligned} R_0f + P_0f &= \sum_k c_{0,k}\varphi_{0,k} + \sum_k d_{0,k}\psi_{0,k} \\ &= \tfrac{1}{\sqrt{2}}\sum_k c_{0,k}[\sum_n h_n\varphi_{1,2k+n}] + \sum_k d_{0,k}[\sum_n g_n\varphi_{1,2k+n}] \\ &= \tfrac{1}{\sqrt{2}}\sum_k[\sum_n h_{k-2n}c_{0,n} + \sum_n g_{k-2n}d_{0,n}]\varphi_{1,k}. \end{aligned}$$

We see that the associated vector $Z = (\sum_n h_{k-2n}c_{0,n} + \sum_n g_{k-2n}d_{0,n})_{k\in\mathbb{Z}}$, satisfies

$$Z = \frac{1}{\sqrt{2}}MX = \frac{1}{2}M\tilde{M}Y = Y, \tag{2.6.19}$$

i.e. $P_0 + R_0 = P_1$, and thus $R_0 = Q_0$ and $S_0 = W_0$. ◇

Remark 2.6.2 *We clearly have similar results concerning the dual scaling function: $\tilde{\psi}_{j,k}$, $k \in \mathbb{Z}$, is a Riesz basis of the complement space $\widetilde{W}_j$ defined as the kernel of P_j^* in $\tilde{V}_{j+1}$, and the projection onto this space can be computed by the inner products against the functions $\psi_{j,k}$, $k \in \mathbb{Z}$*

The dual wavelets provide a multiscale decomposition of V_{j_1} into the (non-orthogonal) sum $V_{j_0} \oplus W_{j_0} \oplus \cdots \oplus W_{j_1-1}$: any function $f_{j_1} \in V_{j_1}$ can be expanded into

$$f_{j_1} = \sum_{k\in\mathbb{Z}} c_{j_1,k}\varphi_{j_1,k} = \sum_{k\in\mathbb{Z}} c_{j_0,k}\varphi_{j_0,k} + \sum_{j=j_0}^{j_1-1}\sum_{k\in\mathbb{Z}} d_{j,k}\psi_{j,k}, \tag{2.6.20}$$

with $d_{j,k} = \langle f_{j_1}, \tilde{\psi}_{j,k}\rangle$ and $c_{j,k} = \langle f_{j_1}, \tilde{\varphi}_{j,k}\rangle$.

From (2.6.13), we see that $\widetilde{W}_j$ is orthogonal to V_j and thus to all V_l, $l \le j$, and W_l, $l < j$. By symmetry, we obtain that W_j is orthogonal to $\tilde{V}_l$ if $l \le j$, and that the spaces W_j and $\widetilde{W}_l$ are orthogonal for all $l \ne j$. In particular, we have the biorthogonality relations

$$\langle \varphi_{j_0,k}, \tilde{\psi}_{j,m}\rangle = \langle \tilde{\varphi}_{j_0,k}, \psi_{j,m}\rangle = 0, \;\; j \ge j_0 \text{ and } \langle \psi_{j,k}, \tilde{\psi}_{l,n}\rangle = \delta_{j,l}\delta_{k,n}.$$

These relations can be used to show that $\{\varphi_{j_0,k}\}_{k\in\mathbb{Z}} \cup \{\psi_{j,k}\}_{j_0\le j<j_1,k\in\mathbb{Z}}$ is a Riesz basis of V_{j_1}: if a function f_{j_1} has the form (2.6.20), we have on the one hand

$$\begin{aligned} \|f_{j_1}\|_{L^2}^2 &\le (j_1 - j_0)[\|\sum_{k\in\mathbb{Z}} c_{j_0,k}\varphi_{j_0,k}\|_{L^2}^2 + \sum_{j=j_0}^{j_1-1}\|\sum_{k\in\mathbb{Z}} d_{j,k}\psi_{j,k}\|_{L^2}^2] \\ &\lesssim (j_1 - j_0)[\sum_{k\in\mathbb{Z}}|c_{j_0,k}|^2 + \sum_{j=j_0}^{j_1-1}\sum_{k\in\mathbb{Z}}|d_{j,k}|^2], \end{aligned}$$

from the stability at each level. On the other hand, we use the biorthogonality relations to derive

$$\begin{aligned}\sum_{k\in\mathbb{Z}}|c_{j_0,k}|^2+\sum_{j=j_0}^{j_1-1}\sum_{k\in\mathbb{Z}}|d_{j,k}|^2 &= \sum_{k\in\mathbb{Z}}|\langle f,\tilde{\varphi}_{j_0,k}\rangle|^2\\ &\quad+\sum_{j=j_0}^{j_1-1}\sum_{k\in\mathbb{Z}}|\langle f,\tilde{\psi}_{j,k}|^2\\ &\lesssim [\|P_{j_0}f\|_{L^2}^2+\sum_{j=j_0}^{j_1-1}\|Q_jf\|_{L^2}^2]\\ &\lesssim (j_1-j_0)\|f_{j_1}\|_{L^2}^2.\end{aligned}$$

If we use Theorem 2.3.3, we derive that any function $f \in L^2$ has a converging expansion of the form

$$f=\lim_{j_1\to+\infty}[\sum_{k\in\mathbb{Z}}c_{j_0,k}\varphi_{j_0,k}+\sum_{j=j_0}^{j_1-1}\sum_{k\in\mathbb{Z}}d_{j,k}\psi_{j,k}], \tag{2.6.21}$$

with $d_{j,k}=\langle f_{j_1},\tilde{\psi}_{j,k}\rangle$ and $c_{j,k}=\langle f_{j_1},\tilde{\varphi}_{j,k}\rangle$.

However, we cannot conclude from the stability of the decomposition in V_{j_1} that $\{\varphi_{j_0,k}\}_{k\in\mathbb{Z}}\cup\{\psi_{j,k}\}_{j\geq j_0,k\in\mathbb{Z}}$ and $\{\tilde{\varphi}_{j_0,k}\}_{k\in\mathbb{Z}}\cup\{\tilde{\psi}_{j,k}\}_{j\geq j_0,k\in\mathbb{Z}}$ are biorthogonal Riesz bases of $L^2(\mathbb{R})$: this would mean that we can control the stability constants independently of j_1, which is not the case here since our (crude) estimate involves (j_1-j_0) in the constant. Note that a particular instance where this global stability always holds is the case of *orthonormal wavelet bases*, i.e. when $\tilde{\varphi}=\varphi$ and $\tilde{\psi}=\psi$. The global stability property for biorthogonal decompositions of L^2 will be proved under mild assumptions on the functions φ and $\tilde{\varphi}$ in §3.8 of Chapter 3.

We already described a two-level decomposition and reconstruction procedure (between level 0 and 1) in the proof of Theorem 2.6.1. The multilevel procedure operates in the same way as for the particular cases that we studied in Chapter 1: on the one hand, one *decomposes* from fine to coarse scale, using the formulae

$$c_{j,k}=\frac{1}{\sqrt{2}}\sum_n\tilde{h}_{n-2k}c_{j+1,n}\text{ and }d_{j,k}=\frac{1}{\sqrt{2}}\sum_n\tilde{g}_{n-2k}c_{j+1,n}, \tag{2.6.22}$$

i.e. applying the matrix $\frac{1}{\sqrt{2}}\tilde{M}$ to the fine scale approximation coefficients to obtain the intertwined sequences of coarser approximation and detail coefficients. On the other hand, one *reconstructs* from coarse to fine scale, using the formula

$$c_{j+1,k}=\frac{1}{\sqrt{2}}[\sum_n h_{k-2n}c_{j,n}+\sum_n g_{k-2n}d_{j,n}], \tag{2.6.23}$$

i.e. applying the inverse matrix $\frac{1}{\sqrt{2}}M$.

As an example, we illustrate the pyramidal structure of these algorithms on Figure 2.6.1, in the case where the support of h_n is $[-1,1]$ and the support of $\tilde{h}_n$ is $[-3,3]$. The full lines correspond to the coefficients $\tilde{h}_n$ (in the decomposition) and h_n (in the reconstruction), while the dashed lines correspond to the coefficients $\tilde{g}_n$ (in the decomposition) and g_n (in the reconstruction). Note that these algorithms are again of complexity $\mathcal{O}(N)$, provided that we know how to deal in a proper way with functions that are defined on a bounded interval such as $[0,1]$. A simple solution is to extend these functions out of $[0,1]$ by periodization: this amounts to using the 1-periodizations ($f \mapsto f_{per} = \sum_k f(\cdot+k)$) of the scaling functions and wavelets at resolution levels $j \geq 0$. The terms $Q_j f$ and $P_j f$ are then characterized by 2^j samples, that can be computed by a straightforward adaptation of (2.6.22) and (2.6.23), However, this solution has the disadvantage of creating an "artificial jump" of f_{per} at 0 and 1 even when f is regular near these points. Better strategies will be discussed in §2.12.

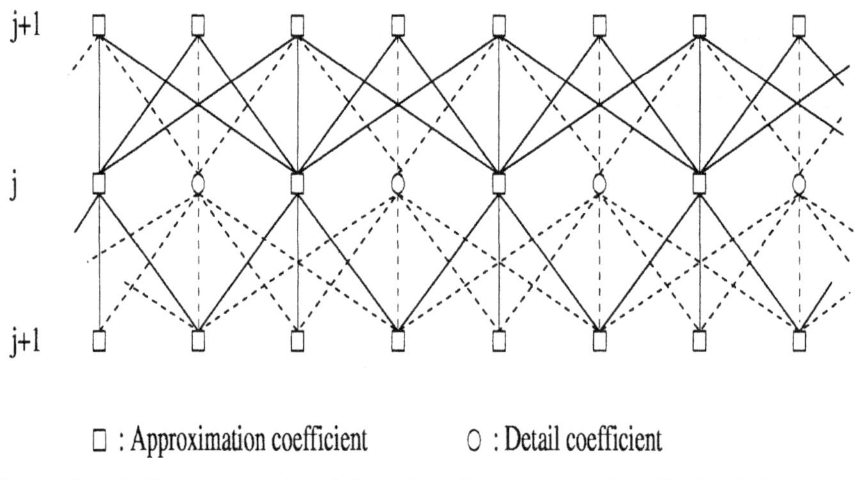

Figure 2.6.1: Decomposition (above) and reconstruction (below) algorithms

We remark that (2.6.22) amounts to a *convolution* of the fine scale approximation coefficients by the discrete sequence $\tilde{h}_{-n}$ and $\tilde{g}_{-n}$ (i.e. an autocorrelation by $\tilde{h}_n$ and $\tilde{g}_n$) followed by a *decimation* of one sample out of two in the result. Similarly, (2.6.23) amounts to convolutions by the discrete sequences h_n and g_n preceded by the insertion of zero values between each sample.

Such an operation is well-known in signal processing as a *subband coding scheme* based on a *two channel filter bank*. Generally speaking, subband coding aims to split a discrete signal $s(n)$, $n \in \mathbb{Z}$ into M components

$s_1(n), \cdots, s_M(n)$, each of them representing a certain range of the frequency content of $s(n)$: the s_i are obtained by filtering the initial signals with M filters having complementary frequency characteristics, and by downsampling the resulting signals by a factor of M in order to preserve the total amount of information. Ideally, if the filtering process amounts to a sharp frequency cut-off, i.e. s_i is obtained by preserving the frequencies $|\omega| \in [\frac{i-1}{M}\pi, \frac{i}{M}\pi]$, in the Fourier series $S(\omega) = \sum s_n e^{in\omega}$, then Shannon's sampling theorem states that s_i is perfectly characterized by retaining one sample out of M. Since in practice such ideal filters are not implementable (due to the slow decay of their impulse response), one approximates them by finite impulse response filters, i.e. convolutions by finite sequences $(f_{1,n})_{n\in\mathbb{Z}}, \cdots, (f_{M,n})_{n\in\mathbb{Z}}$. In order to reconstruct the original signal, the decimated signals are upsampled by inserting $M-1$ zero values between each sample, and filtered by finite sequences $(g_{1,n})_{n\in\mathbb{Z}}, \cdots, (g_{M,n})_{n\in\mathbb{Z}}$. This ensemble of operations is represented by the so-called "block diagram" in Figure 2.6.2: the downpointing and uppointing arrows respectively stand for downsampling and upsampling operations. In order to ensure prefect reconstruction of the signal, particular properties need to be satisfied by the decomposition filters $f_{i,n}$, and reconstruction filters $g_{i,n}$. In the case of a two-channel filter bank, i.e. $M = 2$, these properties are exactly those which are stated in Lemma 2.6.1, i.e. (2.6.8) and (2.6.9).

A general treatment of filter banks and their relation to wavelet bases can be found in the books by VAIDYANATHAN [1992], KOVACEVIC and VETTERLI [1995] and NGUYEN and STRANG [1996].

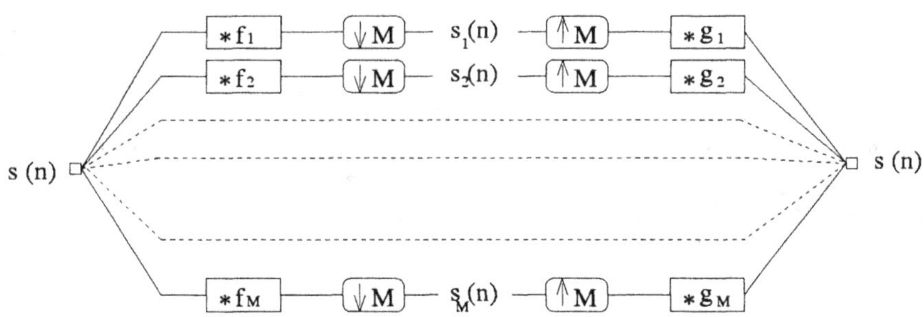

Figure 2.6.2: Subband coding scheme

In our case of interest which is displayed in Figure 2.6.3, we notice that the filters h_n and $\tilde{h}_n$ have low-pass frequency characteristics: according to the sum-rules (2.3.18), their Fourier series $m(\omega)$ and $\tilde{m}(\omega)$ (defined by (2.2.39)) satisfy

$$m(0) = \tilde{m}(0) = 1 \text{ and } m(\pi) = \tilde{m}(\pi) = 0. \quad (2.6.24)$$

In contrast, the filters g_n and $\tilde{g}_n$ have "mirror" high pass characteristics since by their definition, they satisfy

$$\frac{1}{2}\sum_n g_n e^{-in\omega} = e^{-i\omega}\overline{\tilde{m}(\omega+\pi)} \text{ and } \frac{1}{2}\sum_n \tilde{g}_n e^{-in\omega} = e^{-i\omega}\overline{m(\omega+\pi)}.$$

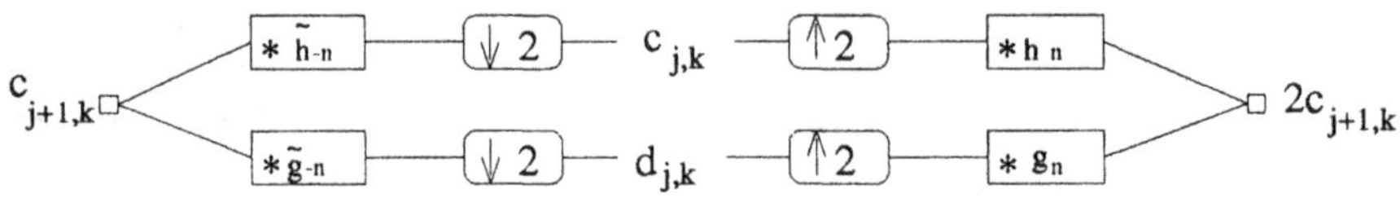

Figure 2.6.3: Block diagram for the wavelet two-scale algorithm (without normalization of filters by $1/\sqrt{2}$)

Note that the oscillation of the sequences g_n and $\tilde{g}_n$, expressed by $\sum g_n = \sum \tilde{g}_n = 0$ implies that ψ and $\tilde{\psi}$ have vanishing integral.

Remark 2.6.3 *As we already noted in Remark 2.2.6, interpolatory scaling functions can be viewed as a "degenerate" case of dual pair with $\tilde{\varphi} = \delta$ the Dirac distribution centered at the origin. In that case the same derivation of wavelets and multiscale algorithm can be made with $\tilde{h}_n = 2\delta_{0,n}$. As in the particular case of the Schauder hierarchical basis studied in §1.3 of Chapter 1, we obtain a wavelet ψ which is simply given by*

$$\psi(x) = \varphi(2x-1), \tag{2.6.25}$$

and a dual wavelet $\tilde{\psi}$ which is a combination of Dirac distributions. The application of $\tilde{\psi}_{j,k}$ corresponds to the computation of the wavelet coefficient $d_{j,k}$ as the difference between the exact value of f at $2^{-j}(k+1/2)$ and its predicted value which is obtained by interpolation of the neighbouring values on the coarser grid $2^{-j}\mathbb{Z}$, using the coefficients h_n.

As a conclusion to this section, it is important to note that the decomposition and reconstruction algorithms that we described only involve the the sequences h_n, g_n, $\tilde{h}_n$ and $\tilde{g}_n$ as discrete filters and are based on their particular algebraic properties described in Lemma 2.6.1. These algorithms are purely discrete transformations, and they do not directly involve the underlying functions φ, $\tilde{\varphi}$, ψ and $\tilde{\psi}$, if these exist. In particular, one can apply these transformations on the discretization of a function which was not obtained by approximation in the corresponding space V_j.

2.7 Smoothness analysis

The goal of this section, as well as the next two sections, is to show how to "guess" from the coefficients h_n, some important properties of the function φ that solves the refinement equation (2.2.13).

Our starting point is thus a finite set of coefficients h_n and $\tilde{h}_n$ that satisfy the sum rules (2.3.18), and the corresponding refinement equations with unknowns φ and $\tilde{\varphi}$. A first remark is that we can use the Fourier transform in order to obtain a more explicit expression of φ: assuming that φ is an L^1 solution of (2.2.13) with the normalization $\int \varphi = 1$, we can iterate the equation (2.2.40) as follows

$$\begin{aligned}\hat{\varphi}(\omega) &= m(\omega/2)\hat{\varphi}(\omega/2)\\ &= m(\omega/2)m(\omega/4)\hat{\varphi}(\omega/4)\\ &= \cdots = \hat{\varphi}(2^{-j}\omega)\prod_{k=1}^{j} m(2^{-k}\omega).\end{aligned}$$

Letting j tend to $+\infty$, and using the normalization $\hat{\varphi}(0) = 1$, we obtain an explicit expression

$$\hat{\varphi}(\omega) = \prod_{k>0} m(2^{-k}\omega). \tag{2.7.1}$$

The convergence of the infinite product in (2.7.1) is uniform on every compact set, since $m(\omega)$ is a trigonometric polynomial (and thus a C^∞ function), so that we have $1 - m(2^{-k}\omega) = \mathcal{O}(2^{-k}\omega)$. It is also not difficult to check that uniform convergence on every compact set also holds for all derivatives.

Conversely, we remark that if $m(\omega)$ is a trigonometric polynomial such that $m(0) = 1$, the infinite product (2.7.1) defines a tempered distribution: $\hat{\varphi}(\omega)$ defined by (2.7.1) is clearly bounded on any compact set, and for any $j > 0$ and $2^j \le |\omega| < 2^{j+1}$, we have

$$|\hat{\varphi}(\omega)| = |\hat{\varphi}(2^{-j}\omega)\prod_{k=1}^{j} |m(2^{-k}\omega)| \le \sup_{1<|\omega|<2} |\hat{\varphi}|[\sup|m|]^j, \tag{2.7.2}$$

and therefore the estimate

$$|\hat{\varphi}(\omega)| \le C(1+|\omega|)^r, \tag{2.7.3}$$

with $C = \sup_{|\omega|<2}|\hat{\varphi}|$ and $r = \log_2(\sup|m|)$, where we use the notation $\log_2(x) = \log(x)/\log(2)$. By (2.2.40), φ is a solution of the refinement equation in the sense of tempered distributions. We can also write

$$\varphi = (*)_{k=1}^{+\infty}[2\sum_n h_n\delta_{2^{-k}n}], \tag{2.7.4}$$

which shows that φ is compactly supported in $[N_1, N_2]$ if $h_n = 0$ when $n \notin [N_1, N_2]$, a fact which was already established in §2.3 and §2.4, in the setting of dual scaling functions.

Note that the estimate (2.7.3) does not a-priori yield any decay estimate in the Fourier domain, since $\sup_\omega |m(\omega)| \geq m(0) = 1$, i.e. $r \geq 0$. Such a decay estimate is nevertheless crucial in order to ensure that φ has some regularity. We shall now see that a more refined analysis allows us to estimate the smoothness of φ from the properties of the coefficients h_n, or equivalently of the function $m(\omega)$.

In order to measure the smoothness of φ through the properties of its Fourier transform, we use the following definition

Definition 2.7.1 *For $1 \leq q \leq \infty$, the L^q-Fourier smoothness exponent of φ is the quantity*

$$s_q(\varphi) = \sup\{s \; ; \; (1 + |\cdot|)^s \hat{\varphi} \in L^q\}. \tag{2.7.5}$$

Note that for $q = 2$ this is exactly the classical Sobolev exponent. Another classical indicator of smoothness is the Hölder exponent $\mu(\varphi)$ defined as follows: if φ is in C^n but not in C^{n+1}, $n \in \mathbb{N}$, them $\mu(\varphi) = n + \nu(\varphi)$ where $\nu(\varphi)$ is the supremum of $\alpha \in]0, 1[$ such that $\varphi^{(n)} \in C^\alpha$, i.e.

$$\nu(\varphi) = \inf_x \left(\lim_{|t| \to 0} \inf \frac{\log |\varphi^{(n)}(x+t) - \varphi^{(n)}(x)|}{\log |t|} \right). \tag{2.7.6}$$

Now if $(1 + |\cdot|)^s \hat{\varphi} \in L^1$, for some $s \in]n, n+1[$, we have

$$\begin{aligned} |\varphi^{(n)}(x) - \varphi^{(n)}(x+t)| &\leq \int |\omega|^n |(1 - e^{it\omega}) \hat{\varphi}(\omega)| d\omega \\ &\leq 2 \int_{|t\omega| \geq 1} |\omega|^n |\hat{\varphi}(\omega)| d\omega + \int_{|t\omega| \leq 1} |\omega|^{n+1} |\hat{\varphi}(\omega)| d\omega \\ &\lesssim |t|^{s-n}, \end{aligned}$$

which shows that $\mu(\varphi) \geq s_1(\varphi)$. It is also not difficult to show that the exact equality $\mu(\varphi) = s_1(\varphi)$ holds when $\hat{\varphi} \geq 0$.

By Hölder's inequality, one can derive

$$s_q(\varphi) \geq s_p(\varphi) - 1/q + 1/p \text{ if } q \leq p, \tag{2.7.7}$$

which shows that $\mu(\varphi) \geq s_1(\varphi) \geq s_\infty - 1$. When φ has compact support, we also have an estimate from above:

$$s_\infty(\varphi) - 1 \leq \mu(\varphi) \leq s_\infty(\varphi). \tag{2.7.8}$$

To check the upper bound, simply remark that if $\varphi \in C^\mu$ with $\mu \in]0, 1[$, then

$$|\hat{\varphi}(\omega)(1 - e^{i\omega t})| \leq \int |\varphi(x) - \varphi(x+t)| dx \lesssim |t|^\mu. \tag{2.7.9}$$

By taking $\omega = \pi/t$, we see that $|\hat{\varphi}(\omega)| \lesssim |\omega|^{-\mu}$. For $\mu > 1$, we obtain the same result using $|\mathcal{F}\varphi^{(n)}(\omega)| = |\omega|^n|\hat{\varphi}(\omega)|$, where $\mathcal{F}$ is the Fourier transform operator. Thus we can also estimate the Hölder (and the Sobolev) exponent of φ from the rate of decay of its Fourier transform at infinity.

In order to analyze the decay at infinity of the product (2.7.1), we remark that this product decays like $|\omega|^{-1}$ in the particular case $m(\omega) = \frac{1+e^{-i\omega}}{2}$ since it corresponds to $\varphi = \chi_{[0,1]}$. More precisely, we have

$$|\prod_{k>0}\Big(\frac{1+e^{-i2^{-k}\omega}}{2}\Big)| = |\hat{\varphi}(\omega)| = |\frac{1-e^{i\omega}}{i\omega}| \lesssim (1+|\omega|)^{-1}, \qquad (2.7.10)$$

which can also be directly derived from the classical formula

$$\prod_{k\geq 0}\cos(2^{-k}\omega) = \frac{\sin(\omega)}{\omega}. \qquad (2.7.11)$$

We can thus hope that $\hat{\varphi}$ has some decay at infinity if $m(\omega)$ has the factorized form

$$m(\omega) = \Big(\frac{1+e^{-i\omega}}{2}\Big)^L p(\omega), \qquad (2.7.12)$$

where $p(\omega)$ is also a trigonometric polynomial. Note that the sum rules (2.3.18) imply that such a factorization should hold at least for $L = 1$. This factorization also means that $\varphi = B_{L-1} * D(\omega)$, where B_{L-1} is the B-spline of degree $L-1$ and D is a distribution defined by

$$\hat{D}(\omega) = \prod_{k>0} p(2^{-k}\omega). \qquad (2.7.13)$$

Applying the estimate (2.7.3) on $\hat{D}$, we thus obtain

$$|\hat{\varphi}(\omega)| = |\hat{B}_{L-1}(\omega)\hat{D}(\omega)| \leq C(1+|\omega|)^{-L+b_1}, \qquad (2.7.14)$$

where $b_1 = \log_2(\sup|p|)$, i.e. a first estimate

$$s_\infty(\varphi) \leq L - b_1. \qquad (2.7.15)$$

It is possible to improve on (2.7.15): if $2^j \leq |\omega| \leq 2^{j+1}$, we can generalize the estimate (2.7.2) by grouping the factors in blocks of size $n > 0$ which leads to

$$|\hat{D}(\omega)| \leq \sup_{|\omega|\leq 1}|\hat{P}|[\sup|p|]^n[\sup|p(\omega)\cdots p(2^{n-1}\omega)|]^{j/n} \lesssim (1+|\omega|)^{b_n},$$

with

$$b_n = \frac{\log_2[\sup_\omega |\prod_{k=0}^{n-1} p(2^k\omega)|]}{n}. \quad (2.7.16)$$

We thus obtain

$$s_\infty(\varphi) \geq L - b_n, \quad (2.7.17)$$

which is clearly an improvement on (2.7.15) since we have $b_n \leq b_1$, with a strict inequality in general.

Remark 2.7.1 *As n goes to $+\infty$, it can be shown (see* COHEN AND RYAN *[1995, chapter 3]) that the sequence b_n converges to $b_\infty = \inf_{n>0} b_n$, and that moreover, when φ is L^2-stable one has the exact equality*

$$s_\infty(\varphi) = L - b_\infty. \quad (2.7.18)$$

Consequently, (2.7.17) tends to become an exact estimate as n grows. However, evaluating the maximum of $|\prod_{k=0}^{n-1} p(2^k\omega)|$ usually becomes a difficult task for large values of n. In order to approach b_∞ faster than with the sequence b_n, it often pays to introduce the decreasing sequence

$$d_n = \log_2[\sup_\omega \min_{j=0,\cdots,n-1} |\prod_{k=0}^{j} p(2^k\omega)|^{1/j}]. \quad (2.7.19)$$

Clearly $d_n \leq b_n$. Moreover, we also have $b_m \leq d_n + \frac{n}{m} b_1$: for a fixed ω, we can group the factors in the product $\prod_{k=0}^{m-1} p(2^k\omega)$ into blocks of various size $1 \leq s \leq n$, such that each of them, except maybe the last one, is bounded by $\left[\sup_\omega \min_{j=0,\cdots,n-1} |\prod_{k=0}^{j} p(2^k\omega)|^{1/j}\right]^s$. This decomposition yields the estimate

$$|\prod_{k=0}^{m-1} p(2^k\omega)| \leq [\sup_\omega \min_{j=0,\cdots,n-1} |\prod_{k=0}^{j} p(2^k\omega)|^{1/j}]^m [\sup |p(\omega)|]^n. \quad (2.7.20)$$

Letting m go to $+\infty$, we obtain $b_\infty \leq d_n \leq b_n$ so that the sequence d_n also converges to b_∞. In many practical examples the convergence turns out to be much faster than for b_n (in particular in the example of §2.10, it can be shown that $d_n = b_\infty$ for $n \geq 2$).

From the evaluation of $s_\infty(\varphi)$, we can derive an estimate for $s_q(\varphi)$ using (2.7.7), i.e $s_q(\varphi) \geq s_i(\varphi) - 1/q$. In general this estimate is not sharp for $q < \infty$. We shall now describe another strategy that allows us to obtain directly a sharp estimate for $s_q(\varphi)$, in particular for the case of the Sobolev exponent, i.e. for $q = 2$.

With any 2π-periodic real valued continuous function $u(\omega)$, we associate a *transfer operator* T_u acting in the space of continuous 2π-periodic (complex valued) functions as follows:

$$T_u f(\omega) = u(\omega/2)f(\omega/2) + u(\omega/2+\pi)f(\omega/2+\pi). \tag{2.7.21}$$

We first notice two properties of T_u: firstly, if $u(\omega) = \sum_{|n|\leq N} u_n e^{in\omega}$, then T_u preserves the corresponding finite dimensional space spanned by $e^{in\omega}$, $|n| \leq N$. In this basis, T_u is represented by a matrix $(t_{i,j})_{i,j=-N,\cdots,N}$, where $t_{i,j} = 2u_{j-2i}$. Secondly, the dual of T_u (viewed as an operator acting in $L^2([-\pi,\pi])$) is given by

$$T_u^* f(\omega) = 2u(\omega)f(2\omega), \tag{2.7.22}$$

since we have for $f, g \in L^2([-\pi,\pi])$,

$$\begin{aligned}\langle T_u^* f, \overline{g}\rangle &= \textstyle\int_{-\pi}^{\pi} T_u g(\omega) f(\omega) d\omega \\ &= 2\textstyle\int_{-\pi/2}^{\pi/2} [u(\omega)g(\omega) + u(\omega+\pi)g(\omega+\pi)] f(2\omega) \\ &= 2\textstyle\int_{-\pi}^{\pi} g(\omega)u(\omega)f(2\omega)d\omega.\end{aligned}$$

Consequently, if f and g are 2π-periodic continuous functions, we have for all $n \geq 0$,

$$\int_{-\pi}^{\pi} g(\omega)(T_u)^n f(\omega) d\omega = \int_{-2^n\pi}^{2^n\pi} g(\omega)[\prod_{k=1}^{n} u(2^{-k}\omega)] f(2^{-n}\omega) d\omega. \tag{2.7.23}$$

Recalling the factorized form $m(\omega) = \left(\frac{1+e^{-i\omega}}{2}\right)^L p(\omega)$, we denote by T_q the transfer operator associated with the function $u(\omega) = |p(\omega)|^q$. When q is an even integer, $u = |p|^q$ is a trigonometric polynomial and we can thus study the action of T_q in a finite dimensional space E_N, according to the first remark.

Theorem 2.7.1 *If q is an even integer and if ρ_q is the spectral radius of T_q restricted to E_N, we have the estimate*

$$s_q(\varphi) \geq L - \frac{\log_2(\rho_q)}{q}. \tag{2.7.24}$$

Proof Applying (2.7.23) to $f = g = 1$, we obtain that for all $\varepsilon > 0$, there exists a constant $C = C(\varepsilon)$ such that

$$\int_{-2^n\pi}^{2^n\pi} \prod_{k=1}^{n} |p(2^{-k}\omega)|^q d\omega = \langle (T_q)^n 1, 1\rangle \lesssim (\rho_q + \varepsilon)^n. \tag{2.7.25}$$

On the interval $[-2^n\pi, 2^n\pi]$, we have

$$|\hat{D}(\omega)| \leq C \prod_{k=1}^{n} |p(2^{-k}\omega)|, \tag{2.7.26}$$

with $C = \sup_{\omega\in[-\pi,\pi]} |\hat{D}(\omega)|$. It follows that

$$\int_{-2^n\pi}^{2^n\pi} |\hat{D}(\omega)|^q \lesssim (\rho_q + \varepsilon)^n. \tag{2.7.27}$$

Defining $\Delta_n = \{2^{n-1}\pi \leq |\omega| \leq 2^n\pi\}$ and using that $\hat{\varphi}(\omega) = \hat{B}_{L-1}(\omega)\hat{D}(\omega)$, we derive from (2.7.27) the estimate

$$\int_{\Delta_n} (1 + |\omega|)^{sq} |\hat{\varphi}(\omega)|^q \leq C(\rho_q + \varepsilon)^n 2^{-Lqn} 2^{snq}, \tag{2.7.28}$$

which shows that $(1 + |\cdot|)^s\hat{\varphi}$ is in L^q for all $s < L - \frac{\log_2(\rho_q)}{q}$, i.e. (2.7.24). $\diamond$

Remark 2.7.2 *Similarly to (2.7.18), one can also prove the equality in (2.7.24) in the case where φ is L^2-stable (see* VILLEMOES *[1994] or* EIROLA *[1992]). This operator-based technique allows thus to obtain a sharp estimate of the Sobolev exponent $s_2(\varphi)$ by checking the spectral radius of a finite matrix representing T_2 in the basis $e^{in\omega}$. Note that, according to the first remark on transition operators, the matrix coefficients of T_2 are given by $t(i,j) = 2r_{j-2i}$, where $(r_k)_{k\in\mathbb{Z}}$ is the (finite) sequence of Fourier coefficients of $|p(\omega)|^2$. We thus have*

$$t_{i,j} = 2\sum_n p_n p_{n+j-2i}, \tag{2.7.29}$$

where $(p_n)_{n\in\mathbb{Z}}$ is the (finite) sequence of Fourier coefficients of $p(\omega)$. We have already encountered this particular form in §2.5 when we introduced the matrix $H = (H_{k,n})$, $H_{k,n} = 1/2\sum_m h_m h_{m+n-2k}$ for the computation of the mass and stiffness matrix elements: up to a normalization, this matrix exactly represents the transition operator associated with $|m(\omega)|^2$.

Remark 2.7.3 *When q is not an even integer, if we want to obtain a result similar to Theorem 2.7.1, we need to address the question of the function space where we look for the spectral radius of T_q. In* COHEN *and* DAUBECHIES *[1996], it is shown that the choice of an appropriate space of periodic analytic functions allows us to recover a sharp estimate (i.e. an equality in (2.7.24)) when φ is L^2-stable. The computation of the spectral radius in this infinite dimensional space is also shown to be feasible through Fredholm determinant techniques.*

Remark 2.7.4 *When $\hat{\varphi}$, $m(\omega)$ and $p(\omega)$ are positive functions, we can use the above technique with $q = 1$ to obtain a sharp estimate of the Hölder exponent $\mu(\varphi) = s_1(\varphi) = L - \log_2(\rho_1)$. In the case where p and m are not positive, a sharp estimate of μ is still possible, using other operator-based techniques: see in particular* RIOUL *[1992],* JIA *[1999]. Finally, a technique introduced in* DAUBECHIES *and* LAGARIAS *[1991,1992], also based on matrix analysis, allows us to estimate the local pointwise Hölder regularity of scaling functions.*

2.8 Polynomial exactness

In the previous section, we have used the factorized form (2.7.12) of $m(\omega)$ in order to derive smoothness estimates. The goal of this section is to give an interpretation of this factorization in terms of ***polynomial reproduction properties*** satisfied by the functions $\varphi_{j,k}$. These properties will be used later on to prove approximation results in the spaces V_j.

We shall also prove in this section that the factorized form (2.7.12) is necessary to have some smoothness properties for φ, in particular if we want $\varphi \in H^L$. Note that, according to the estimates (2.7.18) and (2.7.24), the presence of the factor $p(\omega)$ results in a loss of smoothness, i.e. (2.7.12) is generally not sufficient to ensure regularity of order L. This will be illustrated in the examples of §2.10 where $p(\omega)$ will be chosen in order to obtain interpolating or orthonormal scaling functions for arbitrary values of the parameter L.

We start by a simple remark: if $m(\omega)$ has the factorized form (2.7.12), then the function φ defined by (2.7.1) satisfies

$$(\frac{\partial}{\partial\omega})^q \hat{\varphi}(2n\pi) = 0, \quad n \in \mathbb{Z} - \{0\}, \ |q| \le L - 1. \tag{2.8.1}$$

Indeed, if $n \in \mathbb{Z} - \{0\}$, there exists some $k > 0$, such that $2^{-k}(2n) \in 2\mathbb{Z} + 1$, so that $m(2^{-k}\omega)$ vanishes at $2n\pi$ with order L. It follows from (2.7.1) that $\hat{\varphi}$ vanishes at $2n\pi$ with the same order of flatness.

If φ is L^2-stable, a converse result also holds. Indeed, from the lower inequality in (2.2.24), there exists $n_0 \in \mathbb{Z}$ such that $\hat{\varphi}((2n_0 + 1)\pi) \neq 0$. Applying (2.8.1) to $n = 2n_0 + 1$, we obtain

$$|m(\pi + \omega/2)\hat{\varphi}((2n_0 + 1)\pi + \omega)| = |\hat{\varphi}((4n_0 + 2)\pi + \omega)| = \mathcal{O}(|\omega|^L), \tag{2.8.2}$$

so that necessarily, $|m(\pi + \omega)| = \mathcal{O}(|\omega|^L)$, which is equivalent to (2.7.12).

The cancellations of $\hat{\varphi}$ that are expressed in (2.8.1) are called the *Strang-Fix conditions* of order $L - 1$. These conditions were introduced for more general (not necessarily refinable) functions in STRANG and FIX [1969]. As shown by the following result, the Strang-Fix conditions measure the degree of polynomial exactness for the spaces generated by the functions $\varphi(\cdot - k)$, $k \in \mathbb{Z}$.

Theorem 2.8.1 *Let φ be an L^1 function with compact support such that $\int \varphi = 1$. The following properties are equivalent:*

1. *φ satisfies (2.8.1), i.e. the Strang-Fix conditions of order $L-1$.*
2. *For all $q = 0, \cdots, L-1$, we can expand the polynomial x^q according to*
$$x^q = \sum_{k\in\mathbb{Z}} [k^q + p_{q-1}(k)]\varphi(x-k), \tag{2.8.3}$$
where p_{q-1} is a polynomial of degree $q-1$, or equivalently, for all $q = 0, \cdots, L-1$,
$$\sum_{k\in\mathbb{Z}} k^q \varphi(x-k) = x^q + r_{q-1}(x), \tag{2.8.4}$$
where r_{q-1} is a polynomial of degree $q-1$.

Proof Remark that since φ is in L^1 with compact support, its Fourier transform is a C^∞ function with bounded derivatives. Applying the Fourier transform on (2.8.4) we obtain that

$$\hat{\varphi}(\omega)\sum_{k\in\mathbb{Z}} k^q e^{-ik\omega} = (i)^q \hat{\varphi}(\omega)[\sum_{n\in\mathbb{Z}} \delta_{2\pi n}^{(q)}], \tag{2.8.5}$$

is a singular distribution supported at the origin only. It follows that for all $q = 0, \cdots, L$, and $n \in \mathbb{Z} - \{0\}$, the distributions $\hat{\varphi}\delta_{2\pi n}^{(q)}$ are identically zero, which is equivalent to the Strang-Fix conditions of order $L-1$.

Conversely, if (2.8.1) is satisfied, we remark that the term $\hat{\varphi}\delta_0^{(q)}$ is a combination of derivatives $\delta_0^{(k)}$, $k = 0, \cdots, q$ where the highest term is exactly $\delta_0^{(q)}$, since we have assumed that $\int \varphi = 1$. Thus (2.8.4) is satisfied. ◇

Remark 2.8.1 *The properties (2.8.3) and (2.8.4) express that the polynomials of degree less than or equal to $L-1$ are "contained" in the space V_0 spanned by the functions $\varphi(\cdot - k)$, $k \in \mathbb{Z}$. From the invariance of Π_{L-1} by a change of scale, we see that these polynomials are also contained in all the spaces V_j, $j \in \mathbb{Z}$. If $\tilde{\varphi}$ is a dual function to φ that satisfies (2.2.49), and which is sufficiently well localized so that the oblic projector P_j is well* *defined by (2.2.50) (for example if φ and $\tilde{\varphi}$ are both L^2 with compact support), we then obtain an explicit expression for the coefficients in (2.8.3) according to*

$$k^q + P_{q-1}(k) = \int x^q \tilde{\varphi}(\cdot - k)dx = \int (x+k)^q \tilde{\varphi}(x)dx, \qquad (2.8.6)$$

i.e. in terms of the moments of $\tilde{\varphi}$. We also have that $P_j p = p$ for all $p \in \Pi_{L-1}$.

Remark 2.8.2 *Theorem 2.3.3 has already revealed that the polynomial exactness property for constant polynomials was necessary and sufficient to ensure the density of the space V_j. As it will be shown in the next chapter, polynomial exactness at order $L-1$ turns out to be crucial in order to derive approximation error estimates of the type*

$$\|f - P_j f\|_{L^p} \lesssim 2^{-jL} \|f^{(L)}\|_{L^p}, \qquad (2.8.7)$$

for functions in the Sobolev space $W^{s,p}(\mathbb{R})$. The Strang-Fix conditions are thus the exact measurement of the approximation power of principal shift invariant spaces (see Remark 2.2.2).

In the case where φ is refinable, the Strang-Fix conditions are also unavoidable to have some smoothness as shown by the following result.

Theorem 2.8.2 *If φ is an L^2-stable compactly supported refinable function in H^L for $L \geq 0$, then it satisfies the Strang-Fix conditions of order L.*

Proof We shall first prove that the Strang-Fix conditions of order L hold when $\varphi \in C^L$. For this, we proceed by induction on L.

Assuming that φ is a continuous compactly supported function, we remark that there exists $n \in \mathbb{Z}$ such that $\varphi(n) \neq 0$: if this were not the case, a simple iteration of the refinement equation (2.2.13) shows that φ vanishes at all dyadic numbers $2^{-j}k$, $j,k \in \mathbb{Z}$, and is thus identically zero. Up to a redefinition of φ by an integer shift (which does not modify the polynomial reproduction properties), we can thus assume that $\varphi(0) \neq 0$.

For $j \geq 0$, we expand $\varphi(2^{-j}\cdot)$ in terms of its scaled version, i.e.

$$\varphi(2^{-j}x) = \sum_{k \in \mathbb{Z}} s_{j,k} \varphi(x-k). \qquad (2.8.8)$$

As it was remarked in the proof of Theorem 2.4.1, the coefficients $s_{j,k}$ can be expressed as inner products

$$s_{j,k} = \langle \varphi(2^{-j}\cdot), \tilde{\varphi}(\cdot - k) \rangle, \qquad (2.8.9)$$

where $\tilde{\varphi}$ is a compactly supported dual function. Since $\varphi(2^{-j}x)$ tends to $\varphi(0)$ uniformly on any compact set as $j \to +\infty$, and since $\int \tilde{\varphi} = 1$, the expansion (2.8.8) yields

$$1 = \sum_{k\in\mathbb{Z}} \varphi(x-k), \tag{2.8.10}$$

and thus, the Strang-Fix conditions at order 0.

Assuming then that the result is proved up to the order $L-1$, we assume that $\varphi \in C^L$. Invoking again the refinement equation, we can assume that the derivative $\varphi^{(L)}$ of order L does not vanish at the origin. Remarking that the function

$$g_L(x) := 2^{jl}[\varphi^{(L)}(0)]^{-1}\Big(\varphi(2^{-j}x) - \sum_{l=0}^{L-1}(2^{-j}x)^l\varphi^{(l)}(0)/l!\Big), \tag{2.8.11}$$

can be expanded in terms of the $\varphi(\cdot - k)$ (by the induction hypothesis), and converges uniformly to x^L on any compact set as $j \to +\infty$, we obtain that

$$x^L = \sum_{k\in\mathbb{Z}} c_{L,k}\varphi(x-k), \tag{2.8.12}$$

with $c_{L,k} = \int x^L\tilde{\varphi}(x-k)dx = k^L + p_{l-1}(k)$, $p_{l-1} \in \Pi_{l-1}$.

We have thus proved that a refinable function $\varphi \in C^L$ satisfies the Strang-Fix conditions of order L.

Assume now that $\varphi \in H^L$. It follows that its autocorrelation function $\phi(x) = \int \varphi(t)\varphi(t+x)dt$ is in C^{2L}. Since $\hat{\phi} = |\hat{\varphi}|^2$, the symbol of ϕ is given by $M(\omega) = |m(\omega)|^2$, where $m(\omega)$ is the symbol of φ. Now ϕ is L^2-stable (if (2.2.24) holds for $\hat{\varphi}$, it also hold for $\hat{\phi} = |\hat{\varphi}|^2$), and satisfies the Strang-Fix conditions at order $2L$: according to the first remarks of this section, we thus have $M(\omega+\pi) = \mathcal{O}(|\omega|^{2L+1})$, which immediately implies

$$M(\omega+\pi) = \mathcal{O}(|\omega|^{2L+2}), \tag{2.8.13}$$

since $M(\omega) \geq 0$. Consequently, we have

$$m(\omega+\pi) = \mathcal{O}(|\omega|^{L+1}), \tag{2.8.14}$$

i.e. φ satisfies the Strang-Fix conditions at order L. $\diamond$

Remark 2.8.3 *This result shows that the property (2.3.12) (reproduction of constant functions) is satisfied by any L^2-stable scaling function.*

Remark 2.8.4 *The flatness of $m(\omega)$ at π expressed by (2.8.14) is equivalent to a discrete polynomial exactness property, even when $m(\omega)$ is not related to an L^2-stable refinable function: consider the subdivision scheme associated with the coefficients h_n as was introduced in §2.4, i.e.*

$$s_{j+1,n} = \sum_k h_{n-2k} s_{j,k}. \tag{2.8.15}$$

Assume that $s_{j,k} = k^q$ with $q \leq L$. Then the refined sequence is given at even points by

$$s_{j+1,2n} = \sum_m h_{2m}(n-m)^q = 2^{-q} \sum_m h_{2m}(2n-2m)^q,$$

and at odd points by

$$s_{j+1,2n+1} = \sum_m h_{2m+1}(n-m)^q = 2^{-q} \sum_m h_{2m+1}((2n+1)-(2m+1))^q.$$

Both sequences are polynomial in n and we see that the whole sequence $s_{j+1,n}$ is polynomial if and only if

$$\sum_m (2m)^l h_{2m} = \sum_m (2m+1)^l h_{2m+1}, \quad l = 0, \cdots, q, \tag{2.8.16}$$

i.e. $m(\omega+\pi) = \mathcal{O}(|\omega|^{q+1})$. By linearity, we thus find that (2.8.14) is equivalent to the property that the subdivision scheme maps polynomial sequences of degree less than or equal to L into sequences of the same type.

Remark 2.8.5 *The Strang-Fix conditions of order L can also be expressed in terms of the oscillation of the dual wavelet $\tilde{\psi}$: since we have*

$$\hat{\tilde{\psi}}(\omega) = e^{-i\omega} m(\omega/2+\pi)\hat{\tilde{\varphi}}(\omega/2), \tag{2.8.17}$$

the condition (2.8.14) is equivalent to

$$|\hat{\tilde{\psi}}(\omega)| = \mathcal{O}(|\omega|^{L+1}), \tag{2.8.18}$$

or to the vanishing moment conditions

$$\int x^q \tilde{\psi}(x)dx = 0, \quad q = 0, \cdots, L. \tag{2.8.19}$$

2.9 Duality, orthonormality and interpolation

The properties of orthonormality (2.2.15) or interpolation (2.2.19) of a refinable function φ, or the duality property (2.2.49) of a pair of refinable functions, induce specific constraints on the corresponding scaling coefficients h_n and $\tilde{h}_n$. More precisely, in the case where φ and $\tilde{\varphi}$ are a pair of dual scaling functions, we have established in Lemma 2.6.1 the necessity of the identity

$$\sum_n h_n \tilde{h}_{n+2k} = 2\delta_{0,k}. \tag{2.9.1}$$

For an orthonormal scaling function, i.e. when $\varphi = \tilde{\varphi}$, (2.9.1) takes the form

$$\sum_n h_n h_{n+2k} = 2\delta_{0,k}. \tag{2.9.2}$$

If φ is interpolatory, since $h_n = \varphi(n/2)$, we obtain the constraint

$$h_{2k} = \delta_{0,k}. \tag{2.9.3}$$

Note that (2.9.3) can also be obtained from (2.9.1) with $\tilde{h}_n = 2\delta_{0,n}$ corresponding to $\tilde{\varphi} = \delta$ (see Remark 2.2.6).

In this section we shall prove converse results: if we start from sequences h_n and $\tilde{h}_n$ that satisfy the above conditions, then, under mild conditions, the associated refinable functions φ and $\tilde{\varphi}$ will satisfy the duality, orthonormality or interpolation conditions.

We recall the symbols and their factorized form

$$m(\omega) = \frac{1}{2}\sum_n h_n e^{-in\omega} = \left(\frac{1+e^{-i\omega}}{2}\right)^L p(\omega), \tag{2.9.4}$$

and

$$\tilde{m}(\omega) = \frac{1}{2}\sum_n \tilde{h}_n e^{-in\omega} = \left(\frac{1+e^{-i\omega}}{2}\right)^{\tilde{L}} \tilde{p}(\omega), \tag{2.9.5}$$

with $L, \tilde{L} > 0$ and $p(0) = \tilde{p}(0) = 1$, and we assume that φ and $\tilde{\varphi}$ are defined by the infinite product formula (2.7.1) applied to m and $\tilde{m}$.

We note that the conditions (2.9.1), (2.9.2) and (2.9.3) are respectively equivalent to the following identities

$$\tilde{m}(\omega)\overline{m(\omega)} + \tilde{m}(\omega+\pi)\overline{m(\omega+\pi)} = 1, \tag{2.9.6}$$

$$|m(\omega)|^2 + |m(\omega+\pi)|^2 = 1, \tag{2.9.7}$$

and

$$m(\omega) + m(\omega + \pi) = 1. \tag{2.9.8}$$

We also introduce the band-limited approximations φ_n and $\tilde{\varphi}_n$ for φ and $\tilde{\varphi}$ defined by

$$\hat{\varphi}_n(\omega) = \Big[\prod_{k=1}^{n} m(2^{-k}\omega)\Big]\chi_{[-\pi,\pi]}(2^{-n}\omega), \tag{2.9.9}$$

and

$$\hat{\tilde{\varphi}}_n(\omega) = \Big[\prod_{k=1}^{n} \tilde{m}(2^{-k}\omega)\Big]\chi_{[-\pi,\pi]}(2^{-n}\omega). \tag{2.9.10}$$

Our first result shows that we can recover duality, orthonormality and interpolation properties from the corresponding identities (2.9.1) to (2.9.3), if these approximations converge in some specific way to φ and $\tilde{\varphi}$ as n goes to $+\infty$.

Theorem 2.9.1 *1. Duality: if m and $\tilde{m}$ satisfy (2.9.6), then the functions φ_n and $\tilde{\varphi}_n$ satisfy the duality property (2.2.49) for all $n \geq 0$. The refinable functions φ and $\tilde{\varphi}$ form a dual pair if they are the L^2-limits of the sequences φ_n and $\tilde{\varphi}_n$.*

2. Orthonormality: if m satisfies (2.9.7), then the function φ_n has orthonormal integer translations for all $n \geq 0$. The refinable function φ is also orthonormal if it is the L^2-limit of the sequence φ_n.

3. Interpolation: if m satisfies (2.9.8), then the function φ_n is interpolatory for all $n \geq 0$. The refinable function φ is also interpolatory if it is the uniform limit of the sequence φ_n.

Proof If m and $\tilde{m}$ satisfy (2.9.6), we obtain for all $n \geq 0$,

$$\begin{aligned}
\int \tilde{\varphi}_n\overline{\varphi_n(\cdot - l)} &= \tfrac{1}{2\pi}\int \overline{\hat{\varphi}_n(\omega)}\hat{\tilde{\varphi}}_n(\omega)e^{il\omega}d\omega \\
&= \tfrac{1}{2\pi}\int_{-2^n\pi}^{2^n\pi}\Big[\textstyle\prod_{k=1}^{n}\overline{m}\tilde{m}(2^{-k}\omega)\Big]e^{il\omega}d\omega \\
&= \tfrac{2^n}{2\pi}\int_{-\pi}^{\pi}\Big[\textstyle\prod_{k=0}^{n-1}\overline{m}\tilde{m}(2^{k}\omega)\Big]e^{i2^n l\omega}d\omega \\
&= \tfrac{2^n}{2\pi}\int_{0}^{\pi}(\overline{m}\tilde{m}(\omega) + \overline{m}\tilde{m}(\omega+\pi))\Big[\textstyle\prod_{k=1}^{n-1}\overline{m}\tilde{m}(2^{k}\omega)\Big]e^{i2^n l\omega}d\omega \\
&= \tfrac{1}{2\pi}\int_{-2^{n-1}\pi}^{2^{n-1}\pi}\Big[\textstyle\prod_{k=1}^{n-1}\overline{m}\tilde{m}(2^{-k}\omega)\Big]e^{il\omega}d\omega \\
&= \cdots = \tfrac{1}{2\pi}\int_{-\pi}^{\pi}e^{il\omega}d\omega = \delta_{0,l}.
\end{aligned}$$

We thus obtain that φ_n and $\tilde{\varphi}_n$ satisfy the duality relation (2.2.49) for all $n \geq 0$. By the continuity of the L^2 inner product, this property also holds for φ and $\tilde{\varphi}$ if they are the L^2-limits of their band-limited approximants φ_n and $\tilde{\varphi}_n$.

Similarly, if m satisfies (2.9.7), we obtain for all $n \geq 0$,

$$\begin{aligned}
\int \varphi_n \overline{\varphi_n(\cdot - l)} &= \frac{1}{2\pi} \int |\hat{\varphi}_n(\omega)|^2 e^{il\omega} d\omega \\
&= \frac{1}{2\pi} \int_{-2^n\pi}^{2^n\pi} \left[\prod_{k=1}^n |m(2^{-k}\omega)|^2\right] e^{il\omega} d\omega \\
&= \frac{2^n}{2\pi} \int_{-\pi}^{\pi} \left[\prod_{k=0}^{n-1} |m(2^k\omega)|^2\right] e^{i2^n l\omega} d\omega \\
&= \frac{2^n}{2\pi} \int_0^{\pi} (|m(\omega)|^2 + |m(\omega+\pi)|^2) \left[\prod_{k=1}^{n-1} |m(2^k\omega)|^2\right] e^{i2^n l\omega} d\omega \\
&= \frac{1}{2\pi} \int_{-2^{n-1}\pi}^{2^{n-1}\pi} \left[\prod_{k=1}^{n-1} |m(2^{-k}\omega)|^2\right] e^{il\omega} d\omega \\
&= \cdots = \frac{1}{2\pi} \int_{-\pi}^{\pi} e^{il\omega} d\omega = \delta_{0,l}.
\end{aligned}$$

We thus obtain that φ_n has orthonormal integer translations for all $n \geq 0$. By the continuity of the L^2 inner product with respect to the L^2 norm, this property is also satisfied by φ if it is the L^2-limit of φ_n.

Finally, if m satisfies (2.9.8), we obtain for all $n \geq 0$,

$$\begin{aligned}
\varphi_n(l) &= \frac{1}{2\pi} \int \hat{\varphi}_n(\omega) e^{il\omega} d\omega \\
&= \frac{1}{2\pi} \int_{-2^n\pi}^{2^n\pi} \left[\prod_{k=1}^n m(2^{-k}\omega)\right] e^{il\omega} d\omega \\
&= \frac{2^n}{2\pi} \int_{-\pi}^{\pi} \left[\prod_{k=0}^{n-1} m(2^k\omega)\right] e^{i2^n l\omega} d\omega \\
&= \frac{2^n}{2\pi} \int_0^{\pi} (m(\omega) + m(\omega+\pi)) \left[\prod_{k=1}^{n-1} m(2^k\omega)\right] e^{i2^n l\omega} d\omega \\
&= \frac{1}{2\pi} \int_{-2^{n-1}\pi}^{2^{n-1}\pi} \left[\prod_{k=1}^{n-1} m(2^{-k}\omega)\right] e^{il\omega} d\omega \\
&= \cdots = \frac{1}{2\pi} \int_{-\pi}^{\pi} e^{il\omega} d\omega = \delta_{0,l}.
\end{aligned}$$

We thus obtain that φ_n is interpolatory for all $n \geq 0$. By the continuity of the pointwise values with respect to the uniform convergence, this property is also satisfied by φ if it is the uniform limit of φ_n. ◇

Remark 2.9.1 *One easily check that $\varphi_n(2^{-n}k) = s_{n,k}$, where $s_{n,k}$ is the result of n iterations of the subdivision algorithm described in §2.4. We can thus view φ_n as the band-limited interpolant of the subdivision values after n-iterations. Recall that in §2.4, we proved the uniform convergence of the linear interpolation of the subdivision, under the assumption that the limit function φ was continuous and L^2-stable. For the L^2 or uniform convergence of φ_n to φ, the situation is very similar: convergence is ensured by the L^2-stability and an assumption on the regularity of the limit. However, we are here interested in placing the assumptions on the function $m(\omega)$ (i.e. on the coefficients h_n) rather than on the function φ.*

The following result shows that we can replace the L^2-stability assumption by a criterion on the cancellations of $m(\omega)$.

Theorem 2.9.2 *Assume that $m(\omega)$ does not vanish on $[-\pi/2,\pi/2]$. Then,*

1. *if $\varphi \in L^2$, it is the L^2-limit of φ_n.*
2. *if $\hat{\varphi} \in L^1$, it is the L^1-limit of $\hat{\varphi}_n$ and φ_n converges uniformly to φ.*

Proof If $m(\omega)$ does not vanish on $[-\pi/2,\pi/2]$, the infinite product formula (2.7.1) shows that $\hat{\varphi}(\omega)$ does not vanish on $[-\pi,\pi]$.

We then remark that

$$\hat{\varphi}_n(\omega)=\hat{\varphi}(\omega)/\hat{\varphi}(2^{-n}\omega) \text{ if } \omega\in[-2^n\pi,2^n\pi], \text{ 0 elsewhere,} \tag{2.9.11}$$

so that

$$|\hat{\varphi}_n(\omega)|\le C|\hat{\varphi}(\omega)|, \tag{2.9.12}$$

with $C=\max_{\omega\in[-\pi,\pi]}|\hat{\varphi}(\omega)|^{-1}$. Thus, if $\hat{\varphi}$ is in L^q, $1\le q<\infty$, the L^q-convergence of $\hat{\varphi}_n$ to $\hat{\varphi}$ is ensured by Lebesgue's dominated convergence theorem (using also that $\hat{\varphi}_n$ converges pointwise to $\hat{\varphi}$).

In the L^1 case, the uniform convergence of φ_n to φ follows, applying the inverse Fourier transform. ◇

Remark 2.9.2 *The condition that $m(\omega)\neq 0$ on $[-\pi/2,\pi/2]$ - or equivalently that $\hat{\varphi}(\omega)\neq 0$ on $[-\pi,\pi]$ - is related to the L^2-stability of φ since it clearly implies the lower inequality in (2.2.24), i.e.*

$$\sum_{n\in\mathbb{Z}}|\hat{\varphi}(\omega+2n\pi)|^2\ge C_1>0. \tag{2.9.13}$$

However, (2.9.13) could be achieved without the assumption that $\hat{\varphi}$ does not vanish on $[-\pi,\pi]$. It can be shown (see e.g. Cohen *and* Ryan *[1995] or* Cohen, Daubechies *and* Feauveau *[1992]) that one can use a condition on $m(\omega)$ that is exactly equivalent to (2.9.13), and reach the same conclusion as in Theorem 2.9.2: one assumes that there exists a compact set K such that*

$$0\in \operatorname{int}(K),\ \ |K|=2\pi \text{ and } \cup_{n\in\mathbb{Z}}(K+2n\pi)=\mathbb{R}, \tag{2.9.14}$$

for which we have

$$m(2^{-k}\omega)\neq 0,\ \ k\ge 0,\ \omega\in K. \tag{2.9.15}$$

A compact set satisfying the assumption (2.9.14) is called congruous to $[-\pi,\pi]$ modulo 2π. Note that when $K=[-\pi,\pi]$, we simply obtain the property that $m(\omega)\neq 0$ on $[-\pi/2,\pi/2]$. In the orthonormal case, it is shown in Cohen *and* Ryan *[1995, chapter 2] that a smarter choice of K allows us to assume only that $m(\omega)$ does not vanish on $[-\pi/3,\pi/3]$. However, such a generalization is not much used in practice: in most relevant examples $m(\omega)$ does not vanish on $[-\pi/2,\pi/2]$.*

It remains to discuss the properties that one can impose on $m(\omega)$ and $\tilde{m}(\omega)$ so that $\hat{\varphi}$ and $\hat{\tilde{\varphi}}$ are ensured to be in L^2 or L^1, in order to obtain the construction of dual, orthonormal or interpolatory scaling functions, directly from some specific assumptions on the refinement coefficients.

In the particular case of orthonormal scaling functions, it is interesting to note that the property (2.9.7) is enough to ensure that $\varphi \in L^2$: indeed, we remarked in the proof of Theorem 2.9.1 that (2.9.7) implies that φ_n is orthonormal, and in particular $\|\varphi_n\|_{L^2} = 1$. Since $\hat{\varphi}_n$ converges pointwise to $\hat{\varphi}$, we obtain by Fatou's lemma that $\varphi \in L^2$ with $\varphi \leq 1$. Clearly the same reasoning shows that if $m(\omega)$ satisfies (2.9.8), and $m(\omega) \geq 0$, the limit function $\hat{\varphi}$ is in L^1.

For more general cases, we already discussed in §2.7 the localization properties of $\hat{\varphi}$. In particular, we see that $\hat{\varphi} \in L^q$ whenever $s_q(\varphi) > 0$, so that we can use the estimate (2.7.18) or (2.7.24): $\hat{\varphi}$ is in L^q if

$$L - b_\infty > 1/q \text{ (i.e. } L - b_n > 1/q \text{ for some } n \geq 0), \tag{2.9.16}$$

or if

$$L - \frac{\log_2(\rho_q)}{q} > 0, \tag{2.9.17}$$

where b_∞, d_n and ρ_q are defined as in §2.7. Additionally, one can easily check that these conditions directly imply the L^q-convergence of $\hat{\varphi}_n$ to $\hat{\varphi}$, so that the assumption that $m(\omega)$ does not vanish on $[-\pi/2, \pi/2]$ can be removed. Introducing the analogous quantities $\tilde{b}_\infty$ and $\tilde{\rho}_q$ related to $\tilde{m}(\omega)$, we can thus summarize the construction of dual, orthonormal and interpolatory scaling functions by the following result.

Theorem 2.9.3 *1. Duality: if m and $\tilde{m}$ satisfy (2.9.6), and if*

$$\max\{L - b_\infty - \frac{1}{2}, L - \frac{\log_2(\rho_2)}{2}\}, \max\{\tilde{L} - \tilde{b}_\infty - \frac{1}{2}, \tilde{L} - \frac{\log_2(\tilde{\rho}_2)}{2}\} > 0,$$

then φ and $\tilde{\varphi}$ form a pair of dual refinable functions in L^2.

2. Orthonormality: if m satisfies (2.9.7) and if

$$\max\{L - b_\infty - 1/2, L - \frac{\log_2(\rho_2)}{2}\} > 0,$$

then φ is an orthonormal refinable function.

3. Interpolation: if m satisfies (2.9.8) and if

$$\max\{L - b_\infty - 1, L - \log(\rho_1)\} > 0,$$

then φ is an interpolatory refinable function.

Remark 2.9.3 *The question of finding minimal (i.e. necessary and sufficient) conditions on the functions m and $\tilde{m}$, that ensure duality, orthonormality or interpolation, has been addressed in numerous contributions. In the orthonormal case, it is shown in* COHEN *and* RYAN *[1995, chapter 2] that the condition (2.9.15) (together with (2.9.7)) is minimal. Other sharp criteria (based on the transfer operator introduced in §2.7 or on the arithmetic properties of the zeros of $m(\omega)$) are also discussed there. In the case of dual scaling functions, a minimal criterion is given in* COHEN *and* DAUBECHIES *[1992], based on transition operators: it essentially states that the conditions on $L - \frac{\log_2(\rho_2)}{2}$ and $\tilde{L} - \frac{\log_2(\tilde{\rho}_2)}{2}$ (together with (2.9.6)) are necessary and sufficient.*

2.10 Interpolatory and orthonormal wavelets

We are now ready to describe the construction of an important class of interpolatory and orthonormal scaling functions. This construction will also be used in the next section to obtain scaling functions that are dual to the B-splines.

In order to build regular interpolatory scaling functions, we need to look for trigonometric polynomials $m(\omega)$ that can be put in the factorized form (2.9.4) and that satisfy the condition (2.9.8). In the case where L is even, we can look for some $m(\omega)$ with real and symmetric Fourier coefficients, i.e. of the form

$$m(\omega) = \frac{1}{2}\Big(h_0 + \sum_{n\geq 1} h_n[e^{in\omega} + e^{-in\omega}]\Big). \tag{2.10.1}$$

Since $m(\omega)$ vanishes with order $L = 2N$ at $\omega = \pi$, it should be of the form

$$m(\omega) = [\cos^2(\omega/2)]^N P(\sin^2(\omega/2)), \tag{2.10.2}$$

where P is an algebraic polynomial. Combining this particular form with the constraint (2.9.8), we obtain the equation

$$(1-x)^N P(x) + x^N P(1-x) = 1, \tag{2.10.3}$$

that should be satisfied for all $x \in [0,1]$ (and thus for all $x \in \mathbb{C}$ since P is a polynomial). The existence of a minimal degree polynomial solution to (2.10.2) is ensured by Bezout's theorem. We denote by P_N this minimum degree solution. Note that for $N = 1$, we have $P_N = 1$ and the corresponding interpolatory refinable function is simply the hat function $\varphi = \max\{1 - |x|, 0\}$ that generates linear splines.

For general N, we can relate the minimal degree solution to a simple discrete interpolation scheme: we consider the subdivision scheme associated with the coeffcients h_n, i.e.

$$s_{j+1,k} = \sum_n h_{k-2n} s_{j,n}. \tag{2.10.4}$$

From (2.9.3), we have $h_0 = 1$ and $h_{2n} = 0$ if $n \neq 0$, so that (2.10.4) rewrites as an interpolatory scheme

$$s_{j+1,2k} = s_{j,k} \text{ and } s_{j+1,2k+1} = \sum_{n \neq 0} h_n s_{j,2k+1+n}. \tag{2.10.5}$$

As we noted in Remark 2.8.4 , the cancellation of $m(\omega)$ at $\omega = \pi$ with order $2N$ is equivalent to a discrete polynomial reproduction property (up to order $2N-1$) for the subdivision scheme associated with the h_n. Due to the discrete interpolatory property (2.10.5), we obtain that if $s_{j,k} = q(2^{-j}k)$ for some $q \in \Pi_{2N-1}$, then we should also have $s_{j+1,k} = q(2^{-j-1}k)$. Clearly, the minimally supported sequence h_n allowing this property corresponds to the choice

$$s_{j+1,2k+1} = q_{j,k}(2^{-j-1}k), \tag{2.10.6}$$

where $q_{j,k}$ is the unique Lagrangian polynomial interpolant in Π_{2N-1} of the points $(2^{-j}(k+n), s_{j,k+n})$, $n = -N+1, \cdots, N$.

This *iterative Lagrangian interpolation scheme*, first introduced in the works of DESLAURIERS and DUBUC [1987], can be viewed as a natural generalization of linear data interpolation viewed as an iterative scheme: when $N = 1$, the new points on the fine grid are simply computed as the averages of the two closest neighbours on the coarser grid. Our general scheme simply consists in using the $2N$ closest neighbours and a higher order polynomial rule.

This interpretation of the minimal degree solution allows us to find an explicit expression of the coefficients h_n: $h_{2n} = \delta_{0,n}$ and

$$h_{-2n-1} = h_{2n+1} = \frac{\prod_{k=-N,k\neq n}^{N-1}(2k+1)}{\prod_{k=-N,k\neq n}^{N-1}(2k-2n)}, \tag{2.10.7}$$

for $n = 0, \cdots, N-1$. Since $m(\omega)$ has degree $2N-1$, it follows that the corresponding minimal degree solution P_N to (2.10.3) has degree $N-1$.

The exact expression of P_N can also be found directly by the following method: in the expansion of $1 = [x + (1-x)]^{2N-1}$, we can factorize x^N in half of the terms, so that we obtain (2.10.3) with

$$P_N(x) = \sum_{n=0}^{N-1} \binom{2N-1}{N+n} x^n (1-x)^{N-1-n}. \tag{2.10.8}$$

Since P_N is of degree $N-1$, we conclude that the trigonometric polynomial $m(\omega)$ defined by (2.10.2) with $P = P_N$ given by (2.10.8) coincides with $m(\omega)$ defined from the Lagrangian interpolation scheme, i.e. by the coefficients (2.10.7).

From now on, we shall use the notation

$$m_N(\omega) = [\cos^2(\omega/2)]^N P_N(\sin^2(\omega/2)), \tag{2.10.9}$$

for this minimal solution, and we denote by ϕ_N the associated refinable function defined by the infinite product (2.7.1). Since $m_N(\omega)$ is positive, as it was remarked in the previous section, we are ensured that the function $\hat{\phi}_N(\omega) = \prod_{k>0} m_N(2^{-k}\omega)$ is in L^1, so that ϕ_N is at least continuous. We also know that $\phi_1 = \max\{1-|x|, 0\}$ is $C^{1-\varepsilon}$ for all $\varepsilon > 0$.

We display in Figure 2.10.1 the iteration of the 4-point iterative interpolation scheme based on cubic interpolation, i.e. corresponding to $N = 2$: the new points are given by

$$s_{j+1,2k+1} = \frac{9}{16}(s_{j,k} + s_{j,k+1}) - \frac{1}{16}(s_{j,k-1} + s_{j,k+2}). \tag{2.10.10}$$

We also display the corresponding interpolatory scaling function ϕ_2 in Figure 2.10.2.

Both figures reveal that we have gained some smoothness in comparison with the linear interpolation process corresponding to $N = 1$. Using the transition operator technique of §2.7, (see in particular Remark 2.7.4), one can actually prove that $\phi_2 \in C^{2-\varepsilon}$ for all $\varepsilon > 0$ and that this estimate is sharp for the Hölder exponent, i.e. $\mu(\phi_2) = 2$.

We shall now prove that we can obtain arbitrarily high regularity by increasing the value of N. For this, we shall make use of the following more synthetic formula for P_N, which was derived in DAUBECHIES [1988] using combinatorial arguments:

$$P_N(x) = \sum_{n=0}^{N-1} \binom{N-1+n}{n} x^n. \tag{2.10.11}$$

Using this form, we first prove two preliminary results.

Lemma 2.10.1 *For all $N \geq 1$, $x \in [0,1]$, we have*

$$P_N(x) \leq [G(x)]^{N-1}, \quad \text{with } G(x) = \max\{2, 4x\}. \tag{2.10.12}$$

Proof We first remark that $P(x)$ is increasing on $[0,1]$, and that from (2.10.3), we have $P(1/2) = 2^{N-1}$. The estimate (2.10.12) follows on $[0, 1/2]$.

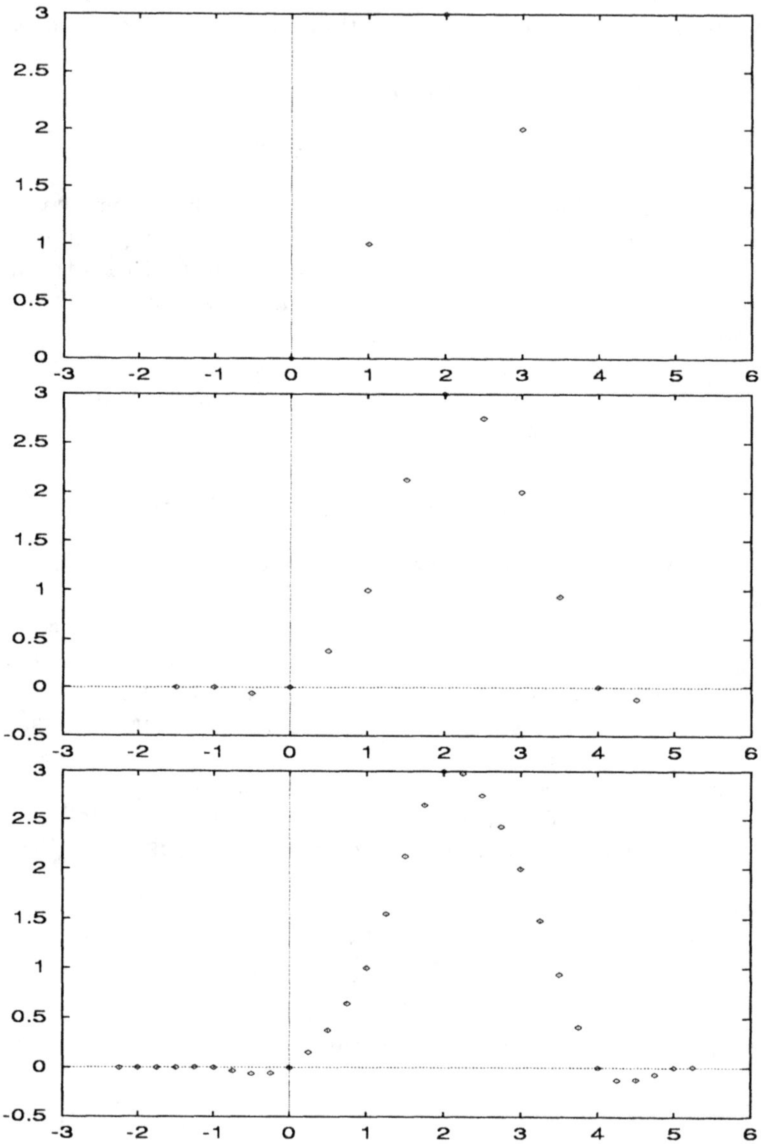

Figure 2.10.1: Iteration of the 4-point interpolation scheme (j=0,1,2)

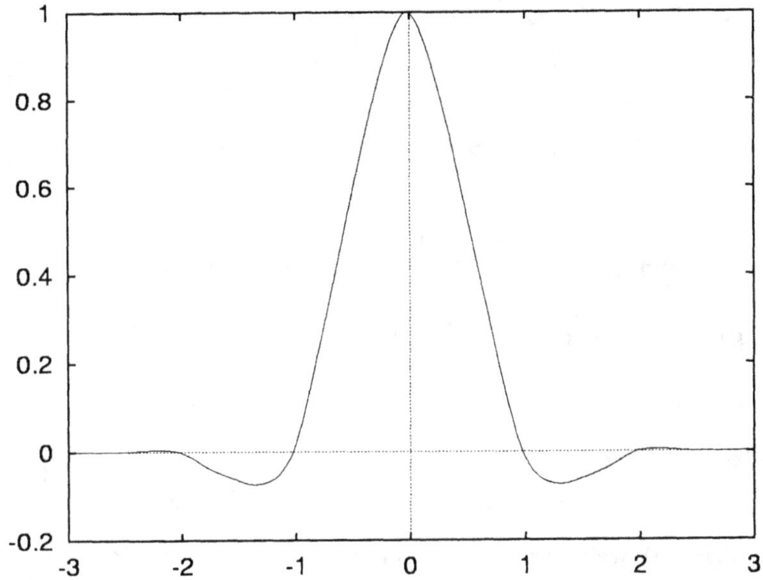

Figure 2.10.2: The limit function ϕ_2

If $x \in [1/2, 1]$, we write

$$\begin{aligned} P_N(x) &= \sum_{n=0}^{N-1} \binom{N-1+n}{n} 2^{-n}(2x)^n \\ &\leq (2x)^{N-1} P_N(1/2) = (4x)^{N-1}. \end{aligned}$$

and thus (2.10.12) holds on the whole of $[0, 1]$. ◇

Lemma 2.10.2 *The function $G(x)$ satisfies*

$$\begin{aligned} &G(x) \leq G(3/4) = 3 \text{ on } [0, 3/4] \text{ and} \\ &G(x)G(4x(1-x)) \leq [G(3/4)]^2 = 9 \text{ on } [3/4, 1]. \end{aligned} \tag{2.10.13}$$

Proof Since G is increasing on $[0, 1]$, we have $G(x) \leq G(3/4)$ on $[0, 3/4]$. Since $3/4$ is left unchanged by $x \mapsto 4x(1-x)$, we have $F(3/4) = [G(3/4)]^2$ where $F(x) = G(x)G(4x(1-x))$.

If $y \in [0, 3/4]$ is such that $4y(1-y) = 1/2$, $F(x) = 64x^2(1-x)$ on $[3/4, y]$ and decreases on this interval, while $F(x) = 8x$ on $[y, 1]$ where it increases. Since $F(1) = 8 \leq [G(3/4)]^2$, we obtain the second part of (2.10.13). ◇

Theorem 2.10.1 *The functions ϕ_N satisfy*

$$s_\infty(\phi_N) \geq 2N(1 - \frac{\log_2 3}{2}) \simeq 0.4\,N. \tag{2.10.14}$$

Consequently, their regularity can be made arbitrarily high by increasing the value of N.

Proof From the results of §2.7, we know that $s_\infty(\phi_N) = 2N - d_2(\phi_N)$ where $d_2(\phi_N)$ is defined through (2.7.19), i.e.

$$d_2(\phi_N) = \log_2\Big(\max_\omega \min\{m_N(\omega), \sqrt{m_N(\omega)m_N(2\omega)}\}\Big). \tag{2.10.15}$$

By a change of variable and using (2.10.12), we obtain

$$\begin{aligned} d_2(\phi_N) &= \log_2[\max_{x\in[0,1]} \min\{P_N(x), \sqrt{P_N(x)P_N(4x(1-x))}\}] \\ &\leq (N-1)\log_2[\max_{x\in[0,1]} \min\{G(x), \sqrt{G(x)G(4x(1-x))}\}]. \end{aligned}$$

From (2.10.13), we know that $\max_{[0,1]} \min\{G(x), \sqrt{G(x)G(4x(1-x))}\} = 3$, and thus

$$d_2(\phi_N) \leq (N-1)\log_2 3, \tag{2.10.16}$$

which yields (2.10.14). ◇

Remark 2.10.1 *For small values of N, one can obtain better estimates for the Hölder or Sobolev regularity, using the operator-based technique described in §2.7 and applying Theorem 2.7.1. Note that this technique becomes however unpractical for large values of N, due to the important size of the matrix representing the transition operator. As N goes to ∞, it can also be shown (see* VOLLKMER *[1991],* COHEN & RYAN *[1995, chapter 3]) that the estimate (2.10.14) is sharp in an asymptotical sense:*

$$\lim_{N\to\infty} \frac{s_\infty(\phi_N)}{2N} = \lim_{N\to\infty} \frac{s_p(\phi_N)}{2N} = \lim_{N\to\infty} \frac{\mu(\phi_N)}{2N} = 1 - \frac{\log_2 3}{2} \simeq 0.2.$$

This behaviour should be compared with the case of B-splines: the functions Φ_N are supported in $S_N = [-2N+1, 2N-1]$, i.e. their regularity grows asymptotically like $|S_N|/10$, while the regularity of the splines B_N is asymptotically equivalent to their support. This loss should be viewed as the price to pay for the interpolation constraint (2.9.8). Note that P_N is not the unique solution to (2.10.3). An open problem is the construction of a family of polynomial solutions of (2.10.3) that would optimize the ratio between the smoothness of the refinable function and its support, either in an asymptotical sense or for a prescribed support size.

We now turn to the construction of orthonormal scaling functions, following the ideas introduced in DAUBECHIES [1988]. It should be noted that most of the techniques that we have used in the previous discussion on the regularity of interpolatory scaling functions were also introduced in this paper, since, as we shall see, both constructions are closely related.

We are now looking for trigonometric polynomials $m^o(\omega)$ having the factorized form

$$m^o(\omega) = \Big(\frac{1+e^{-i\omega}}{2}\Big)^N p(\omega), \tag{2.10.17}$$

and such that the orthonormality constraint (2.9.7) is satisfied. Note that such a polynomial should necessarily satisfy

$$|m^o(\omega)|^2 = [\cos^2(\omega/2)]^N P(\sin^2(\omega/2)), \tag{2.10.18}$$

where $P(x)$ is a polynomial solution of (2.10.3) which is positive on $[0,1]$, or equivalently $|p(\omega)|^2 = P(\sin^2(\omega/2))$.

A natural idea is thus to look for a minimally supported solution

$$m_N^o(\omega) = \Big(\frac{1+e^{-i\omega}}{2}\Big)^N p_N(\omega), \tag{2.10.19}$$

where $p_N(\omega) = \sum_{n=0}^{N-1} c_n e^{-in\omega}$ is such that

$$|p_N(\omega)|^2 = P_N(\sin^2(\omega/2)), \tag{2.10.20}$$

or equivalently,

$$|m_N^o(\omega)|^2 = m_N(\omega). \tag{2.10.21}$$

The existence of p_N is ensured by the following result, originally due to F.Riesz.

Lemma 2.10.3 *Let $A(\omega) = a_0 + \sum_{n=1}^N a_n(e^{in\omega} + e^{-in\omega})$ be a symmetric trigonometric polynomial such that the a_n are real numbers and such that $A(\omega) > 0$. There exists a trigonometric polynomial $B(\omega) = \sum_{n=0}^N b_n e^{-in\omega}$ with real coefficients such that $|B(\omega)|^2 = A(\omega)$.*

Proof We introduce an algebraic polynomial

$$P_A(z) = z^N[a_0 + \sum_{n=1}^N a_n(z^n + z^{-n})], \tag{2.10.22}$$

of degree $2N$, which satisfies $A(\omega) = e^{-iN\omega}P(e^{in\omega})$. Since P_A has real coefficients and since $P_A(z^{-1}) = z^{-n}P_A(z)$, we see that

$$P_A(z) = 0 \Leftrightarrow P_A(\overline{z}) = 0 \Leftrightarrow P_A(z^{-1}) = 0 \Leftrightarrow P_A(\overline{z}^{-1}) = 0. \tag{2.10.23}$$

Since P_A does not vanish on the unit circle (because $m(\omega) > 0$), it can thus be factorized in the following way

$$P_A(z) = C\prod_{k=1}^{N_r}(z-r_k)(z-r_k^{-1})\prod_{k=1}^{N_z}(z-z_k)(z-z_k^{-1})(z-\overline{z}_k)(z-\overline{z}_k^{-1}),$$

where the r_k's are real and the z_k's are complex with non-zero imaginary part.

Since $N = N_r + 2N_z$, we can write

$$A(\omega) = e^{-i(N_r+2N_z)\omega}P(e^{in\omega}) = CA_r(\omega)A_z(\omega),$$

with

$$\begin{aligned} A_r(\omega) &= \prod_{k=1}^{N_r}(e^{i\omega}-r_k)(1-\tfrac{1}{r_k}e^{-i\omega}) \\ &= (-1)^{N_r}[\prod_{k=1}^{N_r} r_k]^{-1}|\prod_{k=1}^{N_r}(e^{-i\omega}-r_k)|^2, \end{aligned}$$

and

$$\begin{aligned} A_z(\omega) &= \prod_{k=1}^{N_z}(e^{i\omega}-z_k)(1-z_k^{-1}e^{-i\omega})(e^{i\omega}-\overline{z}_k)(1-\overline{z}_k^{-1}e^{-i\omega}) \\ &= [\prod_{k=1}^{N_z}|z_k|^2]^{-1}|\prod_{k=1}^{N_z}(e^{-i\omega}-z_k)(e^{-i\omega}-\overline{z}_k)|^2. \end{aligned}$$

It follows that $A(\omega) = |P_B(e^{-i\omega})|^2$ with

$$P_B(z) = D[\prod_{i=1}^{N_r}(z-r_i)\prod_{i=1}^{N_z}(z^2-(z_i+\overline{z}_i)z+|z_i|^2), \tag{2.10.24}$$

and $D = \left(C(-1)^{N_r}[\prod_{i=1}^{N_r} r_i]^{-1}[\prod_{i=1}^{N_z}|z_i|^2]^{-1}\right)^{1/2}$ (by the positivity of $A(\omega)$, we are ensured that the quantity in the square root is also positive). ◇

Note that this lemma is constructive: the coefficients of $m_N^o(\omega)$ can be derived from those of $m_N(\omega)$ through the study of the zeros of the associated algebraic polynomial defined in (2.10.22). We thus have constructed $m_N^o(\omega)$ which satisfies (2.10.21), and the corresponding orthonormal scaling functions φ_N. Note that we have

$$|\hat{\varphi}_N(\omega)|^2 = \hat{\phi}_N(\omega), \tag{2.10.25}$$

or equivalently

$$\phi_N(x) = \int \varphi_N(x+t)\varphi_N(t)dt. \tag{2.10.26}$$

As a consequence, Theorem 2.10.1 also provides an estimate for the regularity of φ_N given by

$$s_\infty(\varphi_N) \geq N(1-\frac{\log_2 3}{2}) \simeq 0.2\,N. \tag{2.10.27}$$

As an example, we display in Figure 2.10.3 and 2.10.4 the graphs of the functions φ_2 (of Hölder regularity $\mu \simeq 0.55$) and φ_5 (of Hölder regularity $\mu \geq 2$) together with their associated wavelets ψ_2 and ψ_5.

2.11 Wavelets and splines

Spline functions are of common use in the approximation of functions. They are particularly attractive because of their numerical simplicity. We shall now discuss different examples of multiscale decompositions into such functions. Here, we shall denote by φ_N the B-spline of degree $N-1$, i.e.

$$\varphi_N(x) = B_{N-1} = (*)^N \chi_{[0,1]}. \tag{2.11.1}$$

and we denote by

$$m_N(\omega) = \Big(\frac{1+e^{-i\omega}}{2}\Big)^N, \tag{2.11.2}$$

its associated symbol. We finally denote by V_j^N the spline approximation space generated by the functions $\{(\varphi_N)_{j,k}\}_{k\in\mathbb{Z}}$.

Example 1. Orthonormal and interpolatory spline wavelets
We recall from Corollary 2.2.2 that we can build an orthonormal scaling function φ_N^o, applying formula (2.2.35) to φ_N. For $N > 1$, the function φ_N^o is not compactly supported but has exponential decay at infinity and satisifies a refinement equation

$$\varphi_N^o(x) = \sum_{n\in\mathbb{Z}} h_n \varphi_N^o(2x-n), \tag{2.11.3}$$

where the coefficients $h_n = 2\langle\varphi_N^o(x), \varphi_N^o(2x-n)\rangle$ also have geometric decay at infinity. One way to estimate the coefficients h_n numerically is by computing the Fourier series $H_N(\omega) = \frac{1}{2}\sum_{n\in\mathbb{Z}} h_n e^{-in\omega}$: combining (2.2.35) together with $\hat\varphi_N(\omega) = m_N(\omega/2)\hat\varphi_n(\omega/2)$ and $\hat\varphi_N^o(\omega) = H_N(\omega/2)\hat\varphi_n^o(\omega/2)$ we obtain

$$H_N(\omega) = \Big(\frac{S_{\varphi_N}(\omega)}{S_{\varphi_N}(2\omega)}\Big)^{1/2} m_N(\omega). \tag{2.11.4}$$

We also have

$$\begin{aligned} S_{\varphi_N}(\omega) &= \textstyle\sum_{n\in\mathbb{Z}} |\hat\varphi_N(\omega+2n\pi)|^2 \\ &= [2\sin^2(\omega/2)]^N \textstyle\sum_{n\in\mathbb{Z}} \frac{1}{|\omega+2n\pi|^{2N}} \\ &= [2\sin^2(\omega/2)]^N R_N(\omega), \end{aligned}$$

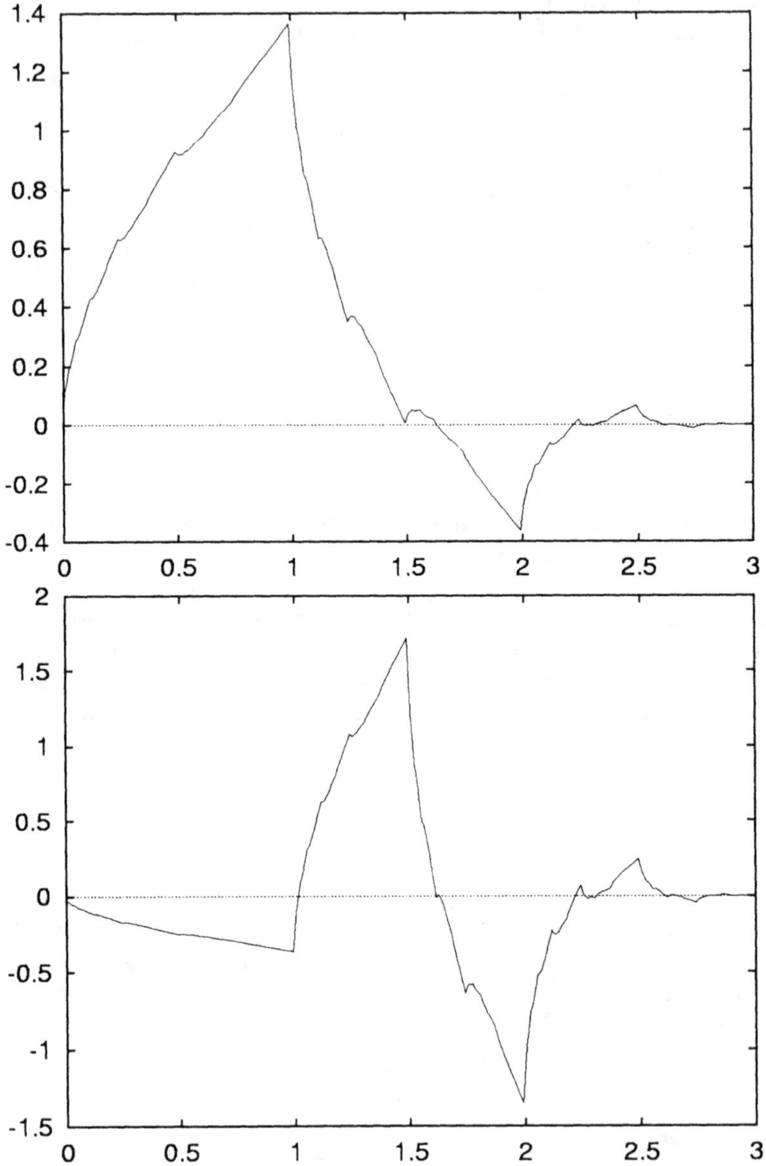

Figure 2.10.3: Orthonormal scaling function and wavelet ($N = 2$)

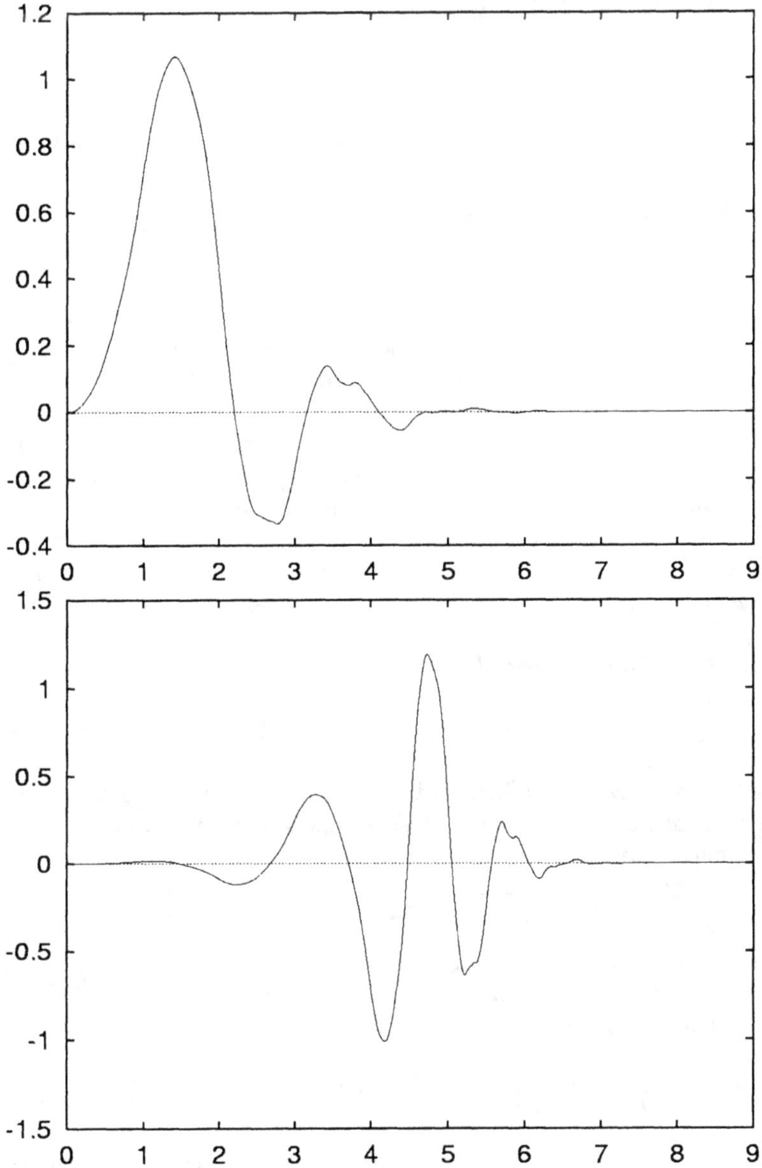

Figure 2.10.4: Orthonormal scaling function and wavelet for $(N = 5)$

where R_N can be computed by induction since $R_1(\omega) = [2\sin^2(\omega/2)]^{-1}$ (due to the orthonormality of φ_1) and

$$R_{N+1}(x) = \frac{R_N''(x)}{2N(2N+1)}. \tag{2.11.5}$$

Orthonormal wavelets ψ_N^o can then be derived according to the formulae of §2.6, which give here

$$\psi_N^o(x) = \sum_{n\in\mathbb{Z}}(-1)^n h_{1-n}\varphi_N^o(2x-n). \tag{2.11.6}$$

This construction was introduced by LEMARIÉ [1988]. Note that for practical computations, the sequence h_n needs to be truncated and the resulting error has to be taken into account.

In a similar way, for even values of N (i.e. for spline of odd order $N-1$), the interpolatory function φ_N^i, defined by applying (2.2.37) to φ_N, also satisfies a refinement equation involving an infinite number of coefficients (when $N > 2$) that can be estimated through the computation of their Fourier series. The corresponding wavelet is then simply given by $\psi_N^i(x) = \varphi_N^i(2x-1)$, as in the case of the Schauder hierarchical basis (and of all wavelets generated from interpolatory scaling functions).

Example 2. Biorthogonal spline wavelets

In order to obtain local multiscale decompositions into spline functions, one needs to give up on orthonormality and work in a biorthogonal setting. The construction of compactly supported refinable functions that are dual to the functions φ_N was addressed in COHEN, DAUBECHIES and FEAUVEAU [1992]. According to the results of §2.9, this construction amounts to finding trigonometric polynomial solutions $\tilde{m}$ to the equation

$$\tilde{m}(\omega)\overline{m_N(\omega)} + \tilde{m}(\omega+\pi)\overline{m_N(\omega+\pi)} = 1, \tag{2.11.7}$$

and to studying the regularity of the corresponding $\tilde{\varphi}$. A family of solutions can be immediately derived using the polynomial P_N defined by (2.10.11): for $L \geq 0$ such that $N + L = 2M \in 2\mathbb{Z}$, we have

$$\cos^{2M}(\omega/2)P_M(\sin^2(\omega/2)) = e^{i\frac{N-L}{2}\omega}\left(\frac{1+e^{-i\omega}}{2}\right)^L P_M(\sin^2(\omega/2))\overline{m_N(\omega)},$$

which implies, since P_N is a solution of (2.10.3), that the trigonometric polynomials

$$\tilde{m}_{N,L} = e^{i\frac{N-L}{2}\omega}\left(\frac{1+e^{-i\omega}}{2}\right)^L P_{\frac{N+L}{2}}(\sin^2(\omega/2)), \tag{2.11.8}$$

are solutions of (2.11.7). We denote by $\tilde{\varphi}_{N,L}$ the corresponding scaling functions. The following result gives a criterion for the choice of L in order to build a regular dual function.

Theorem 2.11.1 *The functions $\tilde{\varphi}_{N,L}$ satisfy*

$$s_\infty(\tilde{\varphi}_{N,M}) \geq L - (N + L - 2)\frac{\log_2 3}{2}) \simeq 0.2\,L - 0.8\,N + 1.6. \qquad (2.11.9)$$

Consequently, their regularity can be made arbitrarily high by increasing the value of L for a fixed value of N. If

$$L > (1 - \frac{\log_2 3}{2})^{-1}(1/2 + (N-2)\frac{\log_2 3}{2}), \qquad (2.11.10)$$

the function $\tilde{\varphi}_{N,L}$ is in L^2 and forms a dual pair together with φ_N.

Proof By a similar reasoning as in the proof of Theorem 2.10.1, we obtain the estimate

$$d_2(\tilde{\varphi}_{N,L}) \leq (\frac{N+L}{2} - 1)\log_2 3, \qquad (2.11.11)$$

which yields (2.11.9) by $s_\infty(\tilde{\varphi}_{N,L}) \geq L - d_2(\tilde{\varphi}_{N,L})$. In particular, (2.11.10) implies that $d_2(\tilde{\varphi}_{N,M}) < L - 1/2$, so that we obtain a dual pair by Theorem 2.9.3. ◇

Remark 2.11.1 *The estimate (2.11.10) of the minimal value of L is crude and can be improved by using a technique based on the spectral analysis of a suitable finite dimensional operator, (see* COHEN *and* DAUBECHIES *[1992]) for small values of N. In particular, the minimal values of L for $N = 2, 3, 4$ are respectively $L = 2, 3, 6$.*

From φ_N and $\tilde{\varphi}_{N,L}$ we can then obtain the wavelets $\psi_{N,L}$ and $\tilde{\psi}_{N,L}$ applying formula (2.6.7). As an example, we display in Figure 2.11.1 the graphs of $\tilde{\varphi}_{2,L}$ dual to the (centred) hat function $\varphi_2 = B_1(x+1)$, and of the wavelets $\psi_{2,2}$ and $\tilde{\psi}_{2,2}$.

Remark 2.11.2 *From the definition of m_N and $\tilde{m}_{N,L}$, we see that both scaling coefficients of φ_N and $\tilde{\varphi}_{N,L}$ are dyadic numbers, i.e. numbers of the type $2^{-j}k$, $j,k \in \mathbb{Z}$. This property can be of particular interest in the practical implementation of the decomposition and reconstruction algorithms, since such dyadic numbers have by definition binary expansions without round-off errors.*

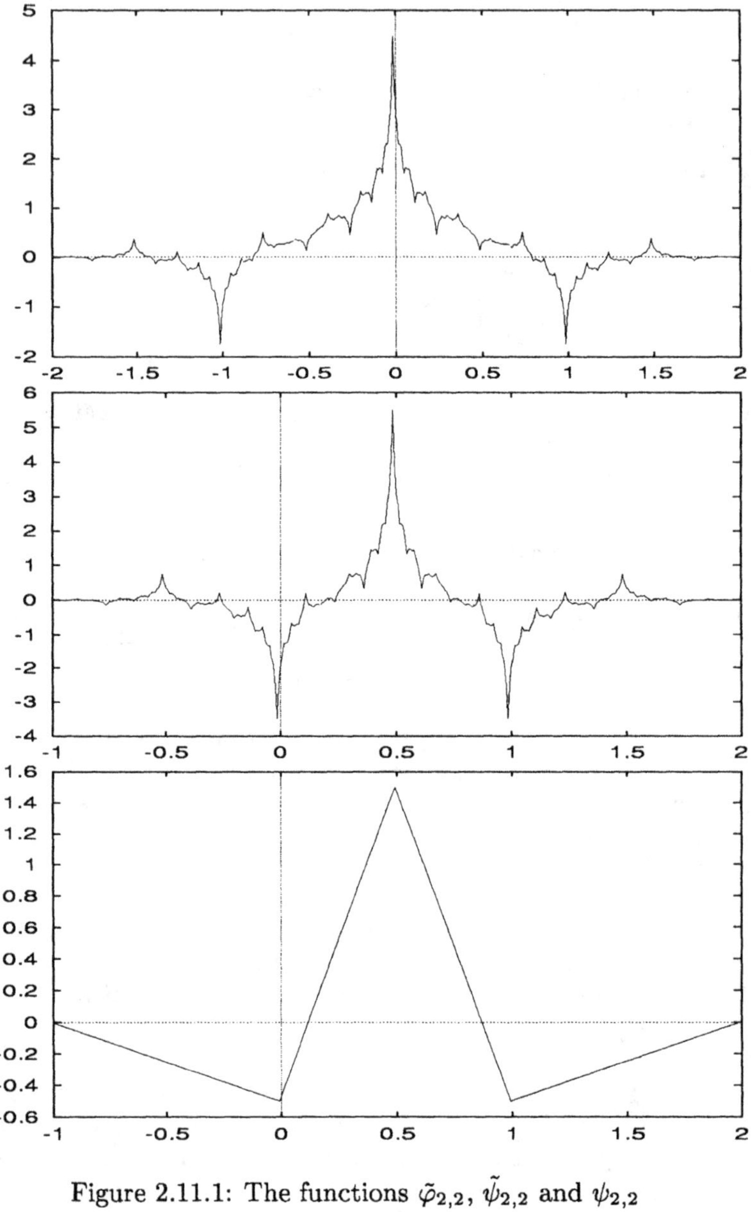

Figure 2.11.1: The functions $\tilde{\varphi}_{2,2}$, $\tilde{\psi}_{2,2}$ and $\psi_{2,2}$

Remark 2.11.3 *From the expression of $\tilde{m}_{N,L}$, we can obtain the recursion formula*

$$\tilde{\varphi}_{N,L} = \tilde{\varphi}_{N,L-1} * \chi_{[0,1]}, \tag{2.11.12}$$

similar to the formula for the B-spline. More generally, if φ and ϕ are refinable functions with symbols $m(\omega)$ and $M(\omega) = \left(\frac{1+e^{-i\omega}}{2}\right)m(\omega)$, we have $\phi = \varphi * \chi_{[0,1]}$ which can also be expressed by a differential relation

$$\phi'(x) = \varphi(x) - \varphi(x-1). \tag{2.11.13}$$

One can then easily check the following facts which were firstly remarked in LEMARIÉ *[1993]: if $\tilde{\phi}$ is a dual function to ϕ with symbol $\tilde{M}(\omega)$, then $\tilde{\varphi}$ defined through the symbol $\tilde{m}(\omega) = \left(\frac{1+e^{i\omega}}{2}\right)\tilde{M}(\omega)$ satisfies*

$$\tilde{\varphi}'(x) = \tilde{\phi}(x+1) - \tilde{\phi}(x), \tag{2.11.14}$$

and it constitutes a dual scaling function to φ. The associated wavelets are also linked by even simpler relations: if ψ, $\tilde{\psi}$, η and $\tilde{\eta}$ denote the wavelets constructed from the functions φ, $\tilde{\varphi}$, ϕ and $\tilde{\phi}$, we have $\psi = \eta'$ and $\tilde{\eta} = -\tilde{\psi}'$. Finally, the associated projectors satisfy commutation formulae of the following type:

$$\Big[\sum_k \langle f, \tilde{\phi}_{j,k}\rangle \phi_{j,k}\Big]' = \sum_k \langle f', \tilde{\varphi}_{j,k}\rangle \varphi_{j,k}, \tag{2.11.15}$$

and

$$\Big[\sum_k \langle f, \tilde{\eta}_{j,k}\rangle \eta_{j,k}\Big]' = \sum_k \langle f', \tilde{\psi}_{j,k}\rangle \psi_{j,k}. \tag{2.11.16}$$

These commutation formulae play a crucial role in certain multivariate constructions where an additional differential constraint is imposed, such as divergence-free wavelets in LEMARIÉ *[1992] and* URBAN *[1995] or multiscale approximation spaces that satisfy the so-called LBB condition in* DAHMEN, KUNOTH *and* URBAN *[1996] for the discretization of velocity and pressure in the mixed formulation of the Stokes problem. Applied to our biorthogonal splines they mean that the derivative of the oblic projection of f onto V_j^N, relative to the dual index L, can be identified with the oblic projection of f' onto V_j^{N-1}, relative to the index $L+1$.*

Figure 2.11.1 reveals that the function $\psi_{2,2}$ can be obtained from the hierarchical basis generator $\psi = \varphi_2(2x)$ by a *coarse grid correction*, i.e. combination of scaling functions at scale $j = 0$. More precisely, we have

$$\psi_{2,2}(x) = 2\psi(x) - [\varphi_2(x) - \varphi_2(x+1)]/2. \tag{2.11.17}$$

More generally, if one is given a set of compactly supported biorthogonal scaling functions and wavelets $(\varphi, \tilde{\varphi}, \psi, \tilde{\psi})$, a new biorthogonal system $(\varphi^n, \tilde{\varphi}^n, \psi^n, \tilde{\psi}^n)$ can be defined (at least in a distributional sense) using such a correction technique. The new wavelet is given by

$$\psi^n(x) = \psi(x) - \sum_{n=N_0}^{N_1} c_n \varphi(x-n), \tag{2.11.18}$$

while the scaling function is left unchanged, i.e. $\varphi^n = \varphi$. On the dual side, this correspond to a modification of the dual scaling function which is now solution of

$$\tilde{\varphi}^n(x) = \sum_n [\tilde{h}_n + \sum_k c_k \tilde{g}_{2k+n}] \tilde{\varphi}^n(2x-n), \tag{2.11.19}$$

where $\tilde{h}_n$ and $\tilde{g}_n$ are the scaling coefficients associated with $\tilde{\varphi}$ and $\tilde{\psi}$. The new dual wavelet is given by the same equation $\tilde{\psi}^n(x) = \sum_n \tilde{g}_n \tilde{\varphi}^n(2x-n)$.

This procedure, known as the ***lifting scheme*** was introduced in SWELDENS [1996]. It ensures that the new dual scaling functions and wavelets have also compact support (since $\tilde{\varphi}^n$ is solution of (2.11.19)). In practice, one often designs the coefficients c_n so that the new wavelet basis has better properties than the old one, e.g. more regularity on the dual side. This is exactly what happens in the case of linear splines that we have discussed: the new dual function $\tilde{\varphi}_{2,2}$ is square integrable (it also has some Sobolev smoothness), in contrast to the Dirac function associated to the decomposition in the hierarchical basis. The main interest of the lifting scheme is its flexibility. In particular, one can easily adapt the above correction process to the hierarchical basis restricted to a bounded interval with some prescribed boundary conditions. Consider for example the spaces V_j^D introduced in §1.3 which are generated by the functions $(\varphi_2)_{j,k}$, $k = 1, \cdots, 2^j - 1$ and incorporate the homogeneous Dirichlet conditions $f(0) = f(1) = 0$. For $j \geq 2$, we can adapt the correction procedure near the boundary to produce a basis for a complement of V_j^D into V_{j+1}^D in terms of the $(\psi_{2,2})_{j,k}$, $k = 1, \cdots, 2^j - 2$ and two boundary adapted functions $2^{j/2}\psi^b(2^j\cdot)$ and $2^{j/2}\psi^b(2^j(1-\cdot))$, where $\psi^b := \psi - \frac{1}{2}\varphi(\cdot - 1)$, which both have one vanishing moment. It is also possible to check that the dual functions corresponding to such a correction process have some positive Sobolev smoothness.

A more general procedure of ***stable completion*** was introduced in CARNICER, DAHMEN and PEÑA [1996] which also authorizes to correct the function ψ with fine grid functions $\varphi(2\cdot - n)$, resulting however in dual functions which are not compactly supported. In §2.13, we shall describe these procedures in a very general framework of discrete multiresolution decompositions which allows us to consider unstructured discretizations, and we shall also use them to build wavelet bases in finite element spaces.

In the next two sections, we discuss examples of wavelets which are built directly as combinations of the fine grid functions $\varphi(2\cdot -n)$ and are instances of stable completion procedures. In particular, these wavelet bases have infinitely compactly supported duals, and the analysis of the properties of these bases are usually more difficult (in particular for Example 4, despite the apparent simplicity of the construction). Note however that an infinite support for the dual function is not a real limitation for several practical applications where only the primal basis is needed. This is the case, in particular, when solving a PDE by a Galerkin method, since one uses the primal basis both for expressing the solution and testing the equation.

Example 3. Semi-orthogonal spline wavelets
A possible approach to the construction of spline wavelet bases consists in looking directly for a compactly supported function $\psi_N \in V_1^N$ orthogonal to V_0^N and such that $\psi_N(\cdot - k)$, $k \in \mathbb{Z}$ forms a Riesz basis of the orthogonal complement W_0^N of V_0^N into V_1^N.

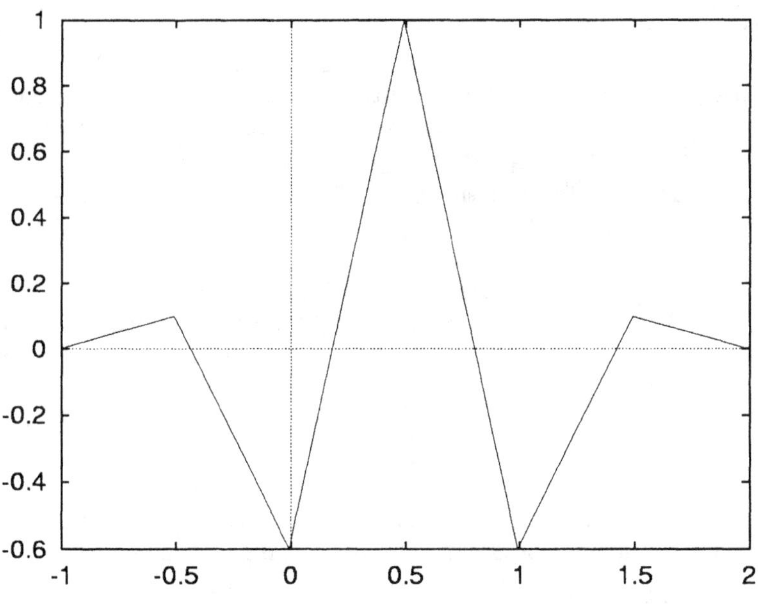

Figure 2.11.2: Semi-orthogonal spline wavelet

In Chui and Wang [1992], it was proved that

$$\psi_N(x) = 2^{1-N} \sum_{k=0}^{2N-2} (-1)^k \varphi_{2N}(k+1)\varphi_{2N}^{(N)}(2x-k), \tag{2.11.20}$$

is the solution of minimal support, among all generators of W_0^N.

Once we are given ψ_N, the system

$$\{\varphi_N(\cdot - k)\}_{k\in\mathbb{Z}} \cup \{(\psi_N)_{j,k}\}_{j\geq 0, k\in\mathbb{Z}}, \tag{2.11.21}$$

is easily proved to be a Riesz basis for the whole of L^2, since the stability across different levels is ensured by the orthogonality of the W_j spaces. The biorthogonal system is then given by

$$\{\varphi_N^d(\cdot - k)\}_{k\in\mathbb{Z}} \cup \{(\psi_N^d)_{j,k}\}_{j\geq 0, k\in\mathbb{Z}}, \tag{2.11.22}$$

where φ_N^d and ψ_N^d are constructed by applying formula (2.2.36) to φ_N and ψ_N. We have thus gained compact support of the basis function and orthogonality across levels, at the cost of having a non-compactly supported dual system.

As an example, we display in Figure 2.11.2 the function ψ_2 which is given by $\psi_2 := \varphi(2\cdot -1) - \frac{3}{5}(\varphi(2\cdot) + \varphi(2\cdot -2)) + \frac{1}{10}(\varphi(2\cdot +1) + \varphi(2\cdot -3))$. Similarly to the previous example, we can adapt this construction near the boundary in the case of multiresolution spaces on a bounded interval with prescribed boundary conditions. Consider again the spaces V_j^D: it is easily seen that for $j \geq 2$ a Riesz basis for the orthogonal complement of V_j^D into V_{j+1}^D is given by the $(\psi_2)_{j,k}$, $k = 1, \cdots, 2^j - 2$ and two boundary adapted functions $2^{j/2}\psi^b(2^j\cdot)$ and $2^{j/2}\psi^b(2^j(1-\cdot))$, where $\psi^b := \frac{9}{10}\varphi(2\cdot -1) - \frac{3}{5}\varphi(2\cdot -2) + \frac{1}{10}\varphi(2\cdot -3)$.

Example 4. Fine grid correction spline wavelets

If we drop the requirement of orthonormality, we can address the more general problem of constructing a compactly supported non-orthogonal wavelet basis by choosing

$$\psi_N(x) = \sum_{n=N_1}^{N_2} a_n \varphi_N(2s - n), \tag{2.11.23}$$

where a_n is an "oscillating" sequence (i.e. $\sum_n a_n = 0$) of finite length.

So far, the general problem of finding a minimal sequence a_n such that the system (2.11.19) (with ψ_N now defined by (2.11.23)) is a Riesz basis of $L^2(\mathbb{R})$ is unsolved. In practice, we have in mind simple sequences such as $(a_0, a_1) = (1, -1)$ or $(a_0, a_1, a_2) = (-1, 2, -1)$, etc. This problem was

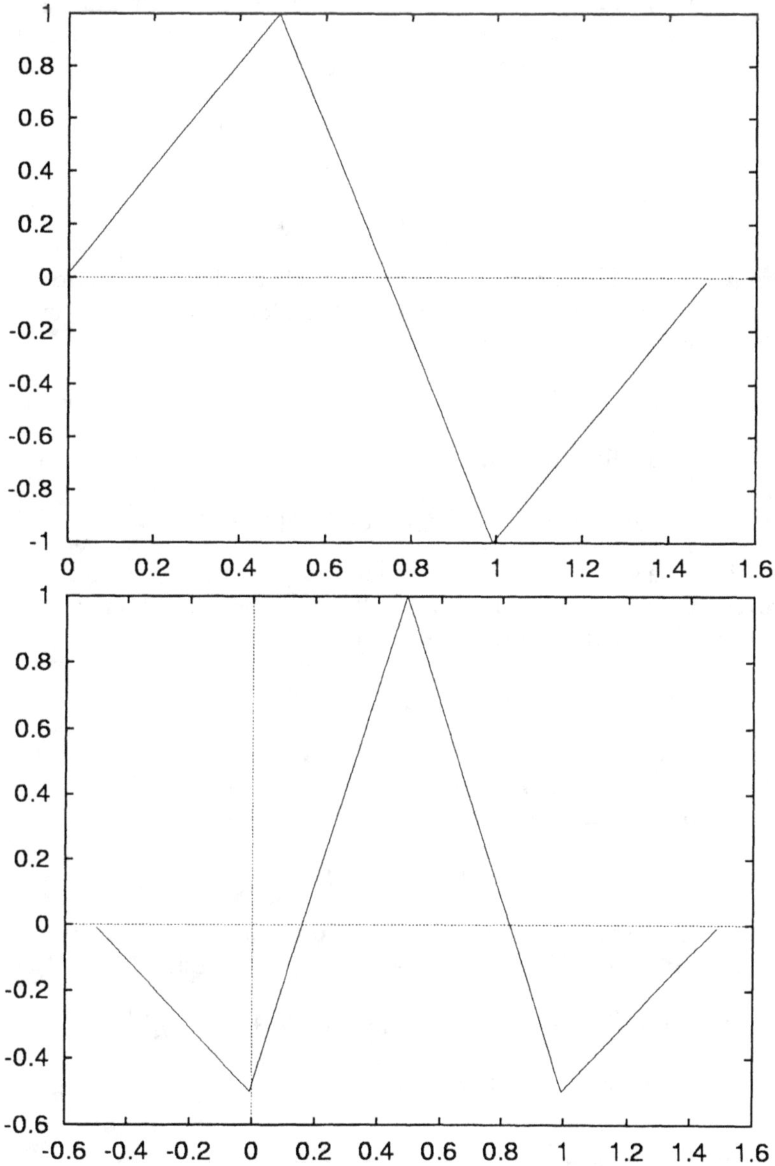

Figure 2.11.3: Non-orthogonal spline wavelets

studied in the case of linear splines, i.e. $N = 2$, in COHEN and DAUBECHIES [1996] and OSWALD and LORENTZ [1996] where it is proved that both choices give rise to a multiscale Riesz basis. As in the previous construction, the biorthogonal system is not compactly supported. We display these non-orthogonal wavelets in Figure 2.11.3.

More generally, in order to study the stability of the multiscale basis generated by ψ_N, we note that (2.11.23) rewrites

$$\hat{\psi}_N(\omega) = a(\omega/2)\hat{\varphi}_N(\omega/2), \tag{2.11.24}$$

where $a(\omega) = \frac{1}{2}\sum_n a_n e^{-in\omega}$. In the case where the construction of $\tilde{\psi}_N$ is related to a dual refinable function (e.g. the biorthogonal spline of Example 2), we know that $a(\omega)$ is related to the dual symbol through

$$a(\omega) = e^{-i\omega}\overline{\tilde{m}(\omega + \pi)}. \tag{2.11.25}$$

The biorthogonality contraint (2.11.7) leads to the equation

$$e^{i\omega}[m_N(\omega)a(\omega + \pi) - m_N(\omega + \pi)a(\omega)] = 1. \tag{2.11.26}$$

that should be solved by $a(\omega)$. This equation is a strong restriction on $a(\omega)$. Hovever, we remark that if $p(\omega)$ is a 2π periodic continuous function that does vanish, then the function $\eta_N(x)$ defined by

$$\hat{\eta}_N(\omega) = p(\omega)\hat{\psi}_N(\omega) = p(\omega)a(\omega/2)\hat{\varphi}_N(\omega/2), \tag{2.11.27}$$

has the property that $(\psi_N)_{j,k}$, $k \in \mathbb{Z}$ is a Riesz basis of a space W_j if and only if $(\eta_N)_{j,k}$, $k \in \mathbb{Z}$ is a Riesz basis for the same space. We can thus relax the equation (2.11.26) by only requiring that it is satisfied by

$$b(\omega) = p(2\omega)a(\omega), \tag{2.11.28}$$

for some non-vanishing continuous function $p(\omega)$. Note that by combining (2.11.28) and (2.11.26), we obtain an explicit expression for $p(\omega)$ as

$$p(2\omega) = e^{-i\omega}[m_N(\omega)a(\omega + \pi) - m_N(\omega + \pi)a(\omega)]^{-1}, \tag{2.11.29}$$

and for the dual symbol $\tilde{m}(\omega)$ as

$$\tilde{m}(\omega) = e^{-i\omega}\overline{b(\omega + \pi)} = \frac{\overline{a(\omega + \pi)}}{\overline{m}_N(\omega)\overline{a}(\omega + \pi) - \overline{m}_N(\omega + \pi)\overline{a}(\omega)}. \tag{2.11.30}$$

The stability of the wavelet basis generated by ψ_N will thus hold if $p(\omega)$ defined by (2.11.29) is bounded, and if the dual symbol $\tilde{m}(\omega)$ defined by

(2.11.30) satisfies together with $m(\omega)$ the conditions of Theorem 2.9.3 for generating a pair of dual refinable functions with $\tilde{\varphi} \in L^2$. Since $\tilde{m}$ is not in general a trigonometric polynomial, the dual system to $(\psi_N)_{j,k}$ is not compactly supported.

Example 5. Multiwavelets
Multiwavelets are obtained from a natural generalization of multiresolution analysis, where one allows V_0 to be generated by more than one scaling function, i.e. there exist $\varphi_1, \cdots, \varphi_r \in V_0$ such that

$$\varphi_q(\cdot - k), \quad k \in \mathbb{Z}, \, q = 1, \cdots, r, \, k \in \mathbb{Z}, \tag{2.11.31}$$

constitutes a Riesz basis of V_0. The sequence V_j is then called a ***multiresolution analysis of multiplicity*** r. Many of the theoretical results that we have described in §2.3 to §2.9 can be extended to this setting: JIA and LEI [1993] describes the appropriate Strang-Fix conditions, PLONKA [1995] gives the generalization of the factorization of the symbol $m(\omega)$ (which is now a matrix valued trigonometric polynomial), COHEN, DAUBECHIES & PLONKA [1997], SHEN [1998] and JIA, RIEMENSCHNEIDER & ZHOU [1998] generalize the smoothness analysis of for the scaling function vector $(\varphi_1, \cdots, \varphi_r)$.

Several interesting examples of multiwavelets involve spline functions.

First, if V_j consists of the functions which are piecewise polynomial of degree N on the intervals $I_{j,k}$, $k \in \mathbb{Z}$, *with no continuity requirement*, a simple orthonormal basis has the form (2.11.31) with

$$\varphi_q = \sqrt{2} L_q(2x - 1)\chi_{[0,1]}, \tag{2.11.32}$$

where L_q is the Legendre polynomial of degree q. In this case, the derivation of orthonormal wavelets (which are also supported on the unit interval) is very simple. These particular wavelets have been introduced in ALPERT [1993] for the purpose of discretizing integral equations that do not require smoothness of the basis functions. One of their interests is their flexibility: since orthogonality is ensured by disjointness of the supports, one can easily adapt them to unstructured discretizations and to triangulations in higher dimensions. Moreover the derivation of orthogonal wavelets reduces to local problems on each support.

More generally, splines with multiple knots (i.e. piecewise polynomials of degree N on which we do not impose up to C^{N-1} smoothness) can be put in this setting. A very simple example is the case of piecewise cubic C^1 functions, for which a natural basis is given by the nodal functions of Hermite interpolation, i.e.

$$\varphi_1(k) = \varphi_2'(k) = \delta_{0,k} \text{ and } \varphi_2(k) = \varphi_1'(k) = 0. \tag{2.11.33}$$

However, the construction of stable dual refinable functions for this type of basis is still an open problem. Recently, the particular case of cubic Hermite splines was treated in DAHMEN, HAN, JIA and KUNOTH [2000] and a general approach to solve this problem was proposed in DYN [2000].

Finally, it was shown in DONOVAN, GERONIMO and HARDIN [1999] that it is possible to construct a set of spline scaling functions $\varphi_1, \cdots, \varphi_q$ with compact support such that (2.11.31) is an orthonormal basis for a spline space V_0^i which is *intertwined* between V_n and V_{n+1} for some $n > 0$ (where V_j is a standard spline multiresolution analysis). Orthonormality can thus be obtained together with compact support, up to a slight modification of the multiresolution approximation spaces.

2.12 Bounded domains and boundary conditions

In this section we shall discuss multiresolution decompositions in the multivariate setting, as well as some commonly used strategies for adapting them to a bounded domain $\Omega \in \mathbb{R}^d$. Another frequently used approach, based on multilevel splitting of finite element spaces, will be discussed in §2.13.

The construction of multiresolution analysis and wavelets that we have described for functions defined on $\mathbb{R}$ can easily be extended to multivariate functions on the whole of $\mathbb{R}^d$. The simplest method by far is the *tensor product* strategy that we already discussed in the setting of the basic examples of the previous chapter (§1.4 of Chapter 1).

In contrast, the adaptation to a bounded domain and the prescription of boundary condition is a delicate task. Several strategies have been proposed to deal with this problem, but so far, none of them has received a full approval in the sense that their efficiency seems to be highly dependent on the type of application.

In a general setting, let us first recall the main principles of the tensor product construction: from a univariate scaling function φ, we define its multivariate version

$$\phi(x) = \varphi(x_1) \cdots \varphi(x_d), \quad \text{for } x = (x_1, \cdots, x_d) \in \mathbb{R}^d. \tag{2.12.1}$$

The multivariate space V_j is then generated by the functions

$$\phi_{j,k}(x) = 2^{dj/2}\phi(2^j - k), \quad k \in \mathbb{Z}^d. \tag{2.12.2}$$

Given a dual function $\tilde{\varphi}$, we define $\tilde{\phi}$ dual to ϕ in a similar way, which yields the L^2-bounded projectors defined by

$$P_j f = \sum_{k \in \mathbb{Z}^d} \langle f, \tilde{\phi}_{j,k} \rangle \phi_{j,k}. \tag{2.12.3}$$

In order to characterize the details, we introduce $2^d - 1$ wavelet functions

$$\psi^\varepsilon(x) = \psi^{\varepsilon_1}(x_1) \cdots \psi^{\varepsilon_d}(x_d), \quad \varepsilon = (\varepsilon_1, \cdots, e_d) \in \{0,1\}^d - \{0\}, \tag{2.12.4}$$

where we have set $\psi^0(x) = \varphi(x)$ and $\psi^1(x) = \psi(x)$. We define similarly the dual wavelets $\tilde{\psi}^\varepsilon$ so that we have

$$Q_j f = P_{j+1} f - P_j f = \sum_{\varepsilon \in \{0,1\}^d - \{0\}} \sum_{k \in \mathbb{Z}^d} \langle f, \tilde{\psi}^\varepsilon_{j,k} \rangle \psi^\varepsilon_{j,k}, \tag{2.12.5}$$

where $\psi^\varepsilon_{j,k}$ and $\tilde{\psi}^\varepsilon_{j,k}$ are defined by a similar rescaling as in (2.12.2). From a computational point of view, the multiscale decomposition and reconstruction algorithms are performed in a separable manner, applying at each stage $j \to j+1$ in the reconstruction (resp. $j+1 \mapsto j$ in the decomposition) the filters h_n and g_n after upsampling (resp $\tilde{h}_n$ and $\tilde{g}_n$ before downsampling) successively in each direction.

Remark 2.12.1 *As we already remarked in §1.4 of Chapter 1, an alternate tensor product multiscale decomposition makes use of the full tensor product system defined by*

$$\psi_{j,k}(x_1, \cdots, x_d) = \psi_{j_1,k_1}(x_1) \cdots \psi_{j_d,k_d}(x_d), \tag{2.12.6}$$

where $j = (j_1, \cdots, j_d)$ *and* $k = (k_1, \cdots, k_d)$. *If* $j_{\min} < j_{\max}$, *using the convention* $\psi_{j,k} = \varphi_{j_{\min},k}$ *for* $j = (j_{\min} - 1, \cdots, j_{\min} - 1)$, *we obtain a multiscale decomposition of a function* $f \in V_{j_{\max}}$ *into*

$$f = \sum_{j_{\min} - 1 \le j_1, \cdots, j_d < j_{\max}} \sum_{k \in \mathbb{Z}^d} d_{j,k} \psi_{j,k}, \tag{2.12.7}$$

where $d_{j,k} = \langle f, \tilde{\psi}_{j,k} \rangle$, *with* $\tilde{\psi}_{j,k}$ *defined in an analogous manner. The associated decomposition and reconstruction algorithms consist in applying the univariate algorithms between the extreme values of the scale, successively in each direction.*

An advantage of the tensor product construction is that it inherits all the properties of the associated univariate multiscale decomposition: smoothness and local support of the basis functions, fast $\mathcal{O}(N)$ transformation algorithms, polynomial reproduction properties (if the univariate φ satisfies the Strang-Fix conditions at the order L, then the multivariate polynomials $x^k = x_1^{q_1} \cdots x_d^{q_d}$, $q_1, \cdots, q_d \leq L$ are reproduced by the projectors P_j).

However, this construction excludes some very classical approximation spaces such as $\mathbb{P}_1$ finite elements defined on a regular grid: if we consider the space V_0 consisting of all bivariate continuous L^2 functions that are affine on the triangles $[(m,n),(m+1,n),(m,n+1)]$ and $[(m,n),(m-1,n),(m,n-1)]$ and its scaled versions $V_j = \{f \, ; \, f(2^{-j}\cdot) \in V_0\}$, we find that the spaces V_j form an embedded sequence of approximation spaces with Riesz bases $\phi_{j,k}$, $k \in \mathbb{Z}^2$, where the generator ϕ is the hat function

$$\phi \in V_0, \quad \phi(m,n) = \delta_{0,m}\delta_{0,n}, \tag{2.12.8}$$

which is not of tensor product type. This very simple example motivates the study of multiresolution spaces generated by non-tensor product functions.

From an algorithmic point of view, this means that we are using non-separable filters in a multivariate generalization of subband decompositions that were described in §2.6, but we are keeping the same type of downsampling and upsampling operations (one sample out of two in each direction) which can be represented by the dilation matrix $2I$ that maps $2^{-j}\mathbb{Z}^d$ to $2^{-j+1}\mathbb{Z}^d$. Another generalization, which was essentially considered in multivariate signal processing, consists in using for these operations a dilation matrix D that differs from $2I$. In order to define proper downsampling and upsampling operators, this matrix should have integer entries (in order to map $\mathbb{Z}^d$) into itself and be *expanding*, i.e. all eigenvalues λ_i of D satisfy $|\lambda_i| > 1$. It is thus natural to introduce the following generalization of definition 10.1.

Definition 2.12.1 *A multiresolution analysis of $L^2(\mathbb{R}^d)$ is a nested and dense sequence of closed approximation subspaces*

$$\{0\} \cdots \subset V_{-1} \subset V_0 \subset V_1 \subset \cdots L^2(\mathbb{R}^d), \tag{2.12.9}$$

that satisfies the scaling relation

$$f \in V_j \Leftrightarrow f(D\cdot) \in V_{j+1}, \tag{2.12.10}$$

for a fixed expanding matrix D with integer entries, and such that V_0 is generated by a Riesz basis of the type $\{\phi(\cdot - k)\}_{k \in \mathbb{Z}^d}$, for some $\phi \in V_0$.

In this more general setting, the refinability property is expressed by a relation of the type

$$\phi(x) = \sum_{n \in \mathbb{Z}^d} h_n \phi(Dx - n), \tag{2.12.11}$$

and one can also require the existence of a dual scaling function $\tilde{\phi}$ with similar refinability properties. A projector P_j of the the type (2.12.3) is then obtained, using the functions

$$\phi_{j,k} = |\det(D)|^{j/2}\phi(D^j \cdot -k), \quad k \in \mathbb{Z}^d \tag{2.12.12}$$

and $\tilde{\phi}_{j,k}$ defined in an analogous manner.

In the attempt to mimic step by step the construction of such refinable functions and related multiscale decompositions that we have described in the previous sections for univariate functions, one is however faced with several difficulties that we want to point out here. Firstly, the derivation of wavelets from the scaling functions is no longer straightforward as in §2.6, except for some particular choice of dilation matrices or refinable functions. Secondly, the relations between the cancellations of the symbol $m(\omega_1, \cdots, \omega_d)$ and the polynomial reproduction properties of the shifts of ϕ, expressed by the Strang-Fix conditions in §2.8, are more intricate. Thirdly, the smoothness analysis cannot be done in the same way as we presented in §2.7 since there is no simple multivariate analog to the factorized form $m(\omega) = \left(\frac{1+e^{-i\omega}}{2}\right)^L p(\omega)$. Finally, for the design of orthonormal scaling function, there is no analogous result to the Riesz factorisation Lemma 2.10.3.

Concerning the second and the third difficulties, the reader will find in COHEN, GRÖCHENIG and VILLEMOES [1999] and JIA [1999], other techniques, involving transition operators, to measure the regularity without the help of a factorization of $m(\omega)$. Nevertheless, these difficulties explain the small number of existing examples of non-separable multiscale decompositions that are of practical use. In the context of numerical simulation, the most relevant nonseparable constructions are related to finite element spaces. They are discussed in the next section. Generally speaking, the tensor product construction remains by far the simplest way to generalize multiscale decomposition to functions defined on $\mathbb{R}^d$.

In most numerical applications, one is faced with the task of decomposing a function which is defined on a *bounded domain* $\Omega \subset \mathbb{R}^d$. We thus need to adapt the ideas of multiscale approximation and decomposition to this setting. A possibility is clearly to extend a function f defined on Ω to the whole of $\mathbb{R}^d$, and apply the techniques that we have described for $L^2(\mathbb{R}^d)$. However, the effectiveness of such a method is strongly dependent on the type of extension which is used: a simple extension by zero, or by periodization in the case of tensor product domains, has the effect of generating an artificial discontinuity on the boundary of Ω even when f is smooth. As a consequence, many wavelets will be needed to resolve this singularity and this might ruin the approximation process. A "smoother" extension process (in the sense that it preserves Hölder continuity of order $\mu < 1$) based on symmetrization can be performed in the case of tensor product domains (see COHEN, DAUBECHIES and VIAL [1993]): such a process is frequently used in image processing where the generation of a singularity in the normal derivative is acceptable. Smoother extensions, although they are known to exist for general Lipschitz domains (STEIN [1970]) are somehow more difficult to perform numerically, making unrealistic this approach for the effective numerical computations of multiscale decompositions.

A more appropriate solution is to build nested approximation spaces $V_j^\Omega \subset V_{j+1}^\Omega \subset L^2(\Omega)$ that reproduce the main features that we have encountered throughout this chapter, in particular the existence of a local basis with a simple structure, satisfying stability, smoothness and polynomial reproduction properties, as well as of a dual basis yielding local projectors P_j and fast decomposition algorithms. An additional difficulty that might arise is the need to incorporate homogeneous boundary conditions in the construction of V_j^Ω.

A very natural strategy for the construction of such spaces is to use a basis that consists of those multivariate functions $\phi_{j,k}$ of (2.12.2) or (2.12.12) that are supported in the interior of Ω and of a set of adapted basis functions that are localized against the boundary. The main interest of this approach is the following: in an interior region Ω_j which contains all points of Ω that are distant at least $C2^{-j}$ from the boundary $\partial\Omega$, for some constant $C > 0$ independent of j, the analysis of a function at resolution 2^{-j} operates exactly as if the function were defined on the whole of $\mathbb{R}^d$. In particular, the quadrature rules and decomposition-reconstruction formulae are the same.

The main problem that we need to address is thus the construction of the basis functions near the boundary. A simple possibility is to generate V_j^Ω by all restrictions to Ω of the basis functions which have a support that is not disjoint with Ω. This choice amounts to defining V_j^Ω as the restriction to Ω of the functions in V_j, i.e.

$$V_j^\Omega = \{f \in L^2(\Omega) \text{ such that } f = g_{|\Omega},\ g \in V_j\}. \tag{2.12.13}$$

This choice, which was first studied in MEYER [1991] in the case of an interval, ensures trivially the nestedness of the spaces V_j.

The restrictions to Ω of the functions $\phi_{j,k}$, $k \in K_j$, where K_j is the set of indices such that $|\Omega \cap \mathrm{Supp}(\phi_{j,k})| \neq 0$, actually form a basis of V_j^Ω. This fact is easy to check in the case of a univariate interval I. One first remark that since the restriction does not affect the inner product of the boundary functions with the interior functions, both group generate independent spaces. Finally, the interior ones already form a basis, while the boundary functions are necessarily independent due to their support properties (we also assume here that the scale level j is larger than a minimal j_0 so that the supports of the left side and right side boundary functions do not overlap). For more general domains, this can be shown for tensor product scaling functions, invoking a stronger argument: the functions $\phi_{j,k}$ are ***locally linearly independent***, i.e. if for some region $R \in \mathbb{R}^d$ of non-zero measure, we have

$$x \in R \Rightarrow \sum_{k \in \mathbb{Z}^d} c_k \phi_{j,k}(x) = 0, \tag{2.12.14}$$

then $c_k = 0$ for all k such that $|R \cap \mathrm{Supp}(\phi_{j,k})| \neq 0$. This remarkable property (which was already known for univariate spline functions) was proved in LEMARIE and MALGOUYRES [1991] in the case of an L^2-stable univariate compactly supported refinable function and can easily be extended to tensor product refinable functions (see Theorem 3.7.1 of §3.7 in the next chapter).

Note that this basis inherits many of the useful features of the basis for $L^2(\mathbb{R}^d)$: the polynomials are reproduced up to the same order, the restricted functions have the same smoothness and they are locally supported. However, the main defect of this construction resides in its lack of stability: the basis functions that overlap Ω on a small portion of Ω, have in general very bad stability constants. This defect will in general result in numerical instabilities in the process of orthonormalizing these functions or of building a proper dual basis. For this reason, the construction of V_j^Ω by (2.12.13) is avoided in most practical computations.

A variant of this construction that yields a more stable basis, was proposed in the case of an interval by COHEN, DAUBECHIES and VIAL [1993]. The basic idea for constructing the boundary functions can be described as follows, in the case where $\Omega = [0, 1]$. Suppose that φ is a refinable function with compact support, i.e. $\mathrm{Supp}(\varphi) = [0, L]$ (which we can always assume up to an integer shift in the definition of φ), and such that the Strang-Fix conditions are satisfied up to order N, i.e.

$$x^q = \sum_{k \in \mathbb{Z}} \langle (\cdot)^q, \tilde{\varphi}(\cdot - k) \rangle \varphi(x - k), \quad q = 0, \cdots N. \tag{2.12.15}$$

For some fixed $M_1 \geq 0$ and $M_2 \geq L$, and for some j large enough so that $2^{-j}(M_1 + M_2 + 2L) \leq 1$, we define $V_j^{[0,1]}$ as the space generated by the interior functions

$$\varphi_{j,k}, \quad M_1 \leq k \leq 2^j - M_2, \tag{2.12.16}$$

and the left and right edge functions $\varphi_{j,q}^0$ and $\varphi_{j,q}^1$, $q = 0, \cdots, N$, defined by

$$\varphi_{j,q}^0(x) = 2^{j(1/2-q)} \sum_{k < M_1} \langle (\cdot)^q, \tilde{\varphi}_{j,k} \rangle \varphi_{j,k}, \tag{2.12.17}$$

and

$$\varphi_{j,q}^1(x) = 2^{j(1/2-q)} \sum_{k>2^j-M_2} \langle (1-\cdot)^q, \tilde{\varphi}_{j,k} \rangle \varphi_{j,k}(x). \quad (2.12.18)$$

Note that we also have

$$\varphi_{j,q}^0(x) = 2^{j(1/2-q)} [x^q - \sum_{k\geq M_1} \langle (\cdot)^q, \tilde{\varphi}_{j,k} \rangle \varphi_{j,k}], \quad (2.12.19)$$

and

$$\varphi_{j,q}^1(x) = 2^{j(1/2-q)} [(1-x)^q - \sum_{k\leq 2^j-M_2} \langle (1-\cdot)^q, \tilde{\varphi}_{j,k} \rangle \varphi_{j,k}(x)]. \quad (2.12.20)$$

Instead of taking all restrictions of the basis functions that are located near the boundary, we are thus selecting the specific linear combinations of these functions that allow us to reproduce polynomials up to degree N. These combinations are also independent, due to the particular structure of the coefficients: we recall from the discussion on the Strang-Fix conditions in §2.8 that

$$2^{j(1/2-q)} \langle (\cdot)^q, \tilde{\varphi}_{j,k} \rangle = k^q + P_{q-1}(k), \quad (2.12.21)$$

where P_{q-1} is a polynomial of degree $q-1$. The linear independence of the left side functions is then an immediate consequence of both linear independence of the polynomials $x^q + P_{q-1}(x)$, $q = 0, \cdots, N$, and of the restrictions of the functions $\varphi_{j,k}$, $k \leq M$. The same holds at the right edge. Note that the scaling factor $2^{j(1/2-q)}$ has been chosen in such a way that $\|\varphi_{j,q}^\varepsilon\|_{L^2}$, $\varepsilon = 0, 1$ is independent of j.

Remark 2.12.2 *There exist variants to the definition of the edge functions by (2.12.17) and (2.12.18) which amount to taking other polynomial bases for the coefficients of $\varphi_{j,k}$. The above choice is particularly well fitted for imposing homogeneous boundary conditions in the construction of $V_j^{[0,1]}$: since the functions $\varphi_{j,q}^0$ (resp. $\varphi_{j,q}^1$) have a smooth extension (up to the order of smoothness of φ) as $2^{j/2}x^q$ on $]-\infty, 0]$ (resp. $2^{j/2}(1-x)^q$ on $[1, +\infty[$), we can define a version of $V_j^{[0,1]}$ that satisfies homogeneous Dirichlet conditions by removing $\varphi_{j,0}^0$ and $\varphi_{j,0}^1$ from the set of generators: since these are the only two functions that do not vanish at the boundary, the resulting space V_j^D satisfies the homogeneous Dirichlet boundary condition (we are assuming that φ is at least continuous). More generally, we can impose cancellations of order r at the endpoints by removing the first r edge functions at 0 and 1, assuming that $\varphi \in C^{r-1}$. This idea has been developed in* PERRIER *and* MONASSE *[1998] for orthonormal scaling functions, and in* MASSON *[1996],* DAHMEN, KUNOTH *and* URBAN *[1997] for more general biorthogonal scaling functions.*

The system defined by (2.12.16), (2.12.17) and (2.12.18) turns out to be better adapted for orthonormalization or construction of a dual basis than the functions obtained by simple restriction, in the sense that these processes do not generate numerical instabilities (note in particular that the edge functions have a larger support than the interior functions which exclude "small tails" that appear by restriction to the end of the support of φ). On the other hand, the nestedness property of the spaces $V_j^{[0,1]}$ is no more straightforward but still holds as shown by the following result.

Theorem 2.12.1 *The spaces $V_j^{[0,1]}$ defined by the scaling functions (2.12.16), (2.12.17) and (2.12.18) are nested, i.e. $V_j^{[0,1]} \subset V_{j+1}^{[0,1]}$.*

Proof If $\varphi_{j,k}$ is an interior scaling function, i.e. $M_1 \leq k \leq 2^j - M_2$, its decomposition as a combination of $\varphi_{j+1,n}$ based the refinable equation, only involves those $\varphi_{j+1,n}$ with support embedded in the support of $\varphi_{j,k}$, and thus n such that $2M_1 \leq n \leq 2^{j+1} - 2M_2$, i.e. interior functions at the finer level. The interior basis functions of $V_j^{[0,1]}$ are thus contained in $V_{j+1}^{[0,1]}$.

If we now consider the left edge basis function $\varphi_{j,q}^0$ for some $q = 0, \cdots, N$, we remark that the difference $D(x) := \varphi_{j,q}^0(x) - \frac{1}{\sqrt{2}}\varphi_{j+1,q}^0(x)$ also writes

$$D(x) = 2^{j(1/2-q)}[\sum_{k\geq M_1} \langle (\cdot)^q, \tilde{\varphi}_{j,k}\rangle \varphi_{j,k}(x) - \sum_{k\geq M_1} \langle (\cdot)^q, \tilde{\varphi}_{j+1,k}\rangle \varphi_{j+1,k}(x)].$$

It is therefore identically zero for $x \geq (M_1 + L)2^{-j}$ since we then have exactly

$$\sum_{k\geq M_1} \langle (\cdot)^q, \tilde{\varphi}_{j,k}\rangle \varphi_{j,k}(x) = \sum_{k\geq M_1} \langle (\cdot)^q, \tilde{\varphi}_{j+1,k}\rangle \varphi_{j+1,k}(x) = x^q. \qquad (2.12.22)$$

Expanding the interior functions $\varphi_{j,k}$, $k \geq M_1$ in terms of the interior functions at the finer level, we thus obtain

$$\varphi_{j,q}^0 = \frac{1}{\sqrt{2}}\varphi_{j+1,q}^0 + \sum_{k=M_1}^{2M_1-1} h_{q,k}^0 \varphi_{j+1,k}, \qquad (2.12.23)$$

for some coefficients $h_{q,k}^0$, i.e. $\varphi_{j,q}^0 \in V_{j+1}^{[0,1]}$. A similar reasoning applies to the right edge functions. ◇

Remark 2.12.3 *In the case where we start with an orthonormal scaling function φ, the left and right edge scaling functions need to be orthonormalized. The resulting orthonormal bases are presented in* COHEN, DAUBECHIES *and* VIAL *[1993], where it is also shown that a natural way to build a wavelet basis is to take some interior $\psi_{j,k}$ and construct "edge wavelets" by an orthonormalization of the functions $\varphi^{\varepsilon}_{j+1,q} - P_j\varphi^{\varepsilon}_{j+1,q}$, $\varepsilon = 0, 1$, $q = 0, \cdots, N$, where P_j is the orthonormal projector onto $V_j^{[0,1]}$.*

Remark 2.12.4 *In the case where we start with a pair of dual scaling functions, there is much more freedom in the construction of the dual system. A basic constraint is still that the parameters (M_1, M_2) and $(\tilde{M}_1, \tilde{M}_2)$ (corresponding to the analogous construction of $\widetilde{V}_j$) should be linked in such a way that V_j and $\widetilde{V}_j$ have the same dimension. The edge functions $\varphi^{\varepsilon}_{j,q}$, $\varepsilon = 0, 1$ and $\tilde{\varphi}^{\varepsilon}_{j,q}$, $\varepsilon = 0, 1$ should then be "biorthogonalized", i.e. recombined in such a way that they satisfy duality properties. Such a process amounts to a matrix inversion at each edge, but in contrast to the orthonormalization process of the previous remark, the matrix is not guaranteed to be invertible. In the specific case of biorthogonal spline functions (example 2 of §2.11), the well-posedness of this approach could be proved in* DAHMEN, KUNOTH *and* URBAN *[1999]. One also has the possibility (by choosing the parameters M_1 and M_2 large enough) of building a dual basis which only consists of interior functions $\tilde{\varphi}_{j,k}$. There is also more flexibility in the derivation of biorthogonal wavelets. In particular, it was shown in* COHEN *and* MASSON *[1998] that a basis of $W_j = V_{j+1} \cup \widetilde{V}_j^{\perp}$ can be obtained by taking those inner wavelets $\psi_{j,k}$ that can be expressed as combinations of the inner functions $\varphi_{j+1,k}$, $M_1 \leq k \leq 2^j - M_2$ at the next level, and the projection by $Q_j := P_{j+1} - P_j$ of an appropriate selection of inner functions $\varphi_{j+1,k}$ located near the edges onto W_j. A similar technique is used for generating $\widetilde{W}_j$ and the resulting wavelets bases are then biorthogonalized near the edges by the same process as for the scaling functions.*

Remark 2.12.5 *The edge functions are also "refinable" in the sense that they can have a local expansion of the type (2.12.23) as a linear combination of the basis functions at the finer level. The coefficients $h^0_{q,k}$ can easily be derived from the standard coefficients h_n of (2.2.13) and the values of the polynomial coefficients in (2.2.19). Similarly, one derives particular $h^1_{q,k}$ at the right edges, and coefficients $\tilde{h}^{\varepsilon}_{q,k}$, $g^{\varepsilon}_{q,k}$ and $\tilde{g}^{\varepsilon}_{q,k}$ associated with dual scaling functions, wavelets and dual wavelets at each edge $\varepsilon = 0, 1$, depending on the construction which is chosen. In the fast decomposition and reconstruction algorithm, these particular coefficients correspond to a modification of the filtering process when aproaching the boundary.*

For a general domain $\Omega \subset \mathbb{R}^d$, the construction of multiscale decompositions is of course a more difficult task, except in the simple case of rectangular domains, i.e.

$$\Omega = I_1 \times \cdots \times I_d. \tag{2.12.24}$$

where we can simply apply the tensor product strategy to the constructions that we described for the interval $[0, 1]$. Note that this strategy also allows us to prescribe specific homogeneous boundary conditions on a given side of the domain, imposing the corresponding univariate boundary condition in some of the directions.

Such a construction can be further used to treat more general domains Ω, provided that these domains can be patched into cube-like subdomains Ω_i, $i = 1, \cdots, n$, which are the images of a reference cube $[0, 1]^d$ by isoparametric maps T_i. Such maps allow us to define successively multiresolution spaces $V_j^{\Omega_i}$ as the image of $V_j^{[0,1]^d}$ (defined by means of tensor product) by T_i, and V_j^{Ω} as

$$V_j^{\Omega} = C^0(\Omega) \cap (\oplus_{i=1}^n V_j^{\Omega_i}). \tag{2.12.25}$$

Since the maps are isoparametric, one can easily build a first basis of V_j^{Ω} by keeping all pullbacks $\varphi_{j,k}(T_i^{-1}\cdot)$ and $\varphi_{j,q}^{\varepsilon}(T_i^{-1}\cdot)$, $q \geq 1$ of those basis functions that vanish on the boundary of $[0, 1]^d$, and by a simple adaptation on the interface: one simply glues together the symmetric pairs of scaling functions on both sides of the interface associated with the edge function $\varphi_{j,0}^{\varepsilon}$, $\varepsilon = 0, 1$ in the normal direction and corresponding to the same function in the tangential direction. A similar process can be performed to construct a basis for $\tilde{V}_j$ defined the same way as V_j.

The biorthogonalization of these bases is made with respect to a pullback inner product defined by

$$\langle f, g \rangle_p = \sum_{i=1}^n \int_{[0,1]^d} f(T_i x)\overline{g(T_i x)}dx. \tag{2.12.26}$$

The resulting bases $(\varphi_\gamma)_{\gamma \in \Gamma_j}$ of V_j and $(\tilde{\varphi}_\gamma)_{\gamma \in \Gamma_j}$ of $\tilde{V}_j$ define an oblic projector of the form

$$P_j f = \sum_{\gamma \in \Gamma_j} \langle f, \tilde{\varphi}_\gamma \rangle_p \varphi_\gamma. \tag{2.12.27}$$

This approach was proposed in DAHMEN and SCHNEIDER [1999b] and further developed in CANUTO, TABACCO and URBAN [1999] and in COHEN and MASSON [1998], where explicit constructions of the biorthogonal scaling functions and wavelets are given at the interface between the sub-domains where they need to be adapted. Similarly, the biorthogonality of the wavelet bases is intended in the sense of the pullback inner product $\langle \cdot, \cdot \rangle_p$.

A serious difficulty in such domain decomposition strategies appears if we require more smoothness than Lipschitz ($C^{1,1}$) continuity for the basis functions overlapping the interfaces. This is a problem when it comes to the characterization of high order smoothness from the properties of multiscale decompositons, by the techniques that will be developed in the next chapter. A solution to this problem was recently proposed in DAHMEN and SCHNEIDER [1998,1999a], based on general results of CIECELSKI and FIGIEL [1983] that allow us to describe smoothness spaces on Ω as the direct sum of the analogous spaces on the subdomains Ω_i subject to specific boundary conditions.

In summary, the construction of multiresolution approximation spaces and wavelet bases on d-dimensional domains can be derived from the constructions on $\mathbb{R}$ by the following steps:

$$\Omega = \mathbb{R} \rightarrow \Omega = [0,1] \rightarrow \Omega = [0,1]^d \rightarrow \Omega = \cup\Omega_i,\ \Omega_i \sim [0,1]^d, \tag{2.12.28}$$

where (1) → (2) is done by restriction and adaptation near the boundary, (2) → (3) uses tensor product and (3) → (4) can be performed when domain decomposition is feasible. This general approach is currently the object of active research. It is far from being understood in its full generality, but its success on model geometries (such as L-shaped domains) is promising.

If one wants to avoid domain decomposition techniques and discretization by tensor product grid in each domain, in particular in the setting of triangular finite element discretization, other strategies are also available for the construction of multiscale decompositions on $\Omega \in \mathbb{R}^d$. They will be discussed in the next section.

Remark 2.12.6 *For a fairly general class of domains, it is also possible to mimic the approach used in dealing with the interval, i.e. do a particular adaptation of the scaling functions near the boundary, based on local linear combinations generalizing (2.12.17) and (2.12.18). A strategy for building these "boundary scaling functions" is proposed in* COHEN, DAHMEN *and* DEVORE *[2000]. It avoids orthonormalization and biorthogonalization near the boundary, by the choice of a dual system which consists of scaling functions supported in the interior of the domain. This approach is theoretically feasible for any domain with Lipschitz boundary. However, one expects practical difficulties in the implementation of such a general framework, due to the possibly complicated structure of the boundary: this structure enters in the computations of quantities such as mass or stiffness matrices and inner products with boundary scaling functions (since the integrals have to be restricted to the domain). It also complicates the prescription of homogeneous and non-homogeneous boundary conditions in the discretization of boundary value problems.*

2.13 Point values, cell averages, finite elements

In this last section, we want to adopt a somewhat different perspective on the construction of multiscale decompositions, in the sense that these decompositions will be based on specific types of discretizations which are commonly used in numerical analysis. We have in mind three types of discretizations which respectively correspond to the three most frequently used numerical methods.

1. **Point values** are the unknowns and data of *finite element methods*: given a set of points Γ, a function f is discretized by its values $f(\gamma)$ for $\gamma \in \Gamma$.

2. **Cell averages** are the unknowns and data of *finite volume methods*: a function f is discretized by its averages $|\Omega_\gamma|^{-1} \int_{\Omega_\gamma} f(x)dx$ on cells Ω_γ, $\gamma \in \Gamma$ which constitute a disjoint partition of the domain of interest.

3. **Finite element spaces** are of common use in the context of *Petrov-Galerkin methods*: given a triangulation $\mathcal{T}$, a function f is approximated by another function which is piecewise polynomial on each triangle of $\mathcal{T}$ and characterized by certain local linear functionals, e.g. nodal values.

Note that in the two first examples we are only specifying the discretization process in terms of a *functional* applied to the function f (the Dirac distribution δ_γ for point values and the indicator function $|\Omega_\gamma|^{-1}\chi_{\Omega_\gamma}$ for cell averages), while in the last example we also focus on a specific *approximation space* whose elements are characterized by such functionals.

We shall discuss several strategies which allow us to build multiscale representations of such discretizations. Such representations constitute a "bridge" between wavelets and more classical numerical methods. They are particularly promising for practical applications since they allow us to exploit the advantages of wavelet discretizations which were evoked in the previous chapter (adaptivity and preconditioning) within a specific numerical method which might be well fitted to the resolution of a given problem.

All these strategies can be thought in terms of a very general *discrete framework* which was introduced in HARTEN [1993,1996]. This framework is based on the concept of interscale operators (which also play an important role in multigrid methods), as expressed by the following definition.

Definition 2.13.1 *A generalized discrete multiresolution approximation is given by a sequence of index sets Γ_j, $j \geq 0$, which are equipped with two types of operator:*

1. *Restriction/projection operators* P_j^{j+1} *from* $\ell^\infty(\Gamma_{j+1})$ *to* $\ell^\infty(\Gamma_j)$,

2. *Prolongation/prediction operators* P_{j+1}^j *from* $\ell^\infty(\Gamma_j)$ *to* $\ell^\infty(\Gamma_{j+1})$.

Such operators should in addition satisfy the property

$$P_j^{j+1}P_{j+1}^j = I, \tag{2.13.1}$$

i.e. the projection operator is a left inverse to the prediction operator.

Such operators should be thought as a generalization of the filtering processes involving $\tilde{h}_n$ and h_n in (2.6.1) and (2.6.2). Note that property (2.13.1) implies that the image of the projection operator P_j^{j+1} is the entire $\ell^\infty(\Gamma_j)$. If U_{j+1} is a discrete vector in $\ell^\infty(\Gamma_{j+1})$, we can define its "approximation"

$$\hat{U}_{j+1} := P_{j+1}^j P_j^{j+1} U_{j+1}, \tag{2.13.2}$$

which is obtained from its projection $U_j := P_j^{j+1}U_{j+1}$ at the coarser resolution. Clearly, U_{j+1} can be characterized in a redundant way by the data of U_j and of the prediction error

$$E_{j+1} := U_{j+1} - \hat{U}_{j+1}. \tag{2.13.3}$$

In order to eliminate the redundancy, we notice that (2.13.1) implies that the prediction error satisfies $P_j^{j+1}E_{j+1} = 0$, i.e. belongs to the null space N_{j+1} of the projection operator. Let us assume that we have at our disposal a basis for this null space in the sense of a family $(\Psi_{j,\lambda})_{\lambda\in\nabla_j}$ of vectors in $\ell^\infty(\Gamma_{j+1})$ such that any vector $V \in N_{j+1}$ has the unique expansion

$$V = \sum_{\lambda\in\nabla_j} v_\lambda \Psi_{j,\lambda} \tag{2.13.4}$$

for some sequence $(v_\lambda) \in \ell^\infty(\nabla_j)$ in the sense that for all $\gamma \in \Gamma_{j+1}$ the series $\sum_{\lambda\in\nabla_j} v_\lambda\Psi_{j,\lambda}(\gamma)$ is absolutely summable and converges to $V(\gamma)$. Note that such a basis always exists in the case where Γ_{j+1} is a finite set, in which case we clearly have $\#(\Gamma_{j+1}) = \#(\Gamma_j) + \#(\nabla_j)$.

Denoting by D_j the coordinate vector of the error E_{j+1} in the basis $(\Psi_{j,\lambda})_{\lambda\in\nabla_j}$, we can define two operators associated with the "details" between resolution levels j and $j+1$: (i) a projection operator Q_j^{j+1} acting from $\ell^\infty(\Gamma_{j+1})$ to $\ell^\infty(\nabla_j)$ which maps U_{j+1} to D_j, and (ii) a prediction operator Q_{j+1}^j acting from $\ell^\infty(\nabla_j)$ to $\ell^\infty(\Gamma_{j+1})$ which maps D_j to E_{j+1}.

Here again, these operators should be viewed as generalizations of the filtering processes involving the filters $\tilde{g}_n$ and g_n. From their definition, it is clear that a property similar to (2.13.1) holds, i.e.

$$Q_j^{j+1} Q_{j+1}^j = I. \tag{2.13.5}$$

In addition, we also remark that

$$U_{j+1} = \hat{U}_{j+1} + E_{j+1} = P_{j+1}^j U_j + Q_{j+1}^j D_j = P_{j+1}^j P_j^{j+1} U_{j+1} + Q_{j+1}^j Q_j^{j+1} U_{j+1},$$

which shows that

$$P_{j+1}^j P_j^{j+1} + Q_{j+1}^j Q_j^{j+1} = I, \tag{2.13.6}$$

and that U_{j+1} is equivalently characterized by (U_j, D_j). By iteration, we obtain a one to one correspondance between a vector U_J in $\ell^\infty(\Gamma_J)$ and its multiscale decomposition

$$\mathcal{M}U_J = (U_0, D_0, D_1, \cdots, D_{J-1}) \in \ell^\infty(\nabla^J), \tag{2.13.7}$$

where we have set

$$\nabla^J := \Gamma_0 \cup \nabla_0 \cup \cdots \cup \nabla_{J-1}. \tag{2.13.8}$$

This generalized wavelet representation can be implemented by a (fine to coarse) decomposition algorithm using P_j^{j+1} and Q_j^{j+1}, and a (coarse to fine) reconstruction algorithm using P_{j+1}^j and Q_{j+1}^j, similar to the algorithms discussed in §2.6.

Of course such a tool is much too crude as such, and in practice we always have in mind a more specific situation. In particular, Γ_j and ∇_j correspond to some spatial discretization at resolution 2^{-j} and in this context we are interested that the above operators have some *local properties* in the following sense: there exists a constant C independent of j and γ such that $P_{j+1}^j U(\gamma)$ (resp. $Q_{j+1}^j U(\gamma)$, $P_j^{j+1} U(\gamma)$, $Q_j^{j+1} U(\gamma)$) depends only on those $\mu \in \Gamma_j$ (resp. $\mu \in \nabla_j$, $\mu \in \Gamma_{j+1}$, $\mu \in \Gamma_{j+1}$) such that $\mathrm{dist}(\gamma, \mu) \le C2^{-j}$. We also have in mind the typical situation in which the Γ_j are nested meshes, i.e. $\Gamma_j \subset \Gamma_{j+1}$, such as e.g. $\Gamma_j = 2^{-j}\mathbb{Z}$ corresponding to the algorithms of §2.6, in which case we can simply define ∇_j as $\Gamma_{j+1} \setminus \Gamma_j$. Such prescriptions will typically hold for the examples that we shall present, associated with our three classical discretizations.

Before going to these examples, we shall explain the *lifting scheme* - which was already introduced in §2.11 - in this general context. This procedure consists of building new projection and prediction operators based on the following remark: given a matrix A_j acting from $\ell^\infty(\nabla_j)$ to $\ell^\infty(\Gamma_j)$, we can define new operators by

$$\tilde{P}_{j+1}^j = P_{j+1}^j, \ \tilde{Q}_{j+1}^j = Q_{j+1}^j - P_{j+1}^j A_j, \tag{2.13.9}$$

and

$$\tilde{P}_j^{j+1} = P_j^{j+1} + A_j Q_j^{j+1}, \ \tilde{Q}_j^{j+1} = Q_j^{j+1}. \tag{2.13.10}$$

It is then easily seen that these new operators also satisfy (2.13.1), (2.13.5) and (2.13.6) and therefore qualify to define a new one to one multiscale transform of the same type as (2.13.7). We notice that if both A_j and the four initial operators are local, the new operators will also be local.

We can give an interpretation of the lifting scheme in matrix terms: if we insert

$$I = \begin{pmatrix} I & -A_j \\ 0 & I \end{pmatrix} \begin{pmatrix} I & A_j \\ 0 & I \end{pmatrix}, \tag{2.13.11}$$

in between the two factors of the identity

$$(P_{j+1}^j, Q_{j+1}^j) \begin{pmatrix} P_j^{j+1} \\ Q_j^{j+1} \end{pmatrix} = I, \tag{2.13.12}$$

which expresses (2.13.6) as well as (2.13.1) and (2.13.5), we obtain a similar identity for the new projectors. In turn, we see that there is a natural alternative version of the lifting scheme since we also have

$$I = \begin{pmatrix} I & 0 \\ -B_j & I \end{pmatrix} \begin{pmatrix} I & 0 \\ B_j & I \end{pmatrix}, \tag{2.13.13}$$

for any operator B_j acting from $\ell^\infty(\Gamma_j)$ to $\ell^\infty(\nabla_j)$. If we use (2.13.13) instead of (2.13.11), the new operators are defined by

$$\tilde{P}_{j+1}^j = P_{j+1}^j - Q_{j+1}^j B_j, \ \tilde{Q}_{j+1}^j = Q_{j+1}^j, \tag{2.13.14}$$

and

$$\tilde{P}_j^{j+1} = P_j^{j+1}, \ \tilde{Q}_j^{j+1} = Q_j^{j+1} + B_j P_j^{j+1}. \tag{2.13.15}$$

If the initial operators express the decomposition and reconstruction of a function in some wavelet basis, (2.13.9) and (2.13.10) correspond to a correction of the primal wavelets by coarse grid primal scaling functions (similar to Example 2 in §2.11), while (2.13.14) and (2.13.15) correspond to a correction of the dual wavelets by coarse grid dual scaling functions (similar to Examples 3 and 4 in §2.11).

We can think of more general modifications of the operators, since (2.13.11) and (2.13.13) are not the only possible factorizations of the identity. In particular, the ***stable completion process*** proposed in DAHMEN, CARNICER and PENA [1996], consists of introducing an additional ***invertible*** operator L_j acting inside $\ell^\infty(\nabla_j)$, and defining the new operators by a more general transformation than (2.13.9) and (2.13.10):

$$\tilde{P}_{j+1}^j = P_{j+1}^j, \ \tilde{Q}_{j+1}^j = Q_{j+1}^j L_j - P_{j+1}^j A_j, \tag{2.13.16}$$

and

$$\tilde{P}_j^{j+1} = P_j^{j+1} + A_j L_j^{-1} Q_j^{j+1}, \ \tilde{Q}_j^{j+1} = L_j^{-1} Q_j^{j+1}. \tag{2.13.17}$$

If L_j is local, the resulting reconstruction operators in (2.13.16) remain local, while the decomposition operators in (2.13.17) are generally global due to the presence of L_j^{-1}. In turn, the new dual wavelet system is not locally supported. This situation was encountered with the fine grid correction spline wavelets presented in §2.11.

Remark 2.13.1 *In the particular case corresponding to the sub-band filtering algorithms of §2.6, the following result was proved in* DAUBECHIES *and* SWELDENS *[1998] which reflects the flexibility of the lifting scheme: any arbitrary biorthogonal filter-bank* $(h_n, \tilde{h}_n, g_n, \tilde{g}_n)$ *can be obtained by applying a finite number of lifting procedures of the two above type to the "lazy wavelet" filters corresponding to* $h_n = \tilde{h}_n = \sqrt{2}\delta_{0,n}$ *and* $g_n = \tilde{g}_n = -\sqrt{2}\delta_{1,n}$.

Remark 2.13.2 *In the general framework that we have described, it is worth mentionning that one can relax the assumption that the prediction operator is linear: some of the prediction operators proposed in* HARTEN *[1996] are nonlinearly data dependent for the purpose of a better adapted prediction near the jumps or singularities of the data which usually generate spurious oscillations (i.e. Gibbs phenomenon). These nonlinear predictions are based on essentially non-oscillatory (ENO) reconstruction techniques and we shall briefly discuss them in the next two examples. However, the corresponding multiresolution transforms can no longer be thought as a change of basis, which results in numerous difficulties when analyzing the stability of these transforms.*

We shall now turn to several concrete examples, corresponding to the three types of discretizations. In all these examples, the discretizations will be nested. Although this restriction is not imposed in the above general framework, it is often encountered in practice and it simplifies the construction of multiresolution decompositions in the sense that the restriction operator has a natural definition.

Example 1. Point value multiresolution

In this first example, $(\Gamma_j)_{j\geq 0}$ is a sequence of grids which discretizes $\mathbb{R}^d$ or some domain $\Omega \in \mathbb{R}^d$ at different resolutions 2^{-j}: if $\mu(\gamma) \in \Gamma_j$ is the closest neighbour of a point $\gamma \in \Gamma_j$, we have

$$|\gamma - \mu(\gamma)| \sim 2^{-j}, \tag{2.13.18}$$

with constants independent of γ and j. We assume furthermore that these grids are nested, i.e.

$$\Gamma_j \subset \Gamma_{j+1}. \tag{2.13.19}$$

In the point value setting, we can interpret a discrete vector $U_j \in \ell^\infty(\Gamma_j)$ as the values of a continuous function u on the grid Γ_j:

$$U_j = (u(\gamma))_{\gamma\in\Gamma_j}. \tag{2.13.20}$$

This suggests taking for P_j^{j+1} the simple decimation operator defined by

$$P_j^{j+1}U_{j+1}(\gamma) = U_{j+1}(\gamma), \quad \gamma \in \Gamma_j, \tag{2.13.21}$$

which is the only choice compatible with (2.13.20) in the sense that we have $U_j = P_j^{j+1}U_{j+1}$. This operator is illustrated in the one dimensional case in Figure 2.13.1.

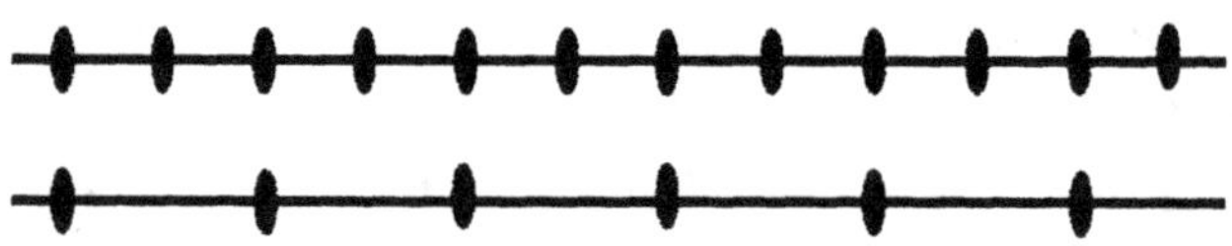

Figure 2.13.1: Restriction for point value discretizations

Let us now turn to the prediction operator. We first notice that by (2.13.1), the vector $\hat{U}_{j+1} = P_{j+1}^j U_j$ should coincide with U_{j+1} on the coarse grid Γ_j. Therefore, building a prediction operator may be viewed as an interpolation problem: from the values $u(\gamma)$, $\gamma \in \Gamma_j$, we want to reconstruct approximations $\hat{u}(\lambda)$ of the exact values $u(\lambda)$ for λ on the grid ∇_j which is simply defined by

$$\nabla_j := \Gamma_{j+1} \setminus \Gamma_j \tag{2.13.22}$$

Given such an interpolation process, it is natural to define the details by the restriction of the interpolation error E_{j+1} on ∇_j (since E_{j+1} vanishes on Γ_j), i.e.

$$D_j = (u(\lambda) - \hat{u}(\lambda))_{\lambda\in\nabla_j}. \tag{2.13.23}$$

In the case of univariate structured meshes $\Gamma_j = 2^{-j}\mathbb{Z}$, such multiresolution transforms are typically associated with ***interpolatory scaling functions and wavelets***. In this particular setting, we already presented in §2.10 a class of local prediction operators based on ***lagrangian polynomial interpolation***, which are accurate up to some prescribed order N in the sense that the prediction is *exact* if u is a polynomial of degree less than $N-1$. We illustrate this prediction operator in Figure 2.13.2 in the case of cubic polynomial interpolation.

In a more general setting, let us explain how one follows the same principle in order to build prediction operators which are exact for polynomials of

Figure 2.13.2: Prediction for point value discretizations

total degree less than $N-1$. A first step is to associate to each $\lambda \in \nabla_j$ a *prediction stencil* $S(\lambda)$ consisting of a set of points $\gamma \in \Gamma_j$ in a neighbourhood of λ such that

$$|\lambda - \gamma| \leq C2^{-j}, \quad \gamma \in S(\lambda). \tag{2.13.24}$$

From the values $(u(\gamma))_{\gamma\in S(\lambda)}$, we want to build a polynomial $p_\lambda \in \Pi_{N-1}$ with the property that if the function u is in Π_{N-1} then $p_\lambda = u$. We then simply define

$$\hat{u}(\lambda) = p_\lambda(\lambda). \tag{2.13.25}$$

The construction of the polynomial p_λ remains to be discussed. Ideally, we would like to define p_λ as the unique polynomial of degree $N-1$ which interpolates the values of u on $S(\lambda)$, i.e.

$$p_\lambda(\gamma) = u(\gamma), \quad \gamma \in S(\lambda). \tag{2.13.26}$$

Example 3 below will discuss a typical instance of such a prediction operator associated with Lagrange finite elements. However, the choice of a stencil for which this problem admits a unique solution is not always possible, in particular for unstructured grids. In such cases, a possibility is to take a sufficiently large stencil so that the application which maps a polynomial $p \in \Pi_{N-1}$ to its values on $S(\lambda)$ is injective, and to define p_λ by solving the least square problem

$$\inf_{p\in\Pi_{N-1}} \sum_{\gamma\in S(\lambda)} |u(\gamma) - p(\gamma)|^2. \tag{2.13.27}$$

Such a reconstruction is clearly exact if u is in Π_{N-1}.

Remark 2.13.3 *Similarly to dyadic lagrangian interpolation, we can analyze the convergence of the subdivision scheme corresponding to the iterative application of the prediction operator from coarse to fine scales. If such a scheme is uniformly convergent, we can define the associated interpolatory* *scaling functions* $\varphi_{j,\gamma}$, $\gamma \in \Gamma_j$, *as the limit of the subdivision scheme applied to the Dirac vector* $U_{j,\gamma}(\mu) = \delta_{\gamma,\mu}$, $\mu \in \Gamma_j$. *It is easy to check that such functions are refinable in the sense that they are linear combinations of the* $\varphi_{j+1,\mu}$, $\mu \in \Gamma_{j+1}$ *(using the same arguments as in §2.4), and that they satisfy the interpolatory condition*

$$\varphi_{j,\gamma}(\mu) = \delta_{\gamma,\mu}, \quad \mu \in \Gamma_j. \tag{2.13.28}$$

Such functions can thus be used to generate an interpolatory multiresolution analysis V_j *and the corresponding hierarchical wavelet basis is simply given by*

$$\{\varphi_{0,\gamma}\}_{\gamma \in \Gamma_0} \cup \{\varphi_{j+1,\lambda}\}_{\lambda \in \nabla_j, j \geq 0}, \tag{2.13.29}$$

i.e. the wavelets coincide with the fine grid scaling functions on the grid ∇_j. *The dual scaling functions are of course the Dirac distributions* $\tilde{\varphi}_{j,\gamma} = \delta_\gamma$.

Remark 2.13.4 *It is possible to apply the lifting scheme on such a construction. Note that while (2.13.14) and (2.13.15) result in a similar "interpolatory decomposition", the application of (2.13.9) and (2.13.10) modifies the projection operator which is no more the decimation operator (2.13.21). As we shall see in the finite element context, one of the interests of such a modification is that it may lead to dual scaling functions which are smoother than the Dirac distribution.*

Remark 2.13.5 *As explained in Remark 2.13.2, the prediction need not be a linear operator. In the point value context, Essentially Non-Oscillatory (ENO) nonlinear prediction schemes of* HARTEN *[1996] can be explained as follows: we consider several prediction stencils* $S_i(\lambda)$, $i = 1, \cdots, N$, *resulting in different polynomials* $p_{i,\lambda}$ *which qualify to predict the value of* u *at the point* λ. *We then select by some prescribed numerical criterion the polynomial which is the least oscillatory in the neighbourhood of* λ. *Such a data dependent process aims to minimize the Gibbs effect and improve the accuracy of the prediction in the regions where* u *is singular. We can then hope for a reduction of the coefficient size near the singularities, and thus better compression properties than with linear reconstruction. However, one should be aware of the difficulties which are specific to such nonlinear transforms: in particular, the convergence and stability of the subdivision process is no longer clearly understood in terms of limit functions, as well as the perturbation induced by thresholding the small detail coefficients.*

Example 2. Cell average multiresolution

In the cell average context, Γ_j represents a partition of $\mathbb{R}^d$ or $\Omega \subset \mathbb{R}^d$ by a set of disjoint *cells* Ω_γ, $\gamma \in \Gamma_j$. Such cells tile the space at various resolutions 2^{-j} in the sense that there exists two constants c and C such that

$$c2^{-j} \leq \inf_{\gamma\in\Gamma_j} r(\Omega_\gamma) \leq \sup_{\gamma\in\Gamma_j} R(\Omega_\gamma) \leq C2^{-j}, \tag{2.13.30}$$

where $r(A)$ (resp. $R(A)$) is the radius of the largest (resp. smallest) circle contained in (resp. containing) a set A. We assume furthermore that these discretizations are nested in the sense that each Ω_γ, $\gamma \in \Gamma_j$ is the union of a finite number of cells Ω_μ, $\mu \in \Gamma_{j+1}$. Note that this number is always bounded by $2^dC/c$ where C and c are the constants in (2.13.30).

In this context, we interpret a discrete vector $U_j \in \ell^\infty(\Gamma_j)$ as the average of a locally integrable function over the cell Ω_γ, i.e.

$$U_j = (u_\gamma)_{\gamma\in\Gamma_j} \text{ with } u_\gamma := |\Omega_\gamma|^{-1} \int_{\Omega_\gamma} u(x)dx. \tag{2.13.31}$$

As in the point value setting, this suggests taking for P_j^{j+1} the averaging operator defined by

$$P_j^{j+1} U_{j+1}(\gamma) = |\Omega_\gamma|^{-1} \sum_{\mu\in\Gamma_{j+1},\Omega_\mu\subset\Omega_\gamma} |\Omega_\mu| U_{j+1}(\mu), \tag{2.13.32}$$

which is compatible with (2.13.31) in the sense that we have $U_j = P_j^{j+1}U_{j+1}$. This operator is illustrated in the one dimensional case in Figure 2.13.3.

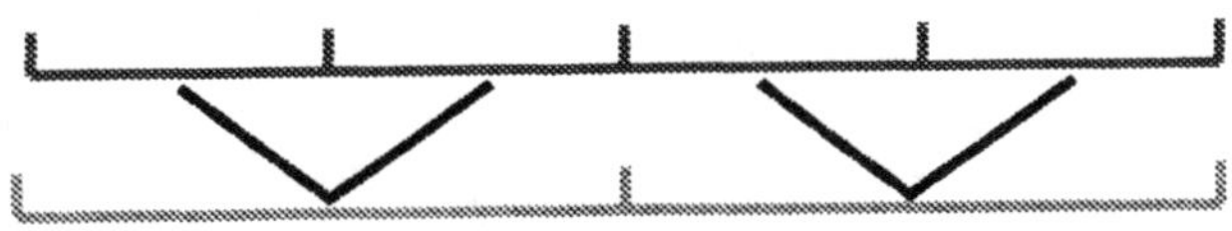

Figure 2.13.3: Restriction for cell average discretizations

Turning to the prediction operator, we first notice that by (2.13.1), the vector $\hat{U}_{j+1} = P_{j+1}^j U_j$ should satisfy the compatibility condition

$$\sum_{\mu\in\Gamma_{j+1},\Omega_\mu\subset\Omega_\gamma} |\Omega_\mu| \hat{U}_{j+1}(\mu) = \sum_{\mu\in\Gamma_{j+1},\Omega_\mu\subset\Omega_\gamma} |\Omega_\mu| U_{j+1}(\mu). \tag{2.13.33}$$

In turn, the prediction error E_{j+1} satisfies the linear constraint

$$\sum_{\mu\in\Gamma_{j+1},\Omega_\mu\subset\Omega_\gamma} |\Omega_\mu| E_{j+1}(\mu) = 0. \tag{2.13.34}$$

Therefore, we can easily represent E_j in a non-redundant fashion, by defining ∇_j as a subset of Γ_{j+1} obtained as follows: for each $\gamma \in \Gamma_j$, we select a $\mu(\gamma) \in \Gamma_{j+1}$ such that $\Omega_\mu \subset \Omega_\gamma$, and we define

$$\nabla_j := \Gamma_{j+1} \setminus \{\mu(\gamma)\ ;\ \gamma \in \Gamma_j\}. \tag{2.13.35}$$

We can then define the detail vector D_j as the restriction of E_{j+1} to ∇_j, i.e.

$$D_j = (u_\lambda - \hat{u}_\lambda)_{\lambda \in \nabla_j}, \tag{2.13.36}$$

where $\hat{u}_\lambda := \hat{U}_{j+1}(\lambda)$.

In the case of univariate structured meshes $\Gamma_j = 2^{-j}\mathbb{Z}$, such multiresolution transforms are typically associated with biorthogonal scaling functions and wavelets for which the dual scaling function is the box function, i.e. $\tilde{\varphi} = \chi_{[0,1]}$. Some examples of primal functions have been given in §2.11: we can take for φ any of the functions $\tilde{\varphi}_{1,L}$ defined by (2.11.8). For such examples, the corresponding prediction algorithm has an interpretation which is very similar to point value dyadic lagrangian interpolation: it consists of reconstructing the unique polynomial $Q_{j,k}$ of degree $L-1$ which admits the same average as u on the intervals $[2^{-j}(k+l), 2^{-j}(k+l+1)]$ for $l = 1-L, \cdots, L-1$ and defining the predicted averages as those of $Q_{j,k}$ on the half intervals $[2^{-j}k, 2^{-j}(k+1/2)]$ and $[2^{-j}(k+1/2), 2^{-j}k]$. We illustrate this prediction operator in Figure 2.13.4 in the case of a quadratic polynomial reconstruction.

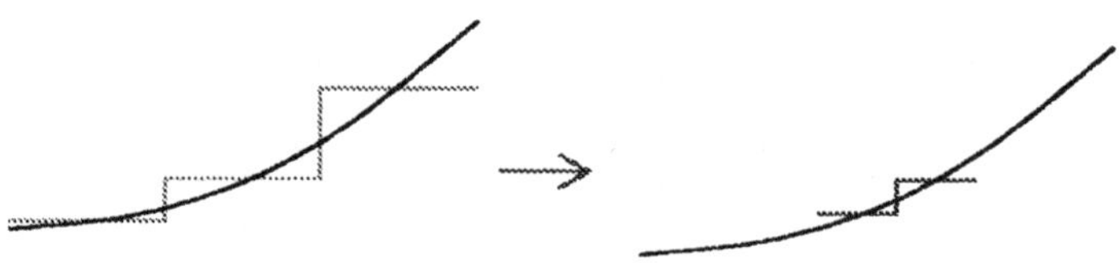

Figure 2.13.4: Prediction for cell average discretizations

Similarly to the point value setting, we can generalize this idea by introducing a stencil of cells $S(\gamma)$ for each $\gamma \in \Gamma_j$, using the averages u_μ, $\mu \in S(\gamma)$ to reconstruct a polynomial p_γ, either by interpolation or by least square minimization. We can also derive the same remarks concerning the existence of underlying scaling functions and wavelets, the application of the lifting scheme, and the possiblity of using ENO-type reconstructions based on data dependent selection of the stencil.

Example 3. Hierarchical finite elements
Finite element spaces are of common use in the numerical discretization and simulation of physical processes, in particular when they take place on multidimensional domains that are geometrically less trivial than tensor products of univariate intervals.

For such domains Ω, the construction of a family of finite element spaces V_h, $h > 0$, relies on a decomposition of Ω into subdomains of simple shape - typically triangles or rectangles in the case of a bidimensional domain - and size of order h. The functions that constitute V_h are then piecewise polynomials of a fixed degree on each subdomain, with certain smoothness properties at the interfaces separating these subdomains.

A general question that one can address is the following:

Given a finite element space, can we find natural multiscale decompositions and wavelet bases for this space ?

In some sense, we have already given an answer to this question in the case of rectangular finite elements that are obtained by tensor products of univariate spline functions: we can combine the construction of spline wavelet bases described in §2.11 together with the tensor product strategy of §2.12 (provided that Ω can be patched by square-shaped domains), to obtain "rectangular" finite element wavelets.

For many domains however, triangulations - and more generally simplicial decompositions in higher dimensions - are more adaptable, in particular for polygons in 2D and polyhedrons in 3D. We thus need to address the construction of multiscale decomposition in this setting.

More general domains Ω are also covered by the subsequent development provided that they can be patched into simplex-like sub-domains Ω_i, which are the images of a reference simplex by isoparametric maps T_i. Such maps allow us to analyze - at least locally - a function f defined on Ω through its pull-back on a reference polygonal or polyhedron geometry, by means of the multiscale decompositions and wavelet bases that we shall now describe. For the sake of simplicity, we consider the case of a bidimensional polygonal domain. It should be understood that the development below can be reformulated in the more general case of simplicial finite elements for higher dimensions.

A natural framework here is a ladder of finite element spaces $V_j = V_{h\simeq 2^{-j}}$, based on *nested triangulations* $\mathcal{T}_j$. The construction of such triangulations, which was already evoked in §1.4 of Chapter 1, is particularly simple in the case of a polygonal domain: one starts from an initial conformal finite triangulation $\mathcal{T}_0$ and obtains $\mathcal{T}_j$ by a uniform subdivision process, as displayed in Figure 2.13.5.

Then, a finite element space $V_j = V_{h\simeq 2^{-j}}$ is defined by choosing a polynomial space Π and associating a suitable set of nodes with each triangle $T \in \mathcal{T}_j$, so that each function $f \in V_j$ coincides with an element of Π_k on each triangle, which is uniquely defined by its values (and the values of some of its partial derivatives) at the associated nodes.

Even in the setting of nested triangulations, arbitrary choices of finite elements do not in general provide the property $V_j \subset V_{j+1}$: classical (and useful!) counter-examples are the Argyris C^1 quintic triangle and the Crouzeix-Raviart non-conforming (mid-point) linear triangle. However, an important class that does satisfy this property are the $\mathbb{P}_k$ Lagrange finite elements, i.e. piecewise Π_k functions (i.e. polynomials of total degree k) that are continuous at the interfaces between each triangle.

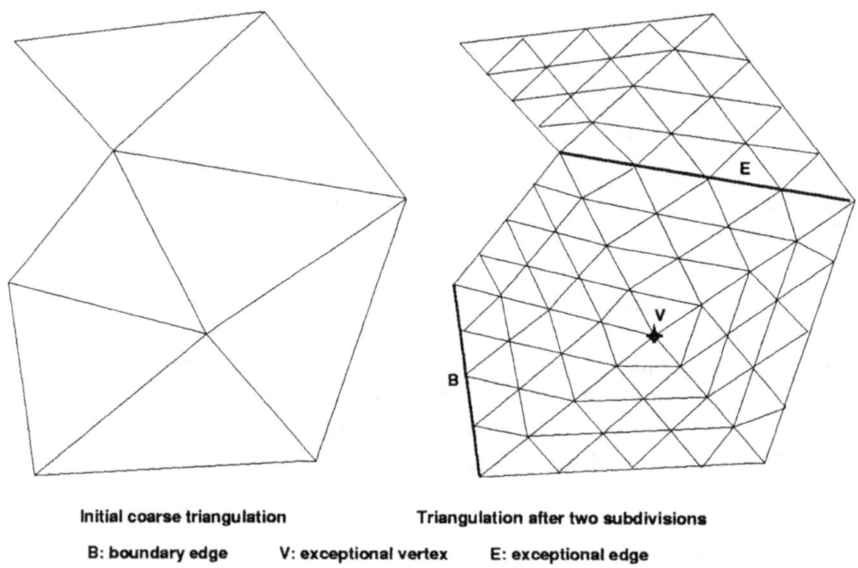

Figure 2.13.5: Triangulation of Ω by uniform subdivison: from $\mathcal{T}_0$ to $\mathcal{T}_2$

For such elements, a natural set of nodes associated with each triangle $T = (a, b, c) \in \mathcal{T}_j$ is given by

$$N_T := \{a + \frac{n}{k}(b-a) + \frac{m}{k}(c-a) \; ; \; 0 \le n + m \le k\}. \qquad (2.13.37)$$

On each triangle $T \in \mathcal{T}_j$, a function of V_j is then uniquely defined by its values at the node points of N_T. We denote by $\Gamma_j = \cup_{T\in\mathcal{T}_j} N_T$ the set of nodes globally associated with V_j. In the case of $\mathbb{P}_1$ finite elements this set coincides with the grid of vertices of $\mathcal{T}_j$. More generally, Γ_j is a superset of this grid.

The *nodal* function $\varphi_{j,\gamma}$, $\gamma \in \Gamma_j$ is then defined to be the unique element of V_j such that

$$\varphi_{j,\gamma}(\mu) = \delta_{\gamma,\mu}, \quad \mu \in \Gamma_j. \tag{2.13.38}$$

A *nodal basis* is then defined as the set $\{\varphi_{j,\gamma}\ ;\ \gamma \in \Gamma_j\}$.

The particular choice of equidistributed nodes (2.13.37) also implies that $\Gamma_j \subset \Gamma_{j+1}$. In this sense, such finite element discretization can be reinterpreted in the setting of point value discretizations of Example 1. In particular, while a natural decimation operator is given by (2.13.21), the choice of a certain finite element also determines a corresponding prediction operator: for $\gamma \in \Gamma_{j+1}$ contained in some triangle $T \in \mathcal{T}_j$ we define its predicted value as

$$\hat{u}(\gamma) = p_T(\gamma), \tag{2.13.39}$$

where p_T is the unique polynomial of degree N which interpolates u on the set N_T of coarse grid nodes.

The corresponding discrete multiscale transform expresses the decomposition of V_J in the *hierarchical basis*: the set of functions $\{\varphi_{j,\gamma}\ ;\ \gamma \in \nabla_j\}$ generates a non-orthogonal complement space W^h_{j-1} of V_{j-1} into V_j, so that the full set

$$\{\varphi_{0,\gamma},\ \gamma \in \Gamma_0\} \cup \{\varphi_{j,\gamma},\ \gamma \in \nabla_j\ j = 0, \cdots, J-1\}, \tag{2.13.40}$$

is a basis for V_J.

As it was remarked in §1.6 of Chapter 1, the hierarchical basis is intimately associated with the interpolation operator onto V_j (defined by $P_j f \in V_j$ such that $P_j f(\gamma) = f(\gamma)$ for all $\gamma \in \Gamma_j$) in the following sense: in the decomposition of a function f into the hierarchical basis, the layer of details at scale j represented by the $\{\varphi_{j,\gamma}\ ;\ \gamma \in \nabla_j\}$ is exactly given by $P_{j+1}f - P_j f$. This situation reveals a serious defect of the hierarchical basis: the interpolation operator is only well defined for continuous functions. This is also reflected by the fact that the dual scaling functions and wavelets are Dirac distributions. This results in a lack of L^2-stability for the hierarchical basis, as well as the impossibility of preconditioning in $\mathcal{O}(1)$ second order elliptic operators in more than one dimension with such bases.

A natural idea to remedy this drawback is to apply the lifting scheme according to (2.13.9) and (2.13.10) or more generally the stable completion procedure, which amounts to *correcting* the hierarchical basis, in order to obtain a more stable multiscale basis: from each hierarchical basis function $\varphi_{j,\gamma}$, $\gamma \in \nabla_j$, we would like to construct a wavelet $\psi_{j-1,\gamma}$, $\gamma \in \nabla_j$ by some local perturbation process using neighbouring nodal basis functions at the corresponding scale. Doing so, we expect to obtain a new complement W_j that is "more orthogonal" to V_j than W_{j-1}^h. We already encountered this idea in the different constructions of spline wavelets that were discussed in §2.11. In this task, we are faced with two problems of different natures for choosing an appropriate perturbation:

1. An *algebra* problem: the choice of the perturbation should preserve the structure of a complement of V_{j-1} into V_j, so that we effectively obtain a multiscale basis.

2. An *analysis* problem: the perturbation should be chosen so that the resulting wavelet basis has some improved stability in comparison with the hierarchical basis.

The first problem can be solved by applying properly the procedures of lifting scheme or stable completion which are precisely meant to derive a proper complement space. The second problem will be solved if we can analyze the stability properties of the new approximation operator P_j that has replaced the initial interpolation operator in the sense that the details at scale j in the new decomposition of a function f are represented by $P_{j+1}f - P_jf$. Such an operator will have the general form

$$P_jf = \sum_{\gamma\in\Gamma_j} \langle f, \tilde{\varphi}_{j,\gamma}\rangle \varphi_{j,\gamma}, \tag{2.13.41}$$

and we thus need to understand the nature (smoothness, support) of the dual functions $\tilde{\varphi}_{j,\gamma}$ which have replaced the Dirac functions δ_γ corresponding to the interpolation operator.

These problems are currently the object of active research. So far, concrete constructions have only been proposed in the case of $\mathbb{P}_1$ finite elements, mostly in 2D. They are particularly well understood in the case of a bi-infinite uniform triangulation (i.e. an hexagonal mesh), in which Fourier methods can be used to solve the two above-mentioned problems. This suggests using these constructions in the interior of each coarse triangle, since at a sufficiently fine scale, the grid coincides with a uniform mesh in the perturbation neigbourhood of a hierarchical basis function. Particular adaptations need to be done at exceptional vertices and edges that belong to the coarse triangulation $\mathcal{T}_0$, as well as near the boundary (see Figure 2.13.5).

We shall now describe three possible constructions in the case of a uniform mesh. These examples should be viewed as the $\mathbb{P}_1$ finite elements analog to the univariate spline wavelets described in Examples 2-3-4 of §2.11.

Example 4. Coarse grid correction wavelets
In this example, we apply the lifting scheme according to (2.13.9) and (2.13.10) which amounts to defining the new wavelets from the hierarchical basis function by a perturbation formula of the type

$$\psi_{j-1,\gamma} = \varphi_{j,\gamma} + \sum_{\mu \in S_j(\gamma)} c_{\gamma,\mu}\varphi_{j-1,\mu}, \quad \gamma \in \nabla_{j-1}, \tag{2.13.42}$$

where the ***stencil*** $S_j(\gamma)$ is a neigbourhood of γ contained in the coarse grid Γ_{j-1}. Note that this choice always ensures that $\{\psi_{j-1,\gamma}\ ;\ \gamma \in \nabla_j\}$ is a basis for a complement space W_{j-1}^c of V_{j-1}.

The main problem is of course the choice of the coefficients $c_{\gamma,\mu}$ in order to stabilize the basis. A first natural requirement is to impose an oscillation property, i.e.

$$\int \psi_{j,\gamma} = 0, \tag{2.13.43}$$

since the presence of low frequencies in $\varphi_{j,\gamma}$ can be viewed as the origin of the L^2 instability of the hierarchical basis.

However, a complete understanding of the new system necessarily comes through the analysis of the new dual functions $\tilde{\varphi}_{j,\gamma}$ that appear in (2.13.41). As we remarked in §2.11, in the case of univariate spline wavelets, and in the beginning of the present section, the lifting scheme yields a new refinable dual scaling function that has also compact support and satisfies a refinement equation with a finite number of coefficients.

We can then use the techniques of §2.7 (adapted to this multivariate setting) to estimate the smoothness of the new dual function. In this case, one can find simple coarse grid corrections that give rise to a compactly supported refinable function $\tilde{\varphi}$ (dual to the hat function φ that generates the $\mathbb{P}_1$ elements on a uniform mesh by its translations and dilations) which is square integrable, and has also some positive Sobolev smoothness. The following choice turns out to coincide with a construction proposed by COHEN and SCHLENKER [1993] for which it is proved that $\tilde{\varphi}$ belongs to the Sobolev space $H^s(\mathbb{R}^2)$ if $s < 0.44$: for $\gamma \in \nabla_j$, $\varphi_{j,\gamma}$ is perturbed by its two adjacent neighbours on the coarse grid Γ_j with weight $-3/16$ and by its two opposite neighbours with weight $1/16$ as shown in Figure 2.13.6. Note that this results in three types of wavelets that can be derived from each other by rotation of $2\pi/3$.

This particular perturbation gives the maximal Sobolev smoothness for the dual scaling function among all possible weights with such a 4-points stencil. The first numerical results seem to indicate that the use of a 4-points stencil with modified coefficients near exceptional vertices and edges

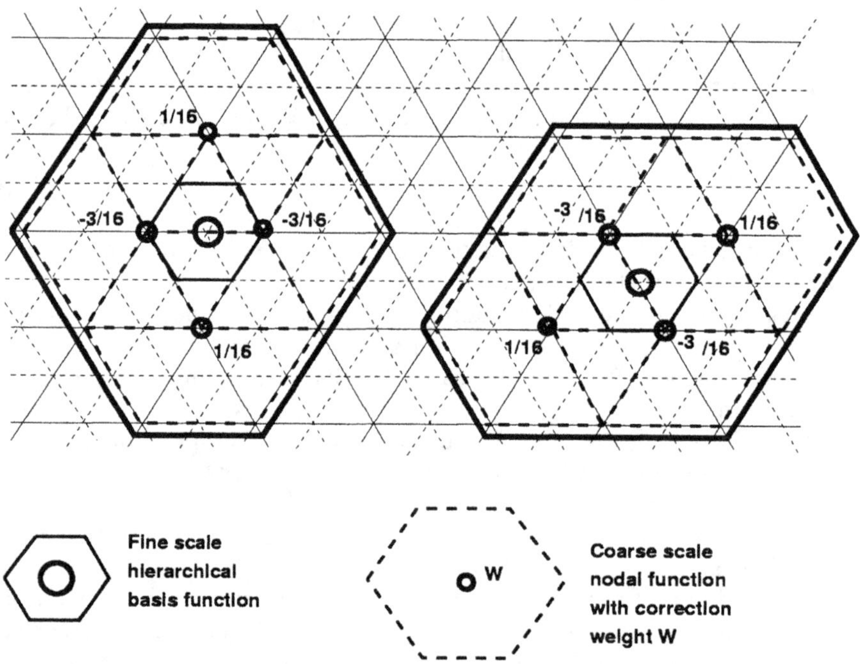

Figure 2.13.6: Two coarse grid correction $\mathbb{P}_1$ wavelets

(in order to preserve (2.13.43)), and of a modified 3-points stencil near the boundary, allows us to preserve the good properties of the resulting multiscale basis. Moreover, the analysis of the corresponding modified dual functions seems feasible.

Example 5. Semi-orthogonal wavelets
In this example, the wavelets are obtained from the hierarchical basis function by a combination formula of the type

$$\psi_{j-1,\gamma} = \sum_{\mu \in T_j(\gamma)} d_{\gamma,\mu} \varphi_{j,\mu}, \quad \gamma \in \nabla_j, \tag{2.13.44}$$

where the stencil $T_j(\gamma)$ is a neigbourhood of γ contained in the fine grid Γ_j, and where the coefficients are chosen in such a way that $\psi_{j-1,\gamma}$ is orthogonal to V_{j-1}. We also have to check that the functions $\psi_{j-1,\gamma}$, $\gamma \in \nabla_j$ constitute a Riesz basis for the orthogonal complement W^o_{j-1} of V_{j-1} into V_j.

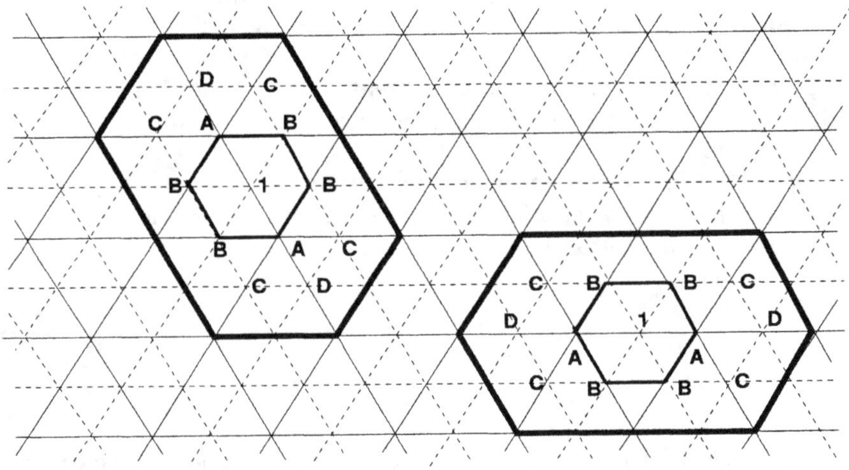

Figure 2.13.7: Two semi-orthogonal $\mathbb{P}_1$ wavelets

Note that while the number of coarse grid functions overlapping $\psi_{j-1,\gamma}$ is roughly proportional to $|T_j(\gamma)|/4$, so that taking the stencil $T_j(\gamma)$ large enough we should be able to find coefficients $d_{\gamma,\mu}$ that ensure the orthogonality. The main problem is thus to check the stability of the wavelet basis *within one level of scale*, i.e. that the functions $\psi_{j-1,\gamma}$, $\gamma \in \nabla_j$ are a Riesz basis for W^o_{j-1}. In the case of a uniform mesh, this can easily be done by the

Fourier transform techniques introduced in §2.2 (adapting Theorem 2.2.1 to this multivariate setting).

We display in Figure 2.13.7 a possible choice of stencil that yields such a semi-orthogonal wavelet basis, whereas in the previous example, we have three types of wavelets that can be derived from each other by rotation of $2\pi/3$.

This construction seems to be adaptable near exceptional vertices and edges, as well as boundaries. First results going in this direction are presented in STEVENSON [1998]. The main advantage of the semi-orthonormal approach is the relative simplicity of the L^2-stability analysis which can be done separately at each scale level due to the orthogonality between the W_j^o spaces. A generalization of this approach was recently proposed in DAHMEN and STEVENSON [1999]: the corrected wavelets are chosen to be orthogonal to another lagrange finite element space $\tilde{V}_j$ which satisfies a minimal angle condition with respect to V_j (for all $f \in V_j$ (resp. $\tilde{V}_j$), $\|f\|_{L^2} = 1$, there exists $g \in \tilde{V}_j$ (resp. V_j), $\|g\|_{L^2} = 1$, such that $\langle f, g\rangle > \alpha$, with $\alpha > 0$ independent of j), and such that the $\tilde{V}_j$ are also nested, i.e. $\tilde{V}_J \subset \tilde{V}_{j+1}$. The wavelets can be constructed by an element by element orthogonalization procedure, which allows a proper treatment of exceptional edges and vertices. In both constructions, the dual basis is globally supported in Ω.

Example 6. Fine grid correction wavelets
In this example, the wavelets are obtained from the hierarchical basis function by a perturbation formula of the type

$$\psi_{j-1,\gamma} = \varphi_{j,\gamma} + \sum_{\mu \in R_j(\gamma)} c_{\gamma,\mu}\varphi_{j,\mu}, \quad \gamma \in \nabla_j, \tag{2.13.45}$$

where the stencil $R_j(\gamma)$ is a neighbourhood of γ contained in the fine grid Γ_j.

In contrast with coarse grid correction, this choice does not always ensure that $\{\psi_{j-1,\gamma}\ ;\ \gamma \in \nabla_j\}$ is a basis for a complement space of V_{j-1} into V_j. In order to solve the algebra problem, we can use a correction with fine scale functions that are centred at points of the coarse grid, i.e. impose that $R_j(\gamma) \in \Gamma_{j-1}$. The following example of such a construction was studied in detail in OSWALD and LORENTZ [1996]: the correction consists of removing the two nearest neighbours of $\varphi_{j,\gamma}$, with weight $1/2$. The wavelets are displayed in Figure 2.13.8 (after renormalization by a factor of two).

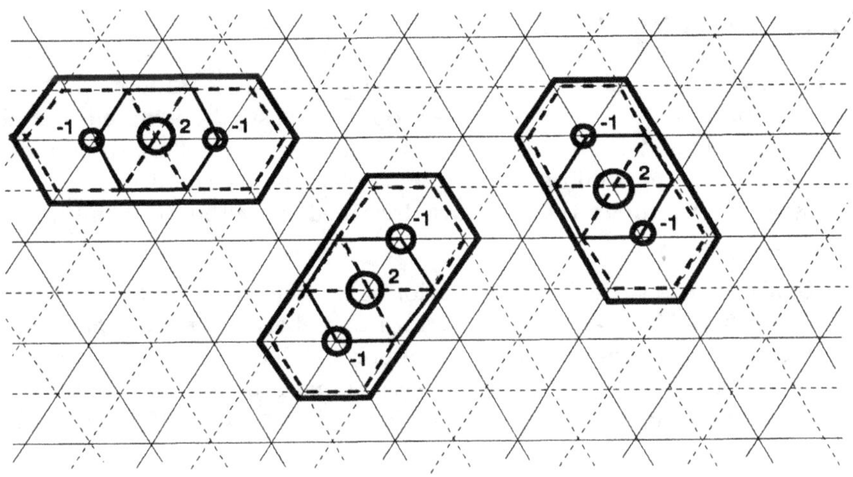

Figure 2.13.8: Three fine grid correction $\mathbb{P}_1$ wavelets

Note that they are orthogonal to the coarse grid nodal functions $\varphi_{j-1,\gamma}$ with respect to the discrete inner product on the fine grid

$$\langle f, g\rangle_{d,j} := \sum_{\gamma\in\Gamma_j} f_\gamma g_\gamma. \tag{2.13.46}$$

However, one cannot only rely on this "discrete semi-orthogonality" to conclude that this multiscale basis is stable: the dual functions still need to be investigated in order to solve the analysis problem. As in Example 4 of §2.11, these dual functions are not compactly supported, but Fourier techniques allow us to estimate their regularity. For the present example, OSWALD and LORENTZ [1996] have proved that they belong to $H^s(\mathbb{R}^2)$ for $s < 0.99$.

This suggests that such a fine grid correction yields a stabilized basis, although a complete analysis in the realistic situation of a bounded domain is still to be done. Such an analysis seems more difficult than in the two previous cases, due to the combination of two difficulties: the lack of orthogonality between levels and the global support of the dual. It is striking to see how this last example, which has the simplest numerical structure and the smallest support, is by far the most complicated to analyze.

2.14 Conclusions

Numerous tools are available to the numerician for multiscale approximation and decomposition of functions. These tools can easily be adapted to multivariate functions defined on $\mathbb{R}^d$, and with a little more effort to bounded domains with the possibility of incorporating boundary conditions.

Most of the examples that we have described share the following fundamental features:

1. **Scaling functions:** approximation takes place in a ladder of embedded spaces V_j, $j \geq 0$, equipped with local bases of ***scaling functions***. More precisely, these bases have supports of diameter $\mathcal{O}(2^{-j})$ and they satisfy a property of ***controlled overlapping***: a point x is contained in at most M supports of basis function at level j with M independent of x and j.

2. **Wavelets:** a (not necessarily orthogonal) complement W_j of V_{j+1} onto V_j is generated by a ***wavelet basis***, with the same local properties. This results in a multiscale basis consisting of scaling functions at a fixed level and wavelets at all finer levels. The local nature of the scaling functions and wavelets yields a fast reconstruction algorithm to obtain the coefficients of $f \in V_j$ in the scaling function basis from those in the multiscale basis.

3. **Dual bases:** the coefficients of the multiscale decomposition of a function in the wavelet basis can be computed by inner products with dual scaling functions and wavelets. The dual scaling functions are not necessarily compactly supported and refinable. However, if they have these two properties, the coefficients of $f \in V_j$ in the multiscale basis can be computed by a fast algorithm from those in the scaling function basis.

4. **Approximation and smoothness:** the scaling functions have certain ***polynomial exactness properties***, which result in vanishing moments for the dual wavelets. In a shift-invariant setting, these properties are expressed by the Strang-Fix conditions. They also have certain smoothness properties. The dual scaling functions can also be built with such properties. In several constructions, both polynomial exactness and smoothness can be prescribed at an arbitrarily high level, but the price to pay is a support size that grows linearly with these parameters.

These properties play a crucial role in the practical use of multiscale decompositions in numerical analysis, as well as in the proof of most theoretical results underlying these applications, as we shall see in the two remaining chapters.

2.15 Historical notes

Although the concepts and constructions that we have described throughout this chapter have mostly been developed since the end of the 1980's, they should be viewed as the result of the merging of several independent developments that took place in different areas of research:

1. **Time-frequency and time-scale analysis:** since the 1950's, several ideas had been proposed to "localize" Fourier analysis in the context of signal processing. For example, Gabor proposed computing inner products $Gf(a,b) = \langle f, g_{a,b}\rangle$ of the analyzed function f against localized waves,

$$g_{a,b}(x) = e^{iax}g(x-b), \quad a,b \in \mathbb{R}, \tag{2.15.1}$$

where g is typically a gaussian function. In the early 1980's, the geophysicist Morlet who was looking for an analysis with arbitrarily high resolution in space (limited in (2.15.1) by the fixed support of g), proposed using

$$\psi_{a,b}(x) = a^{-1}\psi(\frac{x-b}{a}), \quad a > 0, b \in \mathbb{R}, \tag{2.15.2}$$

where ψ is a function which is well localized both in space and frequency. The function ψ should also satisfy the condition

$$2\pi \int |\omega|^{-1}|\hat{\psi}(\omega)|^2 d\omega = C < +\infty, \tag{2.15.3}$$

which implies $\int \psi = 0$. Moreover, he suggested that a function f can be reconstructed from its continuous wavelet transform $Wf(a,b) = \langle f, \psi_{a,b}\rangle$ by the formula

$$f = C^{-1} \int\int Wf(a,b)\psi_{a,b}(x)\frac{da\,db}{a}. \tag{2.15.4}$$

This formula was proved to be true by Grossmann in 1983. From a more numerical point of view, Daubechies proved in 1984 that the sampling

$$\psi_{m,n}(x) = a_0^{m/2}\psi(a_0^m x - b_0 n), \quad m,n \in \mathbb{Z}, \tag{2.15.5}$$

generates a *frame* of $L^2(\mathbb{R})$ if $a_0 > 1$ and $b_0 > 0$ are chosen small enough. The question of the existence of an orthonormal basis with the above structure and ψ well localized both in space and frequency (in contrast to the Haar system), was left open until a construction by Meyer in 1985, in which ψ belongs to the Schwartz class $\mathcal{S}(\mathbb{R})$. A detailed and comprehensive survey of these developments will be found in DAUBECHIES [1992].

2. **Harmonic analysis:** Since the 1930's, Littlewood and Paley had proposed a systematic way of decomposing a function f into "almost orthogonal blocks". One fixes a positive C^∞ function $r(\omega)$ which has compact support in $[-2, 2]$ and equals 1 on $[-1, 1]$. Any tempered distribution f can then be decomposed according to

$$f = S_0 f + \sum_{j\geq 0} \Delta_j f, \tag{2.15.6}$$

with $\mathcal{F}S_j f(\omega) = \hat{f}(\omega) r(2^{-j}\omega)$ and $\Delta_j f = S_{j+1} f - S_j f$. The block $\Delta_j f$ thus represents the frequency content of f between 2^j and 2^{j+1}. Such a decomposition has proved to be a powerful tool for the study of linear and multilinear operators, as well as for characterizing various function spaces, in particular smoothness classes, from the size properties of the blocks Δ_j (see FRAZIER, JAWERTH AND WEISS [1991] for a review on Littlewood-Paley theory and its applications). Clearly, wavelet decompositions can be viewed as a variant of (2.15.6) which is more adapted for numerical computations, since the $\Delta_j f$ are replaced by combinations of local functions.

3. **Approximation theory:** the study of spline functions began with the work of Schonberg in the 1940's (SCHONBERG [1946]), who is also at the origin of the Strang-Fix conditions described in §2.8. There is a huge amount of existing literature on splines, as well as on finite elements (we already mentioned some general references in this chapter). Let us say that an important step in the understanding of spline approximation was the introduction of local quasi-interpolant operators (which have the same structure as our local projectors P_j, see Remark 2.3.2) in the 1970's by de Boor and Fix. The general relations between spline functions and subdivision algorithms were analyzed in the 1980's (DAHMEN and MICCHELLI [1984]).

4. **Multiresolution image processing:** many algorithms developed in image processing, as well as in computer-aided geometric design,

since the 1960's, are based on a multiresolution processing of the visual information. Beside the subdivision algorithms (that have been discussed in §2.4) for fast curve or surface generation, and the iterated filterbanks (see KOVACEVIC and VETTERLI [1995] for a detailed review) that are principally used for image compression and coding, one should also mention the ***laplacian pyramids*** introduced in BURT and ADELSON [1983] which first made the connexion between techniques of multiscale refinement and image coding ideas, and can be viewed as a discrete implementation of the quasi-interpolant operators that were evoked in Remark 2.3.2.

The introduction of the concept of multiresolution approximation by Mallat in 1986, and the resulting constructions of compactly supported wavelets by Daubechies in 1988, can thus be viewed as a beautiful combination of these ideas.

Chapter 3

Approximation and smoothness

3.1 Introduction

Approximation theory is the branch of mathematics which studies the process of approximating general functions by simple functions such as polynomials, finite elements or Fourier series. It therefore plays a central role in the analysis of numerical methods, in particular approximation of PDE's. Numerous problems of approximation theory have in common the following general setting: we are given a family of subspaces $(S_N)_{N\geq 0}$ of a normed space X, and for $f \in X$, we consider the *best approximation error*

$$\sigma_N(f) := \text{dist}_X(f, S_N) = \inf_{g\in S_N} \|f - g\|_X. \tag{3.1.1}$$

Typically, N represents the number of parameters which are needed to describe an element in S_N, and in most cases of interest, $\sigma_N(f)$ goes to zero as this number tends to infinity. If in addition $\sigma_N(f) \lesssim N^{-s}$ for some $s > 0$, we say that f is approximated at rate s.

Given such a setting, the central problem of approximation theory is to *characterize* by some analytic (typically smoothness) condition those functions f which are approximated at some prescribed rate $s > 0$. Another important problem is how to design simple approximation procedures $f \mapsto f_N \in \Sigma_N$ from f which avoid solving the minimization problem (3.1.1), while remaining near-optimal in the sense that

$$\|f - f_N\|_X \leq C\sigma_N(f), \tag{3.1.2}$$

for some constant C independent of N and f. In the case where the Σ_N are linear vector spaces, near-optimal approximation procedures can be obtained if we can find a sequence of finite element projectors $P_N : X \mapsto S_N$ such that $\|P_N\|_{X\to X} \leq K$ with K independent of N: in this case, according to Lemma 2.3.1, we simply take $f_N = P_N f$ and obtain

$$\|f - f_N\|_X \leq (1 + K)\sigma_N(f). \tag{3.1.3}$$

Perhaps the most well known result for the numerician deals with finite element approximation in Sobolev spaces: consider approximation by finite element spaces V_h defined from regular conforming partitions $\mathcal{T}_h$ of a domain $\Omega \subset \mathbb{R}^d$ into simplices with uniform mesh size h. The classical finite element approximation theory (see e.g. CIARLET [1978]) can be summarized in the following way. If $W^{t,p}$ denotes the classical Sobolev space, consisting of those functions $f \in L^p$ such that $D^\alpha f \in L^p$ for $|\alpha| \leq t$, we typically have

$$f \in W^{t+r,p} \Rightarrow \inf_{g \in V_h} \|f - g\|_{W^{t,p}} \lesssim h^r \tag{3.1.4}$$

provided that V_h is contained in $W^{t,p}$ and that V_h has approximation order larger than $t + r$, i.e. contains all polynomials of degree strictly less than $t + r$. Such classical results also hold for fractional smoothness.We can express (3.1.4) in terms of the multiresolution spaces $V_j \sim V_h$ with $h \sim 2^{-j}$ and write

$$f \in W^{t+r,p} \Rightarrow \inf_{g \in V_j} \|f - g\|_{W^{t,p}} \lesssim 2^{-rj} \tag{3.1.5}$$

We can also express (3.1.4) in terms of the number of parameters, remarking that $N := \dim(V_h) \sim h^{-d}$, so that if we set $X = W^{t,p}$ and $S_N := V_h$ with $h := N^{-1/d}$, we have obtained

$$f \in W^{t+r,p} \Rightarrow \sigma_N(f) \lesssim N^{-r/d}. \tag{3.1.6}$$

We have thus identified an analytic condition which ensures the rate $s = r/d$ in $W^{s,p}$. This rate is governed by the approximation order of the V_h spaces, the dimension d and the level of smoothness of f measured in L^p. Note that this is not a characterization (we only have an implication), i.e. these results express that a smoothness property implies an approximation rate. If the approximation spaces are themselves constituted of smooth functions, it is reasonable to think of a converse: a certain rate of approximation for a function f might imply that f has some smoothness.

Surprisingly enough, it turns out that a large number of smoothness classes, including L^2-Sobolev spaces, can be *exactly characterized* from the rate of decay of the approximation error in the spaces V_j, or also from the summability and decay properties of the wavelet coefficients. Such an instance was already evoked without proof in §1.6 of Chapter 1, in relation to multiscale preconditioning. The goal of this chapter is to prove this type of result in a general context.

As a first step, we introduce in §3.2 the various function spaces that will be considered in this chapter, in particular Hölder, Sobolev and Besov smoothness classes. We then introduce, in §3.3 and §3.4, two important types of inequalities that are essential to build a more general theory: *direct (or Jackson type) inequalities* that are exactly of the type (3.1.5), and *inverse (or Bernstein type) inequalities* that take into account the smoothness properties of the V_j spaces. In the setting of finite element spaces, such inequalities are classically derived by using the properties of the affine mapping between each element and a reference domain (e.g. the unit square in the case of rectangular elements) and basic results of polynomial approximation in this reference domain. Since we have in mind approximation spaces V_j which are not necessarily of finite element types we shall proceed in a slightly different way. Our proofs are based on local estimates on the dyadic cubes

$$I_{j,k} = 2^{-j}(k + [0,1]^d), \quad k \in \mathbb{Z}^d, \tag{3.1.7}$$

and involve the scaling functions generating V_j. They should still be viewed as variants to the proofs of similar results in the finite element context.

In §3.5, we describe a general mechanism, combining direct and inverse estimates with interpolation of function spaces, that yields the characterization of smoothness classes by the rate of approximation.

We apply these ideas in §3.6 to multiscale approximation, and obtain characterization results for L^2-Sobolev, Hölder and more general Besov spaces. In §3.7, we study the possibility of characterizing L^p-Sobolev and Besov spaces by L^p-unstable approximation operators (in particular, we address the case of quasi-normed Besov spaces for $p < 1$ that will be of crucial importance for the nonlinear theory of Chapter 4). We also deal in §3.8 with the characterization of negative smoothness classes and of L^p spaces.

For the sake of simplicity, all these results are at first presented for function spaces defined on the whole of $\mathbb{R}^d$, using multiresolution analysis and wavelets obtained by the tensor product strategy described in §2.12 of Chapter 2. While the generalization of our results to other types of multiscale decompositions using other basis functions (non-tensor product, multiwavelets, finite element bases) is straightforward, their adaptation to function spaces on bounded domains with specific boundary conditions - which is crucial for practical applications - is a more elaborate task, due to the lack of translation invariance of the multiresolution spaces and the specific adaptations needed near the boundary of the domain. We address these issues in §3.9 and §3.10. These two sections are fairly technical and basically reach the same type of norm equivalences that are more easily obtained on $\mathbb{R}^d$.

We end by discussing in §3.11 the application of these results to multilevel preconditioning of elliptic operators. The analysis of such techniques only involves the characterization of L^2-Sobolev spaces (possibly with boundary conditions). The characterization of more general Sobolev and Besov spaces will be useful in the analysis of adaptive approximations that are the object of the next chapter.

Notations. In this chapter, as well as in the next one, we shall frequently adopt notations similar to those used in §2.13, which aim to simplify the representation of a function in a wavelet basis, especially when working on a multivariate domain Ω: for all the constructions in the previous chapter, we can write

$$P_j f := \sum_{\gamma\in\Gamma_j} \langle f, \tilde{\varphi}_\gamma\rangle \varphi_\gamma, \tag{3.1.8}$$

where $(\varphi_\gamma)_{\gamma\in\Gamma_j}$ is the scaling function basis of V_j and $(\tilde{\varphi}_\gamma)_{\gamma\in\Gamma_j}$ the corresponding dual basis of $\widetilde{V}_j$. For the sake of notational simplicity, we denote by $(\psi_\lambda)_{\lambda\in\nabla_{j_0-1}}$ the scaling function basis $(\varphi_\gamma)_{\gamma\in\Gamma_{j_0}}$ at the coarsest level j_0 with a similar convention for the dual functions, so that we can write

$$f = \sum_{j\geq -1}\sum_{\lambda\in\nabla_j} \langle f, \tilde{\psi}_\lambda\rangle \psi_\lambda. \tag{3.1.9}$$

For sake of simplicity, except in certain specified cases, we shall assume that $j_0 = 0$. We also define

$$\nabla^j := \cup_{-1\leq l<j}\nabla_l \text{ and } \nabla = \cup_{j\geq -1}\nabla_j \tag{3.1.10}$$

so that we have

$$f = \sum_{\lambda\in\nabla} \langle f, \tilde{\psi}_\lambda\rangle \psi_\lambda \tag{3.1.11}$$

If an index λ belongs to ∇_j or Γ_j, we set $|\lambda| := j$. Thus,

$$\Phi_j := \{\varphi_\gamma\}_{\gamma\in\Gamma_j} = \{\varphi_\gamma\}_{|\gamma|=j}, \tag{3.1.12}$$

and

$$\Psi_j := \{\psi_\lambda\}_{\lambda\in\nabla^j} = \{\psi_\lambda\}_{|\lambda|<j}, \tag{3.1.13}$$

are respectively the nodal and multiscale basis of V_j.

We recall some size properties of the functions ψ_λ (the same remarks apply to the functions φ_λ, $\tilde{\psi}_\lambda$ and $\tilde{\varphi}_\lambda$). We assume that the basis functions are normalized in L^2. This means that we have scaling factors when we use another norm to measure these functions. For example, we have

$$\|\psi_\lambda\|_{W^{m,p}} \sim 2^{m+(d/2-d/p)|\lambda|}. \tag{3.1.14}$$

The supports of ψ_λ are contained in cubes I_λ of sidelength $C2^{-|\lambda|}$, and do not overlap too much in the following sense: there exists a constant K such that for all $\lambda \in \nabla_j$,

$$\#\{\mu \in \nabla_j \text{ s.t. } \mathrm{Supp}(\psi_\lambda) \cap \mathrm{Supp}(\psi_\mu) \neq \emptyset\} \leq K. \tag{3.1.15}$$

3.2 Function spaces

There exist many different ways of measuring the smoothness of a function f. The most natural one is certainly the order of differentiability, i.e. the maximal index m such that $f^{(m)} = (\frac{d}{dx})^m f$, and more generally $\partial^\alpha f$ with $|\alpha| := \alpha_1 + \cdots + \alpha_d \leq m$ for a multivariate function, is continuous. With this particular measure of smoothness, one associates a class of *function spaces*: for $\Omega \subset \mathbb{R}^d$, we define $C^m(\Omega)$ to be the space of continuous functions which have bounded and continuous partial derivatives $\partial^\alpha f$, $|\alpha| \leq m$. This space is equipped with the norm

$$\|f\|_{C^m(\Omega)} := \sup_{x\in\Omega} |f(x)| + \sum_{|\alpha|=m} \sup_{x\in\Omega} |\partial^\alpha f(x)|, \tag{3.2.1}$$

for which it is a Banach space.

In order to measure the smoothness properties of a function in an average sense, it is also natural to introduce the *Sobolev spaces* $W^{m,p}(\Omega)$ consisting of all functions $f \in L^p$ with partial derivatives up to order m in L^p. Here p is a fixed index in $[1, +\infty]$. This space is also a Banach space, when equipped with the norm

$$\|f\|_{W^{m,p}} := \|f\|_{L^p} + |f|_{W^{m,p}}, \quad |f|_{W^{m,p}} := \sum_{|\alpha|=m} \|\partial^\alpha f\|_{L^p}. \tag{3.2.2}$$

Here, we have used the notation $|\cdot|$ to denote the corresponding semi-norm. Note that the norm (3.2.1) for C^m spaces coincides with the $W^{m,\infty}$ norm although these spaces are different.

All the above spaces share the common feature that the regularity index is an integer.

In many instances of theoretical and numerical analysis of PDE's, one is interested in describing the regularity of a function in a more precise way through fractional degrees of smoothness. The question thus arises of *how to fill the gaps between integer smoothness classes.* There are at least two instances where such a generalization is very natural: firstly, in the case of the L^2-Sobolev spaces $H^m := W^{m,2}$ and when $\Omega = \mathbb{R}^d$, we can define an equivalent norm based on the Fourier transform, since by the Parseval formula we have

$$\|f\|_{H^m}^2 \sim \int_{\mathbb{R}^d} (1+|\omega|)^{2m} |\hat{f}(\omega)|^2 d\omega. \tag{3.2.3}$$

For a non-integer $s \geq 0$, it is thus natural to define the space H^s as the set of all L^2 functions such that

$$\|f\|_{H^s}^2 := \int_{\mathbb{R}^d} (1+|\omega|)^{2s} |\hat{f}(\omega)|^2 d\omega, \tag{3.2.4}$$

is finite.

Secondly, in the case of C^m spaces, we note that for any $h \in \mathbb{R}^d$

$$\sup_{x\in\Omega} |f(x+h) - f(x)| \leq [\sup |f'|]|h| \tag{3.2.5}$$

if $f \in C^1$, whereas for an arbitrary $f \in C^0$, the quantity $\sup_{x\in\Omega} |f(x+h) - f(x)|$ might go to zero arbitrarily slowly as $|h| \to 0$. This motivates the definition of the Hölder space C^s, $0 < s < 1$ consisting of those $f \in C^0$ such that

$$\sup_{x\in\Omega} |f(x+h) - f(x)| \leq C|h|^s. \tag{3.2.6}$$

If $m < s < m+1$, a natural definition of C^s is given by $f \in C^m$ and $\partial^\alpha f \in C^{s-m}$, $|\alpha| = m$. It is not difficult to prove that this property can also be expressed by

$$\sup_{x\in\Omega} |\Delta_h^n f(x)| \leq C|h|^s, \tag{3.2.7}$$

where $n > s$ and Δ_h^n is the n-th order finite difference operator defined recursively by $\Delta_h^1 f(x) = f(x+h) - f(x)$ and $\Delta_h^n f(x) = \Delta_h^1(\Delta_h^{n-1})f(x)$ (for example $\Delta_h^2 f(x) = f(x+2h) - 2f(x+h) + f(x)$). When s is not an integer, the spaces C^s that we have defined are also denoted by $W^{s,\infty}$.

The definition of "s order of smoothness in L^p" for s non-integer and p different from 2 and ∞ is more subject to arbitrary choices. Among others, we find three well-known classes.

For $m < s < m+1$, Sobolev spaces $W^{s,p}$ consist of those distributions f such that $\|f\|_{W^{s,p}} = \|f\|_{L^p} + |f|_{W^{s,p}}$ is finite with

$$|f|_{W^{s,p}} := \sum_{|\alpha|=m} \int_{\Omega\times\Omega} \frac{|\partial^\alpha f(x) - \partial^\alpha f(y)|^p}{|x-y|^{(s-m)p+d}} dx dy. \tag{3.2.8}$$

These spaces coincide with those defined by means of Fourier transform when $p = 2$. They are also a particular case of the Besov spaces defined below when s is not an integer. We refer to ADAMS [1975] for a general introduction.

Potential (or Liouville) spaces $H^{s,p}$ (sometimes also denoted by $L^{p,s}$) are defined by means of the Fourier transform operator $\mathcal{F}$ according to

$$\|f\|_{H^{s,p}} = \|f\|_{L^p} + \|\mathcal{F}^{-1}(1 + |\cdot|^s)\mathcal{F}f\|_{L^p}. \tag{3.2.9}$$

These spaces coincide with the Sobolev spaces $W^{m,p}$ when m is an integer and $1 < p < +\infty$ (see TRIEBEL [1983, p.38]), but their definition by (3.2.9) requires a-priori that $\Omega = \mathbb{R}^d$ in order to apply the Fourier transform. These spaces can of course be defined on arbitrary domains by restriction, but they have no simple inner description for non-integer s. In certain references, they are also referred to as fractional Sobolev spaces. However they differ from the $W^{s,p}$ spaces when $p \neq 2$ and s is not an integer.

Besov spaces $B^s_{p,q}$ involve an extra parameter q and can be defined through finite differences. These spaces include most of those that we have listed so far as particular cases for certain ranges of indices. As it will appear in the forthcoming section §3.5, these spaces are also produced by general "interpolation techniques" between function spaces of integer smoothness, and they can be exactly characterized by the rate of multiresolution approximation error, as well as from the size properties of the wavelet coefficients. For these reasons, we shall insist a bit more on their definitions and properties.

We define the n-th order L^p modulus of smoothness of f by

$$\omega_n(f, t, \Omega)_p = \sup_{|h| \le t} \|\Delta_h^n f\|_{L^p(\Omega_{h,n})}, \tag{3.2.10}$$

(h is a vector in $\mathbb{R}^d$ of Euclidean norm less than t) where

$$\Omega_{h,n} := \{x \in \Omega\ ;\ x + kh \in \Omega,\ k = 0, \cdots, n\}. \tag{3.2.11}$$

We use the simpler notation $\omega_n(f,t)_p$ if Ω is the whole domain where the function is considered. A basic property of this quantity is its monotonicity with respect to t: $\omega_n(f,t)_p \le \omega_n(f,s)_p$ if $t < s$. It is also easily checked that if m is a positive integer, we have $\omega_n(f, mt, \Omega)_p \le m^n \omega_n(f, t, \Omega)_p$,

For $p, q \ge 1$, $s > 0$, the Besov spaces $B^s_{p,q}(\Omega)$ consist of those functions $f \in L^p(\Omega)$ such that

$$(2^{sj}\omega_n(f, 2^{-j})_p)_{j \ge 0} \in \ell^q. \tag{3.2.12}$$

where n is an integer such that $s < n$. A natural norm for such a space is then given by

$$\|f\|_{B^s_{p,q}} := \|f\|_{L^p} + |f|_{B^s_{p,q}},\quad |f|_{B^s_{p,q}} := \|(2^{sj}\omega_n(f, 2^{-j})_p)_{j \ge 0}\|_{\ell^q}. \tag{3.2.13}$$

Remark 3.2.1 *The Besov semi-norm is often instead defined in its integral form*

$$|f|_{B^s_{p,q}} := \|t^{-s}\omega_n(f,t)_p\|_{L^q([0,1],dt/t)}, \tag{3.2.14}$$

i.e. $\left(\int_0^1 (t^{-s}\omega_n(f,t)_p)^q \frac{dt}{t}\right)^{1/q}$ *if* $q < +\infty$. *The equivalence between the above expression and its discrete analog of (3.2.13) is easily derived from the monotonicity of the modulus of smoothness* ($\omega_n(f,t)_p \leq \omega_n(f,t')_p$ *if* $t < t'$). *Also note that* $|f|_{B^s_{p,q}} \sim \|(2^{sj}\omega_n(f,2^{-j})_p)_{j\geq J}\|_{\ell^q}$ *with constants independent of* $J \geq 0$ *such that* $\mathrm{diam}(\Omega) \leq 2^{-J}$.

Remark 3.2.2 *The definition of* $B^s_{p,q}$ *is independent of* n *in the sense that two integers* $n, n' > s$ *yield equivalent norms by (3.2.13). On the one hand, if* $n \geq n'$, *one clearly has* $\omega_n(f,t)_p \lesssim \omega_{n'}(f,t)_p$, *and we get a similar inequality for the corresponding expressions in (3.2.13). To prove the converse part of the norm equivalence, one classically uses the Marchaud type (discrete) inequality*

$$\omega_{n'}(f,2^{-j})_p \lesssim 2^{-n'j}[\|f\|_{L^p} + \|(2^{n'(j+l)}\omega_n(f,2^{-(j+l)})_p)_{l\geq 0}\|_{\ell^q}], \tag{3.2.15}$$

and concludes by using discrete Hardy's inequalities (see for example Butzer *and* Behrens *[1967, p.199] or* DeVore *and* Popov *[1988, Corollary 4.9], or also* DeVore *and* Lorentz *[1993, p.55]).*

The space $B^s_{p,q}$ represents "s order of smoothness measured in L^p", with the parameter q allowing a finer tuning on the degree of smoothness - one has $B^s_{p,q_1} \subset B^s_{p,q_2}$ if $q_1 \leq q_2$ - but plays a minor role in comparison with s since clearly

$$B^{s_1}_{p,q_1} \subset B^{s_2}_{p,q_2}, \text{ if } s_1 \geq s_2, \tag{3.2.16}$$

regardless of the values of q_1 and q_2. Roughly speaking, smoothness of order s in L^p is thus expressed here by the fact that, for n large enough, $\omega_n(f,t)_p$ goes to 0 like $\mathcal{O}(t^s)$ as $t \to 0$. Clearly $C^s = B^s_{\infty,\infty}$ when s is not an integer.

Remark 3.2.3 *More generally, it can be proved (*Triebel *[1983]) that* $W^{s,p} = B^s_{p,p}$ *when* s *is not an integer. One should be aware of the fact that the spaces* $W^{m,p}$ *are not Besov spaces for* $m \in \mathbb{N}$ *and* $p \neq 2$. *In particular, it is well known that* $W^{1,p}$ *can be characterized by the property* $\omega_1(f,t)_p \lesssim t$ *if* $1 < p < +\infty$, *i.e.* $W^{1,p} := \mathrm{Lip1}(L^p)$, *but this space differs from* $B^1_{p,\infty}$ *which involves the second order differences in its definition.*

Remark 3.2.4 *An alternative definition of Besov space is given by the Littlewood-Paley decomposition: one fixes a compactly supported, C^∞ function $S(\omega) \geq 0$ such that $S(0) \neq 0$, and decomposes an arbitrary tempered distribution $f \in \mathcal{S}'(\mathbb{R}^d)$ according to*

$$f = S_0 f + \sum_{j \geq 0} \Delta_j f, \tag{3.2.17}$$

where by definition

$$\Delta_j = S_{j+1} - S_j, \quad \mathcal{F}(S_j f)(\omega) = S(2^{-j}\omega)\hat{f}(\omega). \tag{3.2.18}$$

The Besov space $B^s_{p,q}$ is then defined as the space of all tempered distributions such that

$$\|S_0 f\|_{L^p} + \|(2^{sj}\|\Delta_j f\|_{L^p})_{j \geq 0}\|_{\ell^q}, \tag{3.2.19}$$

is finite. One easily checks that this property is independent of the choice of the function S. This is actually the definition of Besov spaces which is chosen in some classical references on Besov spaces, e.g. in TRIEBEL *[1983], allowing us to consider also negative values of s, and all values $p, q > 0$. It can be shown (see for example* TRIEBEL *[1983]) that (3.2.19) and (3.2.13) are equivalent norms when $s > 0$ and $\Omega = \mathbb{R}^d$ (assuming here that $p, q \geq 1$). Our results of §3.5 and §3.6 will actually provide a proof of this fact. For more general domains, the use of (3.2.19) obliges us to define Besov spaces by restriction, while (3.2.13) offers an inner description. Nevertheless, these two definitions still coincide for fairly general domains (see §3.7). An immediate consequence of the Littlewood-Paley characterization is that $H^s = W^{s,2} = B^s_{2,2}$ for all $s > 0$.*

Sobolev, Besov and potential spaces satisfy two simple embedding relations: for fixed p (and for arbitrary q in the case of Besov spaces, see (3.2.16)), these spaces get larger as s decreases, and in the case where Ω is a bounded domain, for fixed s (and fixed q in the case of Besov spaces), these spaces get larger as p decrease, since $\|f\|_{L^{p_1}} \lesssim \|f\|_{L^{p_2}}$ if $p_1 \leq p_2$.

A less trivial type of result is the so-called *Sobolev embedding* theorem. It states that

$$W^{s_1,p_1} \subset W^{s_2,p_2} \text{ if } s_1 - s_2 \geq d(1/p_1 - 1/p_2), \tag{3.2.20}$$

except in the case where $p_2 = +\infty$ and $s_1 - d(1/p_1 - 1/p_2)$ is an integer, for which one needs to assume that $s_1 - s_2 > d(1/p_1 - 1/p_2)$. In the case of Besov spaces, a similar embedding relation is given by

$$B^{s_1}_{p_1,p_1} \subset B^{s_2}_{p_2,p_2} \text{ if } s_1 - s_2 \geq d(1/p_1 - 1/p_2), \tag{3.2.21}$$

with no restriction on the indices $s_1, s_2 \geq 0$ and $p_1, p_2 \geq 1$. The proof of these embedding theorems can be found in ADAMS [1975] for Sobolev spaces and TRIEBEL [1983] for Besov spaces. We shall also give a proof of the less classical Besov embedding in §3.7.

The Besov spaces can also be defined for p and q less than 1. This extension, which will be of particular importance to us in the study of nonlinear and adaptive approximation in Chapter 4, is the source of additional difficulties. First, the expressions (3.2.13) or (3.2.19) do not define norms but only *quasi-norms*: they fail to satisfy the triangle inquality, but still do satisfy

$$\|f+g\| \leq C(\|f\|+\|g\|) \text{ and } \|f+g\|^\mu \leq \|f\|^\mu + \|g\|^\mu, \tag{3.2.22}$$

for C large enough and $\mu > 0$ small enough. The second property - sometimes called the μ-triangle inequality - clearly holds with $\mu = \min\{p,q\}$ and implies the first property with $C = C(\mu)$. One still has the completeness property, i.e. these spaces are quasi-Banach spaces (in particular $\sum_j f_j$ converges if $\sum_j \|f_j\|^\mu < +\infty$). Secondly, for p less than 1, the L^p boundedness of f does not even ensure that it is a distribution, so that the evaluation of inner products $\langle f, g\rangle$ with test functions g is a-priori meaningless. One way of turning around this difficulty is to replace L^p-spaces by the Hardy H^p-spaces in the definition of the Besov spaces, as proposed in PEETRE [1976, chapter 11]. Finally, the equivalence between the expressions (3.2.13) and (3.2.19) is only valid under the assumption $s/d > \max\{1/p - 1, 0\}$ (see TRIEBEL [1983, p.110]). In all the following (in particular in §3.6 and in the next chapter), we shall mainly consider Besov spaces in this range of indices. By the embedding property (3.2.21), which also holds for $p_1, p_2 < 1$ (see §3.7), this assumption also ensures that the corresponding space is embedded in L^1. In these ranges of indices, the definition of Besov spaces by (3.2.13) is thus sufficient to obtain a proper space of distributions.

In the next sections up to §3.7, we shall only consider Besov and Sobolev spaces for $p \geq 1$.

We end this brief presentation by recalling without proof two important results dealing with local polynomial approximation. We denote by Π_m the set of polynomials of total degree m, i.e. linear combinations of the basis functions $x_1^{k_1} \cdots x_d^{k_d}$ for $k_1 + \cdots k_d \leq m$.

The first result is a simple version of the Deny-Lions theorem, firstly proved in DENY and LIONS [1954], which is of common use in the numerical analysis of the finite element method (see also CIARLET [1989, Theorem 14.1]). A more elaborate version of this result will be proved in §3.10 in order to treat the case of Sobolev spaces with homogeneous boundary conditions.

Theorem 3.2.1 *For $1 \le p \le +\infty$, m a positive integer and $\Omega \subset \mathbb{R}^d$ a connected bounded domain, we have the estimate*

$$\inf_{g\in\Pi_m} \|f-g\|_{L^p(\Omega)} \le C|f|_{W^{m+1,p}(\Omega)}, \tag{3.2.23}$$

where the constant C only depends on d, m, p and Ω.

The second result is a local variant of a theorem by Whitney, involving the modulus of smoothness. In addition to the previous result, it covers the case $0 < p < 1$, but to our knowledge it is only proved for cubic domains (we shall prove it in §3.9 for more general domains). We refer to BRUDNYI [1970] for a proof of this result in the case $p \ge 1$ and to OSWALD and STOROSHENKO [1978] for the generalization to all $p > 0$.

Theorem 3.2.2 *For $p > 0$ and $\Omega = [0,1]^d$, one has*

$$\inf_{g\in\Pi_m} \|f-g\|_{L^p(\Omega)} \le C\omega_{m+1}(f,1,\Omega)_p \tag{3.2.24}$$

where the constant C only depends on d, m and p.

These results have direct consequences on local polynomial approximation error, by a rescaling argument: if I_h is any cube of sidelength $h > 0$, we obtain

$$\inf_{g\in\Pi_m} \|f-g\|_{L^p(I_h)} \lesssim h^{m+1}|f|_{W^{m+1,p}(I_h)}, \tag{3.2.25}$$

and

$$\inf_{g\in\Pi_m} \|f-g\|_{L^p(I_h)} \lesssim \omega_{m+1}(f,h,I_h)_p \tag{3.2.26}$$

by applying (3.2.23) and (3.2.24) to $f_h := f(T_h\cdot)$, where T_h is an affine transformation mapping Q onto I_h.

3.3 Direct estimates

In order to prove the direct estimate for the V_j spaces, we shall combine two ingredients: (i) the local polynomial approximation results of §3.2 and (ii) the polynomial reproduction properties of the V_j spaces.

In this section, as well as in §3.4, §3.6 and §3.7, we assume that $(V_j, \widetilde{V}_j)$ is a pair of biorthogonal multiresolution spaces generated by tensor product type, compactly supported, dual scaling functions φ and $\tilde{\varphi}$ as constructed in the previous chapter. We thus have

$$V_j = \mathrm{Span}\{\varphi_{j,k} = 2^{dj/2}\varphi(2^j\cdot -k)\ ;\ k \in \mathbb{Z}^d\}, \tag{3.3.1}$$

and similarly for $\tilde{V}_j$. We also denote by P_j the corresponding projector defined by

$$P_j f = \sum_{k \in \mathbb{Z}^d} \langle f, \tilde{\varphi}_{j,k} \rangle \varphi_{j,k}. \tag{3.3.2}$$

As in the previous chapters, we also use the notation $Q_j = P_{j+1} - P_j$ for the projectors onto the detail spaces.

We recall that polynomial exactness up to some order n in the V_j spaces is ensured by designing φ so that the Strang-Fix conditions are fulfilled at order n.

As we already mentioned, we are only concerned here (and up to §3.7) with L^p Sobolev and Besov spaces with $p \geq 1$. In the proof of our direct estimates, we shall make use of certain L^p-stability properties of the projector P_j (we already noticed the L^2-stability of this projector in §2.3 of Chapter 2). More precisely, we have the following result.

Theorem 3.3.1 *Let $1 \leq p \leq \infty$. Assume that $\varphi \in L^p$ and $\tilde{\varphi} \in L^{p'}$ where $1/p + 1/p' = 1$. Then the projectors P_j are uniformly bounded in L^p. Moreover the basis $\varphi_{j,k}$ is L^p-stable, in the sense that the equivalence*

$$\|\sum_{k \in \mathbb{Z}^d} c_k \varphi_{j,k}\|_{L^p} \sim 2^{dj(1/2-1/p)} \|(c_k)_{k \in \mathbb{Z}^d}\|_{\ell^p}, \tag{3.3.3}$$

holds with constants that do not depend on j.

Proof We first prove the upper inequality of (3.3.3), i.e.

$$\|\sum_{k \in \mathbb{Z}^d} c_k \varphi_{j,k}\|_{L^p} \lesssim 2^{dj(1/2-1/p)} \|(c_k)_{k \in \mathbb{Z}^d}\|_{\ell^p}, \tag{3.3.4}$$

which is a straightforward consequence of the compact support of φ. On each dyadic cube $I_{j,l}$ defined by (3.1.7), we have the local estimate

$$\begin{aligned} \|\textstyle\sum_{k \in \mathbb{Z}^d} c_k \varphi_{j,k}\|_{L^p(I_{j,l})} &\lesssim \sup_{l-k \in \mathrm{Supp}(\varphi)} |c_k| \|\varphi_{j,0}\|_{L^p} \\ &\lesssim 2^{dj(1/2-1/p)} \|(c_k)_{l-k \in \mathrm{Supp}(\varphi)}\|_{\ell^p}, \end{aligned}$$

where we have used the equivalence of all norms in finite dimensions. Taking the p-th power and summing on l (or taking the supremum on l if $p = \infty$) yields the upper inequality.

For the L^p-stability of P_j, we use the Hölder estimate to derive

$$|\langle f, \tilde{\varphi}_{j,k} \rangle| \leq \|f\|_{L^p(\mathrm{Supp}(\tilde{\varphi}_{j,k}))} \|\tilde{\varphi}_{j,0}\|_{L^{p'}} \lesssim 2^{dj(1/2-1/p')} \|f\|_{L^p(\mathrm{Supp}(\tilde{\varphi}_{j,k}))}.$$

Taking the p-th power and summing on k (or taking the supremum on k if $p = \infty$), we thus obtain

$$\|(\langle f, \tilde{\varphi}_{j,k}\rangle)_{k\in\mathbb{Z}^d}\|_{\ell^p} \lesssim 2^{dj(1/2-1/p')}\|f\|_{L^p}, \tag{3.3.5}$$

which combined with (3.3.4) gives

$$\|P_j f\|_{L^p} \leq C\|f\|_{L^p}, \tag{3.3.6}$$

where C does not depend on j.

In order to prove the lower inequality in the equivalence (3.3.3), we simply apply (3.3.5) to $f = \sum_{k\in\mathbb{Z}^d} c_k\varphi_{j,k}$ and remark that $c_k = \langle f, \tilde{\varphi}_{j,k}\rangle$. $\diamond$

Remark 3.3.1 *The L^p-stability is local, in the sense that*

$$\|P_j f\|_{L^p(I_{j,l})} \leq C\|f\|_{L^p(\tilde{I}_{j,l})}, \tag{3.3.7}$$

where

$$\tilde{I}_{j,l} := \cup_{k\in F_l}\mathrm{Supp}(\tilde{\varphi}_{j,k}),\;\; F_l = \{k \in \mathbb{Z}^d \;;\; |\mathrm{Supp}(\varphi_{j,k}) \cap I_{j,l}| \neq 0\}, \tag{3.3.8}$$

(i.e. $\tilde{I}_{j,l} := I_{j,l} + 2^{-j}[\mathrm{Supp}(\tilde{\varphi}) - \mathrm{Supp}(\varphi)]$) is the larger cube on which the value of f influences the value of $P_j f$ on $I_{j,l}$.

Remark 3.3.2 *If φ continuous and $\tilde{\varphi}$ is a compactly supported Radon measure, the same reasoning can be used to prove that P_j maps $C(\mathbb{R}^d)$ into itself with an L^∞ bound that does not depend on j. This applies for example to the interpolation operator corresponding to a pair $(\varphi, \tilde{\varphi})$ where φ is interpolatory and $\tilde{\varphi} = \delta_0$.*

We now turn to the direct estimate. Note that if P_j is L^p-stable uniformly in j, then according to Lemma 2.3.1 of Chapter 2, we have

$$\inf_{g\in V_j} \|f - g\|_{L^p} \sim \|f - P_j f\|_{L^p}, \tag{3.3.9}$$

i.e. the error estimate $\|f - P_j f\|_{L^p}$ is optimal in V_j. We now establish the direct estimate for this particular approximation process.

Theorem 3.3.2 *Under the same assumptions as in Theorem 3.3.1, we have*

$$\|f - P_j f\|_{L^p} \lesssim 2^{-nj}|f|_{W^{n,p}}, \tag{3.3.10}$$

where $n - 1$ is the order of polynomial exactness in V_j.

Proof On each dyadic cube $\tilde{I}_{j,k}$ as defined by (3.3.8), we consider a polynomial $p_{j,k} \in \Pi_{n-1}$ such that

$$\|f - p_{j,k}\|_{L^p(\tilde{I}_{j,k})} \leq 2 \inf_{g\in\Pi_{n-1}} \|f - g\|_{L^p(\tilde{I}_{j,k})} \lesssim 2^{-nj}|f|_{W^{n,p}(\tilde{I}_{j,k})}, \quad (3.3.11)$$

where the second inequality follows from Theorem 3.2.1.

We now estimate the local error by

$$\begin{aligned} \|f - P_j f\|_{L^p(I_{j,k})} &\leq \|f - p_{j,k}\|_{L^p(I_{j,k})} + \|P_j f - p_{j,k}\|_{L^p(I_{j,k})} \\ &= \|f - p_{j,k}\|_{L^p(I_{j,k})} + \|P_j(f - p_{j,k})\|_{L^p(I_{j,k})} \\ &\lesssim \|f - p_{j,k}\|_{L^p(\tilde{I}_{j,k})} \\ &\lesssim 2^{-nj}|f|_{W^{n,p}(\tilde{I}_{j,k})}, \end{aligned}$$

where we have successively used the polynomial reproduction properties of P_j, the local stability estimate (3.3.7) and the polynomial approximation estimate (3.3.11).

The global estimate (3.3.10) follows by taking the p-th power and summing on k, or by taking the supremum in k when $p = \infty$. ◇

Remark 3.3.3 *In the proof of the direct estimate, as well as for the L^p-stability of P_j we make no use of the duality relation $\langle\varphi, \tilde{\varphi}(\cdot - k)\rangle = \delta_{0,k}$. The conclusion would thus be the same for a general operator P_j of the form (3.3.2) with the sole assumption that P_j reproduces polynomials up to the order n. This is for example the case with the quasi-interpolant operators introduced by* DE BOOR *and* FIX *[1973a] for proving the approximation properties of spline functions. In that particular case, an interesting additional property is that the quasi-interpolant is designed for splines with non-equally spaced knots, while its L^p-norm remains bounded independently of the choice of the knots. In contrast, we have used the existence of dual basis with compact support in order to prove the L^p-stability estimate from below in (3.3.3). Note however that the refinability of $\tilde{\varphi}$ was not used here, i.e. we could have used any other dual basis with a shift invariant structure. The stability property will be useful for proving the inverse estimate of the next section.*

An important variant of the direct estimate (3.3.10) is the Whitney estimate, which involves the modulus of smoothness.

Theorem 3.3.3 *Under the same assumptions as in Theorem 3.3.1, we have*

$$\|f - P_j f\|_{L^p} \lesssim \omega_n(f, 2^{-j})_p, \quad (3.3.12)$$

where $n - 1$ is the order of polynomial exactness in V_j.

Proof We proceed as in the proof of Theorem 3.3.2 by choosing on each dyadic cube $\widetilde{I}_{j,k}$ a polynomial $p_{j,k} \in \Pi_{n-1}$ such that

$$\|f - p_{j,k}\|_{L^p(\widetilde{I}_{j,k})} \leq 2 \inf_{g \in \Pi_{n-1}} \|f - g\|_{L^p(\widetilde{I}_{j,k})} \lesssim \omega_n(f, 2^{-j}, \widetilde{I}_{j,k}), \quad (3.3.13)$$

where the second inequality is easily obtained from (3.2.26) and the fact that $\omega_n(f, mt, \Omega)_p \leq m^n \omega_n(f, t, \Omega)_p$. The local projection error follows by the same arguments as in the proof of Theorem 3.3.2, i.e.

$$\begin{aligned} \|f - P_j f\|_{L^p(I_{j,k})} &\leq \|f - p_{j,k}\|_{L^p(I_{j,k})} + \|P_j f - p_{j,k}\|_{L^p(I_{j,k})} \\ &= \|f - p_{j,k}\|_{L^p(I_{j,k})} + \|P_j(f - p_{j,k})\|_{L^p(I_{j,k})} \\ &\lesssim \|f - p_{j,k}\|_{L^p(\widetilde{I}_{j,k})} \\ &\lesssim \omega_n(f, 2^{-j}, \widetilde{I}_{j,k}). \end{aligned}$$

In order to turn this into a global estimate, we encounter a difficulty since it is not clear that we have

$$\|(\omega_n(f, 2^{-j}, \widetilde{I}_{j,k})_p)_{k \in \mathbb{Z}^d}\|_{\ell^p} \lesssim \omega_n(f, 2^{-j})_p, \quad (3.3.14)$$

except for the case $p = \infty$. At this point, it is useful to introduce, for $p < +\infty$, the following variant of the modulus of smoothness:

$$\tilde{\omega}_n(f, t, \Omega)_p := \Big(t^{-d} \int_{[-t,t]^d} \int_{\Omega_{h,n}} |\Delta_h^n f(x)|^p dx\, dh\Big)^{1/p}. \quad (3.3.15)$$

We clearly have $\tilde{\omega}_n(f, t, \Omega)_p \leq 2\omega_n(f, t, \Omega)_p$. A less trivial result is that, assuming that Ω is a cube and $t < K\text{diam}(\Omega)$ with K a fixed constant, one actually has a full equivalence

$$\tilde{\omega}_n(f, t, \Omega)_p \sim \omega_n(f, t, \Omega)_p. \quad (3.3.16)$$

in the sense that

$$\omega_n(f, t, \Omega)_p \leq C\tilde{\omega}_n(f, t, \Omega)_p, \quad (3.3.17)$$

with a constant C that only depends on K, p and n. A proof of this result can be found in DEVORE and LORENTZ [1993, p.185] in the univariate case, i.e. when Ω is an interval. We adapt this proof below to the case of a multivariate cube. Since the modified modulus clearly satisfies

$$\|(\tilde{\omega}_n(f, 2^{-j}, \widetilde{I}_{j,k})_p)_{k \in \mathbb{Z}^d}\|_{\ell^p} \lesssim \tilde{\omega}_n(f, 2^{-j})_p, \quad (3.3.18)$$

it follows that, if (3.3.17) is satisfied, (3.3.14) also holds for all values of p, allowing us to conclude the proof of (3.3.12).

It remains to prove (3.3.17). By a rescaling argument, it suffices to prove this for $\Omega = [-1,1]^d$. From the property $\omega_n(f,mt,\Omega)_p \leq m^n\omega_n(f,t,\Omega)_p$, it is also sufficient to prove (3.3.17) for t less than some arbitrary constant. Here, we assume $t \leq (2n)^{-1}$. For $|h| \leq t$, we want to prove that $\|\Delta_h^n f\|_{L^p(\Omega_{h,n})}$ is less than $C\tilde{\omega}_n(f,t,\Omega)_p$ with a constant C that does not depend on h and t. To do so, we split the cube Ω into 2^d subcubes of half-side length. One of these is $J := [-1,0]^d$. We shall now estimate $\|\Delta_h^n f\|_{L^p(\Omega_{h,n}\cap J)}$. Our starting point is the identity

$$\Delta_h^n f(x) = \sum_{k=1}^{n}(-1)^k\binom{n}{k}[\Delta_{ky}^n f(x+kh) - \Delta_{h+ky}^n f(x)], \tag{3.3.19}$$

which can be proved by simply remarking that

$$\begin{aligned} 0 &= \textstyle\sum_{k=0}^n \sum_{j=0}^n (-1)^{k+j}\binom{n}{k}\binom{n}{j}[f(x+kh+jky) - f(x+jh+jky)] \\ &= \textstyle\sum_{k=0}^n (-1)^k\binom{n}{k}[\Delta_{ky}^n f(x+kh) - \Delta_{h+ky}^n f(x)], \end{aligned}$$

and remarking that the term corresponding to $k=0$ in the above sum is exactly $-\Delta_h^n f(x)$. From (3.3.19), we derive

$$\|\Delta_h^n f\|_{L^p(\Omega_{h,n}\cap J)} \leq \sum_{k=1}^{n}\binom{n}{k}[\|\Delta_{ky}^n f(\cdot+kh)\|_{L^p(\Omega_{h,n}\cap J)} + \|\Delta_{h+ky}^n f\|_{L^p(\Omega_{h,n}\cap J)}].$$

If we take the p-th power of the right-hand side terms and integrate in y over the domain $[0,t/n]^d$, we obtain by changes of variable

$$\begin{aligned} \textstyle\int_{[0,t/n]^d}\|\Delta_{ky}^n f(\cdot+kh)\|_{L^p(\Omega_{h,n}\cap J)}^p dy &= \textstyle\int_{[0,t/n]^d}\|\Delta_{ky}^n f\|_{L^p(kh+\Omega_{h,n}\cap J)}^p dy \\ &\lesssim \textstyle\int_{[0,t]^d}\int_{\Omega_{y,n}}|\Delta_y^n f(x)|^p dx\, dy, \end{aligned}$$

and

$$\begin{aligned} \textstyle\int_{[0,t/n]^d}\|\Delta_{h+ky}^n f\|_{L^p(\Omega_{h,n}\cap J)}^p dy &\lesssim \textstyle\int_{[h,t+h]^d}\|\Delta_y^n f\|_{L^p(\Omega_{h,n}\cap J)}^p dy \\ &\lesssim \textstyle\int_{[-t,2t]^d}\int_{\Omega_{y,n}}|\Delta_y^n f(x)|^p dx\, dy. \end{aligned}$$

It follows that we have the estimate

$$(t/n)^d\|\Delta_h^n f\|_{L^p(\Omega_{h,n}\cap J)}^p \lesssim \int_{[-2t,2t]^d}\int_{\Omega_{y,n}}|\Delta_y^n f(x)|^p dx\, dy. \tag{3.3.20}$$

We can proceed in a similar way to obtain estimates of the above type for the 2^d-1 other subcubes of Ω. Summing up these estimates yields

$$\omega_n(f,2t,\Omega)_p \leq 2\omega_n(f,t,\Omega)_p \lesssim \tilde{\omega}_n(f,2t,\Omega)_p, \tag{3.3.21}$$

and thus (3.3.17). ◇

Remark 3.3.4 *One can easily check that*

$$\omega_n(f,t)_p \lesssim t^n |f|_{W^{n,p}}. \tag{3.3.22}$$

The Whitney estimate of Theorem 3.3.3 is thus a stronger result than the Jackson estimate of Theorem 3.3.2.

We end this section with a direct estimate for general Besov spaces, which is a simple corollary of the Whitney estimate.

Corollary 3.3.1 *Under the same assumptions as in Theorem 3.3.1, we have*

$$\|f - P_j f\|_{L^p} \lesssim 2^{-js}|f|_{B^s_{p,q}}, \tag{3.3.23}$$

for $0 < s < n$, where $n-1$ is the order of polynomial exactness in V_j.

Proof If $s < n$, it follows from the definition of the Besov semi-norm that

$$\omega_n(f,2^{-j})_p \leq 2^{-js}|f|_{B^s_{p,q}}. \tag{3.3.24}$$

Combining this with the Whitney estimate (3.3.12), we obtain (3.3.19). ◇

Remark 3.3.5 *From the equivalence (3.3.16) between ω and $\tilde{\omega}$ and from (3.3.18), we can also derive*

$$\|(|f|_{B^s_{p,p}(\tilde{I}_{j,k})})_{k\in\mathbb{Z}^d}\|_{\ell^p} \lesssim |f|_{B^s_{p,p}(\mathbb{R}^d)}, \tag{3.3.25}$$

which was not obvious in view of the initial definition of $B^s_{p,p}$ involving only ω. In the case where s is not an integer, (3.3.16) can also be used to prove the equivalence between the $B^s_{p,p}$ and $W^{s,p}$ semi-norms. Note that it is straightforward from (3.2.8) that the $W^{s,p}$ semi-norm satisfies a super-additivity property of the type (3.3.25).

3.4 Inverse estimates

Inverse (or Bernstein-type) estimates aim to take into account the smoothness properties of the approximation spces V_j. Our first basic estimate relates the L^p and $W^{n,p}$ norm of a function $f \in V_j$.

Theorem 3.4.1 *Under the assumptions of Theorem 3.3.1, and assuming that $\varphi \in W^{n,p}$, one has*

$$\|f\|_{W^{n,p}} \leq C 2^{nj}\|f\|_{L^p} \text{ if } f \in V_j, \tag{3.4.1}$$

with a constant C that does not depend on j.

Proof We only need to estimate $|f|_{W^{n,p}}$. For this, we use the decomposition into the local basis $f = \sum_{k\in\mathbb{Z}^d} c_k\varphi_{j,k}$. For any $m = (m_1, \cdots, m_d)$ such that $|m| = m_1 + \cdots + m_d = n$, we obtain

$$\begin{aligned}\|\partial^m f\|_{L^p} &= \|\sum_{k\in\mathbb{Z}^d} c_k\partial^m\varphi_{j,k}\|_{L^p}\\ &= 2^{nj}\|\sum_{k\in\mathbb{Z}^d} c_k 2^{dj/2}(\partial^m\varphi)(2^j\cdot -k)\|_{L^p}\\ &\lesssim 2^{nj}2^{jd(1/2-1/p)}\|(c_k)_{k\in\mathbb{Z}^d}\|_{\ell^p},\end{aligned}$$

where we have used the same computation as in the proof of (3.3.4), replacing φ by $\partial^m\varphi$ (which is also a compactly supported L^p function). We then conclude by using the lower inequality in the equivalence (3.3.3). $\diamond$.

Another type of inverse estimate involves the modulus of smoothness.

Theorem 3.4.2 *Under the assumptions of Theorem 3.4.1, one has*

$$\omega_n(f,t)_p \leq C[\min\{1, 2^j t\}]^n\|f\|_{L^p} \text{ if } f \in V_j, \tag{3.4.2}$$

with a constant C that does not depend on j.

Proof This estimate is trivial for $t \geq 2^{-j}$. From the monotonicity in t of $\omega_n(f,t)_p$, it suffices to prove it for $t = 2^{-l}$, $l > j$. For this we proceed as in Theorem 3.4.1, using the decomposition of f into the scaling function basis, which gives

$$\omega_n(f,t)_p \lesssim \omega_n(\varphi_{j,0},t)_p\|(c_k)_{k\in\mathbb{Z}^d}\|_{\ell^p}. \tag{3.4.3}$$

We then remark that, by (3.3.22), for $l > j$,

$$\omega_n(\varphi_{j,0}, 2^{-l})_p = 2^{jd(1/2-1/p)}\omega_n(\varphi, 2^{j-l})_p \lesssim |\varphi|_{W^{n,p}}2^{jd(1/2-1/p)}2^{n(j-l)}.$$

We conclude the proof, using the lower inequality in the equivalence (3.3.3). $\diamond$

Our third inverse estimates deals with general Besov spaces of integer or fractional order.

Theorem 3.4.3 *Under the assumptions of Theorem 3.3.1, and assuming that $\varphi \in B^s_{p,q}$ one has*

$$\|f\|_{B^s_{p,q}} \leq C2^{sj}\|f\|_{L^p} \text{ if } f \in V_j, \tag{3.4.4}$$

with a constant C that does not depend on j.

Proof For $f = \sum_{k\in\mathbb{Z}^d} c_k\varphi_{j,k} \in V_j$, we first apply (3.4.3) with some $n > s$, and the definition of the Besov semi-norm, to obtain

$$|f|_{B^s_{p,q}} \lesssim \|(c_k)_{k\in\mathbb{Z}^d}\|_{\ell^p}|\varphi_{j,0}|_{B^s_{p,q}}. \tag{3.4.5}$$

It remains to evaluate $|\varphi_{j,0}|_{B^s_{p,q}}$. For $l \le j$, we use the crude estimate

$$\omega_n(\varphi_{j,0}, 2^{-l})_p \lesssim \|\varphi_{j,0}\|_{L^p} \lesssim 2^{dj(1/2-1/p)}. \tag{3.4.6}$$

For $l > j$, we exploit the fact that $\varphi \in B^s_{p,q}$ to obtain the estimate

$$\omega_n(\varphi_{j,0}, 2^{-l})_p = 2^{dj(1/2-1/p)}\omega_n(\varphi, 2^{j-l})_p \le 2^{s(j-l)}2^{dj(1/2-1/p)}\varepsilon_{l-j}, \tag{3.4.7}$$

where $(\varepsilon_n)_{n\ge 0}$ is an ℓ^q sequence and

$$\|(\varepsilon_n)_{n\ge 0}\|_{\ell^q} \le |\varphi|_{B^s_{p,q}}. \tag{3.4.8}$$

Combining (3.4.6) and (3.4.7), we then obtain

$$\begin{aligned} |\varphi_{j,0}|_{B^s_{p,q}} &= \|(2^{ls}\omega_n(\varphi_{j,0}, 2^{-l})_{L^p})_{l\ge 0}\|_{\ell^p} \\ &\le \|(2^{ls}\omega_n(\varphi_{j,0}, 2^{-l})_{L^p})_{0\le l\le j}\|_{\ell^p} + \|(2^{ls}\omega_n(\varphi_{j,0}, 2^{-l})_{L^p})_{l>j}\|_{\ell^p} \\ &\lesssim 2^{js}2^{dj(1/2-1/p)}. \end{aligned}$$

We conclude the proof by injecting this estimate in (3.4.5) and using the lower inequality in the equivalence (3.3.3). ◇

We end this section by mentioning more general direct and inverse estimates involving Sobolev and Besov norms on both sides of the inequalities. They are corollaries of the direct and inverse estimates that we have obtained so far.

Corollary 3.4.1 *Let $1 \le p, q_1, q_2 \le \infty$ and $0 < s < t$, and assume that $\varphi \in L^p$ and $\tilde{\varphi} \in L^{p'}$. If $\varphi \in B^s_{p,q_1}$ and $t < n$ where $n-1$ is the degree of polynomial reproduction in V_j, one has the direct estimate*

$$\|f - P_jf\|_{B^s_{p,q_1}} \lesssim 2^{-j(t-s)}|f|_{B^t_{p,q_2}}. \tag{3.4.9}$$

When s and/or t are integers, these estimates also hold with the classical Sobolev spaces $W^{s,p}$ and/or $W^{t,p}$ and t up to n.

Proof We use the multiscale decomposition of f to estimate the approximation error by

$$\|f - P_jf\|_{B^s_{p,q_1}} \le \sum_{l\ge j}\|P_{l+1}f - P_lf\|_{B^s_{p,q_1}}. \tag{3.4.10}$$

Combining the inverse estimate of Theorem 3.4.3 and the direct estimate of Corollary 3.3.1, we obtain

$$\|P_{l+1}f - P_l f\|_{B^s_{p,q_1}} \lesssim 2^{sl}\|P_{l+1}f - P_l f\|_{L^p} \lesssim 2^{-l(t-s)}|P_{l+1}f - P_l f|_{B^t_{p,q_2}},$$

where we have used $\|P_{l+1}f - P_l f\| \leq \|f - P_l f\| + \|f - P_{l+1}f\|$. This together with (3.4.10), yields the direct estimate (3.4.9). The case of classical Sobolev spaces is treated in the same way, using the direct and inverse estimates of Theorems 3.3.2 and 3.4.1. ◇

Corollary 3.4.2 *Let $1 \leq p, q_1, q_2 \leq \infty$ and $0 < s < t$, and assume that $\varphi \in L^p$ and $\tilde{\varphi} \in L^{p'}$. If $\varphi \in B^t_{p,q_2}$ and $s < n$ where $n-1$ is the degree of polynomial reproduction in V_j, one has the inverse estimate*

$$\|f\|_{B^t_{p,q_2}} \lesssim 2^{j(t-s)}\|f\|_{B^s_{p,q_1}} \ \text{ if } \ f \in V_j. \tag{3.4.11}$$

When s and/or t are integers, these estimates also hold with the classical Sobolev spaces $W^{s,p}$ and/or $W^{t,p}$ and s up to n.

Proof We use the multiscale decomposition of f to estimate its B^t_{p,q_2} norm as follows

$$\|f\|_{B^t_{p,q_2}} \leq \|P_0 f\|_{B^t_{p,q_2}} + \sum_{l=0}^{j-1}\|P_{l+1}f - P_l f\|_{B^t_{p,q_2}}. \tag{3.4.12}$$

We clearly have $\|P_0 f\|_{B^t_{p,q_2}} \lesssim \|f\|_{L^p} \lesssim \|f\|_{B^s_{p,q_1}}$. For the remaining terms, we combine the inverse estimate of Theorem 3.4.3 and the direct estimate of Corollary 3.3.1 and obtain

$$\|P_{l+1}f - P_l f\|_{B^t_{p,q_2}} \lesssim 2^{tl}\|P_{l+1}f - P_l f\|_{L^p} \lesssim 2^{l(t-s)}|f|_{B^s_{p,q_1}}. \tag{3.4.13}$$

This together with (3.4.12), yields the inverse estimate (3.4.11). The case of classical Sobolev spaces is treated in the same way, using the direct and inverse estimates of Theorem 3.3.2 and 27.1. ◇

3.5 Interpolation and approximation spaces

The goal of this section is to describe a general mechanism that allows us to connect approximation and smoothness properties.

Given a sequence of approximation spaces V_j, we would like to relate the property "$\text{dist}_{L^p}(f, V_j) \leq \mathcal{O}(2^{-sj})$" to some classical notion of smoothness satisfied by f. We shall actually be more precise in the description of the approximation rate for the function f, as shown by the following definition.

Definition 3.5.1 *Let X be a Banach space and $(V_j)_{j\geq 0}$ a nested sequence of subspaces of X such that $\cup_{j\geq 0}V_j$ is dense in X. For $s>0$ and $1\leq q\leq\infty$, we define the approximation space $\mathcal{A}^s_q(X)$ related to the sequence V_j by*

$$\mathcal{A}^s_q(X):=\{f\in X \text{ such that } (2^{sj}\text{dist}_X(f,V_j))_{j\geq 0}\in\ell^q\}. \tag{3.5.1}$$

In the case where X is an L^p space, we use the notation $\mathcal{A}^s_{p,q}:=\mathcal{A}^s_q(L^p)$.

Roughly speaking, the space $\mathcal{A}^s_q(X)$ describes those functions such that $\text{dist}_X(f,V_j)\leq\mathcal{O}(2^{-sj})$, with refined information provided by the extra parameter q. One easily checks that it constitutes a proper subspace of X, and that it is a Banach space when equipped with the norm

$$\|f\|_{\mathcal{A}^s_q(X)}:=\|f\|_X+|f|_{\mathcal{A}^s_q(X)}, \tag{3.5.2}$$

with

$$|f|_{\mathcal{A}^s_q(X)}:=\|(2^{sj}\text{dist}_X(f,V_j))_{j\geq 0}\|_{\ell^q}. \tag{3.5.3}$$

Note also that if we set $V_0=\{0\}$, then one can remove the term $\|f\|_X$ in the definition of this norm since it is equal to $\text{dist}_X(f,V_0)$. In the next section, we shall prove that under specific assumptions on the multiresolution approximation spaces, the identity

$$\mathcal{A}^s_{p,q}=B^s_{p,q}, \tag{3.5.4}$$

holds, together with the norm equivalences

$$\|f\|_{B^s_{p,q}}\sim\|f\|_{\mathcal{A}^s_{p,q}} \text{ and } |f|_{B^s_{p,q}}\sim|f|_{\mathcal{A}^s_{p,q}}. \tag{3.5.5}$$

As we shall see in §3.6, these equivalences can be proved directly, combining the definition of the Besov spaces together with the inverse and direct estimates involving the modulus of smoothness (Theorems 3.3.3 and 3.4.2).

In the present section we shall describe a more general mechanism that allows us to identify the approximation spaces $\mathcal{A}^s_{p,q}(X)$ with spaces obtained by *interpolation theory*. Although this mechanism can be avoided when proving (3.5.4) and (3.5.5), its usefulness will appear in other instances of relating approximation and smoothness properties, in particular in the nonlinear context of Chapter 4.

Interpolation theory is a useful tool in functional analysis, operators theory and partial differential equations. Perhaps, the most well known result of this theory is the oldest one: if T is an operator that is bounded in $L^1(\Omega)$ and in $L^\infty(\Omega)$, then the *Riesz-Thorin theorem* states that T is also bounded in $L^p(\Omega)$ for all $p\in]1,+\infty[$.

More generally the goal of interpolation theory is to build from an initial pair of Banach spaces (X, Y), some "intermediate spaces" Z with similar properties with respect to the continuity of operators. Of course, such a process will be of particular interest if for some specific X and Y these intermediate spaces coincide with some known function spaces. The two most well known strategies for interpolation of function spaces - both introduced in the 1950's - are the real method of Lions and Peetre, and the complex method of Calderon. The first one turns out to be directly related to the concept of approximation spaces. Therefore, we shall focus on this approach and we give below some of its main features. A detailed treatment can be found in the books of BERGH and LÖFSTRÖM [1976], and of BENNETT and SHARPLEY [1988].

Let X and Y be a pair of Banach function spaces. With such a pair, we associate the so-called *K-functional* defined for $f \in X + Y$ and $t \geq 0$, by

$$K(f,t) = K(f,t,X,Y) := \inf_{a\in X, b\in Y, a+b=f} [\|a\|_X + t\|b\|_Y]. \tag{3.5.6}$$

This functional has some elementary properties. First, it is continuous, non-decreasing and concave with respect to t. The first two properties are direct consequences of the definition, while the last property can be checked by remarking that

$$[K(f,t) + K(f,s)]/2 \leq \|a\|_X + \frac{t+s}{2}\|b\|_Y \tag{3.5.7}$$

if $a + b = f$, and then by minimizing on a and b in this inequality. Secondly, if $X \cap Y$ is dense in Y, then $K(f,0) := 0$. Similarly, if $X \cap Y$ is dense in X, then $\lim_{t\to+\infty} K(f,t)/t = 0$.

For $\theta \in]0,1[$ and $1 \leq q \leq +\infty$, we define a family of intermediate spaces $X \cap Y \subset [X,Y]_{\theta,q} \subset X + Y$ as follows: $[X,Y]_{\theta,q}$ consists of those functions such that

$$\|f\|_{[X,Y]_{\theta,q}} := \|t^{-\theta} K(f,t)\|_{L^q(]0,+\infty[,dt/t)}, \tag{3.5.8}$$

is finite. One easily checks that the above defined intermediate spaces inherit the Banach spaces structure of X and Y.

A first remark is that this process has the expected property with respect to linear operators: assuming that T is continuous from X_1 to X_2 and from Y_1 to Y_2, we obtain that for all $t \geq 0$,

$$\begin{aligned} K(Tf,t,X_2,Y_2) &= \inf_{a\in X_2, b\in Y_2, a+b=Tf} [\|a\|_{X_2} + t\|b\|_{Y_2}] \\ &\leq \inf_{a\in X_1, b\in Y_1, a+b=f} [\|Ta\|_{X_2} + t\|Tb\|_{Y_2}] \\ &\leq CK(f,t,X_1,Y_1), \end{aligned}$$

with $C = \max\{\|T\|_{X_1\to X_2}, \|T\|_{Y_1\to Y_2}\}$. It follows that T is bounded from $[X_1,Y_1]_{\theta,q}$ to $[X_2,Y_2]_{\theta,q}$ for all θ and q.

One of the main results of interpolation theory is the *reiteration theorem* that we state below without proof (see e.g. BERGH and LÖFSTRÖM [1976]). This theorem confirms the visual intuition of the space $[X,Y]_{\theta,q}$ as barycenter between the "end-points" X and Y, with weight θ in Y and $1-\theta$ in X. It also shows that the parameter q plays a minor role in the definition of intermediate spaces.

Theorem 3.5.1 *If $Z_1 = [X,Y]_{\theta_1,q_1}$ and $Z_2 = [X,Y]_{\theta_2,q_2}$ for some pair of Banach spaces (X,Y), and some $0 < \theta_1 < \theta_2 < 1$ and $q_1, q_2 \in [1,\infty]$, then for $0 < \theta < 1$ and $1 \le q \le \infty$, one has the identities*

$$[X,Z_1]_{\theta,q} = [X,Y]_{\theta_1\theta,q}, \tag{3.5.9}$$

$$[Z_1,Y]_{\theta,q} = [X,Y]_{\theta+\theta_1(1-\theta),q}, \tag{3.5.10}$$

and

$$[Z_1,Z_2]_{\theta,q} = [X,Y]_{\theta_1(1-\theta)+\theta_2\theta,q}. \tag{3.5.11}$$

As we already noted, the main interest of such a theory is the ability to identify classical function spaces as the results of an interpolation process between two other classical spaces. To do so, one needs to understand more deeply the nature of the K-functional related to the pair of spaces which is considered. An important and well understood instance is interpolation between Lebesgue spaces, for which the K-functional can be expressed in terms of the decreasing rearrangement of f, and for which it is known that

$$[L^p,L^q]_{\theta,r} = L^r, \;\; 1/r = \theta/q + (1-\theta)/p, \tag{3.5.12}$$

if $1 \le p \le q \le \infty$.

In the context of this chapter, we shall be interested in interpolation between spaces representing various degrees of smoothness. In particular, we shall always work in the situation where $Y \subset X$ with a continuous embedding and Y is dense in X. A typical example is $X = L^p$ and $Y = W^{m,p}$.

In this specific situation, we write

$$K(f,t) = \inf_{g\in Y} \|f-g\|_X + t\|g\|_Y, \tag{3.5.13}$$

and make a few additional remarks.

Firstly, the K-functional is bounded at infinity since $K(f,t) \le \|f\|_X$. Therefore, the finiteness of (3.5.8) is equivalent to

$$\|t^{-\theta}K(f,t)\|_{L^q(]0,A[,dt/t)} < +\infty \tag{3.5.14}$$

for some fixed $A > 0$ and we can use this modified expression as an equivalent norm for $[X,Y]_{\theta,q}$.

Secondly, due to the monotonicity of $K(f,t)$ in t, we also have an equivalent discrete norm given by

$$\|f\|_{[X,Y]_{\theta,q}} := \|(\rho^{j\theta}K(f,\rho^{-j}))_{j\geq 0}\|_{\ell^q}, \tag{3.5.15}$$

for any fixed $\rho > 1$.

Finally, if $\|f\|_Y = \|f\|_X + |f|_Y$ where $|\cdot|_Y$ is a semi-norm, one can build the same intermediate spaces by using the modified K-functional

$$\widetilde{K}(f,t) = \inf_{g\in Y} \|f-g\|_X + t|g|_Y, \tag{3.5.16}$$

and defining the intermediate semi-norm by

$$|f|_{[X,Y]_{\theta,q}} := \|(\rho^{j\theta}\widetilde{K}(f,\rho^{-j}))_{j\geq 0}\|_{\ell^q}. \tag{3.5.17}$$

An equivalent norm for $[X,Y]_{\theta,q}$ is then given by $\|f\|_X + |f|_{[X,Y]_{\theta,q}}$.

It should be noted that all the above concepts (including the reiteration theorem) carry over to the case where $q \in]0,1[$ and/or X is a quasi-Banach space with $\|.\|_X$ satisfying the μ-triangle inequality of (3.2.22). The resulting intermediate spaces are of course quasi-Banach spaces.

We our now ready to state the main result of this section, connecting approximation spaces and interpolation spaces through direct and inverse estimates.

Theorem 3.5.2 *Assume that V_j is a sequence of approximation spaces*

$$V_j \subset V_{j+1} \subset \cdots \subset Y \subset X, \tag{3.5.18}$$

such that for some $m > 0$, one has a Jackson-type estimate

$$\text{dist}_X(f,V_j) = \inf_{g\in V_j} \|f-g\|_X \lesssim 2^{-mj}\|f\|_Y, \tag{3.5.19}$$

and a Bernstein-type estimate

$$\|f\|_Y \lesssim 2^{mj}\|f\|_X \ \text{ if } \ f \in V_j. \tag{3.5.20}$$

Then, for $s \in]0,m[$, one has the norm equivalence

$$\|(2^{js}K(f,2^{-mj}))_{j\geq 0}\|_{\ell^q} \sim \|f\|_X + \|(2^{js}\text{dist}_X(f,V_j))_{j\geq 0}\|_{\ell^q}, \tag{3.5.21}$$

and thus $[X,Y]_{\theta,q} = \mathcal{A}^s_q(X)$ for $s = \theta m$.

Proof We need to compare the K-functional $K(f,2^{-mj})$ and the error of best approximation $\text{dist}_X(f,V_j)$. In one direction, this comparison is simple: for all $f\in X$, $g\in Y$ and $g_j\in V_j$, we have

$$\text{dist}_X(f,V_j)\le\|f-g_j\|_X\le\|f-g\|_X+\|g-g_j\|_X, \tag{3.5.22}$$

Minimizing $\|g-g_j\|_X$ over $g_j\in V_j$ and using (3.5.19), we obtain

$$\text{dist}_X(f,V_j)\lesssim\|f-g\|_X+2^{-mj}\|g\|_Y. \tag{3.5.23}$$

Finally, we minimize over $g\in Y$ to obtain

$$\text{dist}_X(f,V_j)\lesssim K(f,2^{-mj}). \tag{3.5.24}$$

Since $\|f\|_X\lesssim K(f,1)$ (by the continuous embedding of Y into X and the triangle inequality), we have thus proved that $\|f\|_{\mathcal{A}^s_q(X)}\lesssim\|f\|_{[X,Y]_{\theta,q}}$.

In the other direction, we let $f_j\in V_j$ be such that

$$\|f-f_j\|_X\le 2\,\text{dist}_X(f,V_j), \tag{3.5.25}$$

and we write

$$\begin{aligned}K(f,2^{-mj}) &\le\|f-f_j\|_X+2^{-mj}\|f_j\|_Y\\ &\le\|f-f_j\|_X+2^{-mj}[\|f_0\|_Y+\textstyle\sum_{l=0}^{j-1}2^{ml}\|f_{l+1}-f_l\|_Y]\\ &\lesssim\|f-f_j\|_X+2^{-mj}[\|f_0\|_X+\textstyle\sum_{l=0}^{j-1}2^{ml}\|f_{l+1}-f_l\|_X]\\ &\lesssim 2^{-mj}\|f_0\|_X+2^{-mj}[\textstyle\sum_{l=0}^{j}2^{ml}\text{dist}_X(f,V_l)],\end{aligned}$$

where we have used the inverse inequality (3.5.20) (together with the fact that $f_{l+1}-f_l\in V_{l+1}$) and the crude $\|f_0\|_X\le\|f\|_X+2\text{dist}_X(f,V_0)\le 3\|f\|_X$. We thus do not have the exact converse inequality to (3.5.24).

In order to conclude the proof, we first remark that the term $2^{-mj}\|f\|_X$ satisfies

$$\|(2^{sj}2^{-mj}\|f\|_X)_{j\ge 0}\|_{\ell^q}\lesssim\|f\|_X, \tag{3.5.26}$$

and we concentrate on the second term. We use here a discrete Hardy inequality: if $(a_j)_{j\ge 0}$ is a positive sequence and $b_j:=2^{-mj}\sum_{l=0}^{j}2^{ml}a_l$ with $0<s<m$, one has

$$\|(2^{sj}b_j)_{j\ge 0}\|_{\ell^q}\lesssim\|(2^{sj}a_j)_{j\ge 0}\|_{\ell^q}, \tag{3.5.27}$$

for all $q\in[1,\infty]$. Applying this inequality to $a_j=\text{dist}_X(f,V_j)$ allows us to estimate the weighted ℓ^q norm of the second term and to conclude that $\|f\|_{[X,Y]_{\theta,q}}\lesssim\|f\|_{\mathcal{A}^s_q(X)}$.

We end with the proof of the Hardy inequality. In the case where $q = \infty$, assuming that $a_j \leq C_a 2^{-sj}$, we clearly have

$$b_j \leq C_a 2^{mj} \sum_{l=0}^{j} 2^{(m-s)l} \lesssim C_a 2^{-sj}. \tag{3.5.28}$$

For $q < \infty$, we define q' such that $1/q + 1/q' = 1$ and $\varepsilon = (m-s)/2 > 0$. Using Hölder's inequality, we obtain

$$\begin{aligned}
\sum_{j\geq 0}(2^{sj}b_j)^q &= \sum_{j\geq 0} 2^{(s-m)qj}(\sum_{l=0}^{j} 2^{(m)l}a_l)^q \\
&\leq \sum_{j\geq 0} 2^{(s-m)qj}[\sum_{l=0}^{j}(2^{(m-\varepsilon)l}a_l)^q][\sum_{l=0}^{j}(2^{\varepsilon l})^{q'}]^{q/q'} \\
&\lesssim \sum_{j\geq 0} 2^{-\varepsilon qj}[\sum_{l=0}^{j}(2^{(m-\varepsilon)l}a_l)^q] \\
&= \sum_{l\geq 0}(2^{(m-\varepsilon)l}a_l)^q \sum_{j\geq l} 2^{-\varepsilon qj} \\
&\lesssim \sum_{l\geq 0}(2^{ml}a_l)^q,
\end{aligned}$$

which proves the Hardy inequality. ◇

In practice, we are interested in similar norm equivalences involving specific approximation operators, rather than the error of best approximation. This is treated in the following variant of the previous theorem.

Theorem 3.5.3 *Under the assumptions of Theorem 3.5.2, suppose that we have*

$$\|P_j f - f\| \lesssim 2^{-mj}\|f\|_Y, \tag{3.5.29}$$

for a family of linear operators $P_j : X \mapsto V_j$ which is uniformly bounded in X. Then, for $s = \theta m \in]0, m[$, the $\mathcal{A}^s_q(X)$ and $[X, Y]_{\theta,q}$ norms are equivalent to

$$\|P_0 f\|_X + \|(2^{sj}\|f - P_j f\|_X)_{j\geq 0}\|_{\ell^q}, \tag{3.5.30}$$

and to

$$\|P_0 f\|_X + \|(2^{sj}\|Q_j f\|_X)_{j\geq 0}\|_{\ell^q}, \tag{3.5.31}$$

where $Q_j = P_{j+1} - P_j$.

Proof We first consider (3.5.30). In one direction, since

$$\operatorname{dist}_X(f, V_j) \leq \|f - P_j f\|_X, \tag{3.5.32}$$

and

$$\|f\|_X \leq \|P_0 f\|_X + \|f - P_0 f\|_X, \tag{3.5.33}$$

we clearly have that the $\mathcal{A}^s_q(X)$ is controlled by (3.5.30). In the other direction, we operate as in the second part of the proof of the previous theorem: replacing f_j by P_jf proves that the $[X,Y]_{\theta,q}$ norm is controlled by (3.5.30).

We then turn to (3.5.31). In one direction, we have

$$\|Q_jf\|_X \leq \|f - P_{j+1}f\|_X + \|f - P_jf\|_X, \tag{3.5.34}$$

which shows that (3.5.31) is controlled by (3.5.30). In the other direction, we write

$$\|f - P_jf\|_X \leq \sum_{l\geq j} \|Q_lf\|_X, \tag{3.5.35}$$

and we conclude by using a discrete Hardy inequality: if $(a_j)_{j\geq 0}$ is a positive sequence and $b_j := \sum_{l\geq j} a_l$, then one has

$$\|(2^{sj}b_j)_{j\geq 0}\|_{\ell^q} \lesssim \|(2^{sj}a_j)_{j\geq 0}\|_{\ell^q}, \tag{3.5.36}$$

for all $s > 0$ and $q \in [1,\infty]$. Applying this with $a_j = \|Q_jf\|_X$ allows us to conclude that (3.5.30) is controlled by (3.5.31).

We end by proving the discrete Hardy inequality. For $q = \infty$, it is clear that $a_j \leq C_a 2^{-sj}$ implies $b_j \lesssim C_a 2^{-sj}$. For $q < +\infty$, we define q' such that $1/q + 1/q' = 1$ and $s' = s/2$. Using Hölder's inequality, we obtain

$$\begin{aligned}
\textstyle\sum_{j\geq 0}(2^{sj}b_j)^q & \leq \textstyle\sum_{j\geq 0} 2^{sqj}[\sum_{l\geq j}(2^{s'l}a_l)^q][\sum_{l\geq j} 2^{-s'lq'}]^{q/q'} \\
& \lesssim \textstyle\sum_{j\geq 0} 2^{s'qj}[\sum_{l\geq j}(2^{s'l}a_l)^q] \\
& = \textstyle\sum_{l\geq 0}(2^{s'l}a_l)^q[\sum_{j=0}^{l} 2^{s'qj}] \\
& \lesssim \textstyle\sum_{l\geq 0}(2^{sl}a_l)^q,
\end{aligned}$$

which proves the Hardy inequality. ◇

Remark 3.5.1 *The spaces $\mathcal{A}^s_q(X)$ are always interpolation spaces in the following sense: from their definition, we clearly have the direct estimate*

$$\operatorname{dist}_X(f,V_j) \lesssim 2^{-sj}\|f\|_{\mathcal{A}^s_\infty(X)}. \tag{3.5.37}$$

If $f \in V_j$, we also have $\operatorname{dist}_X(f,V_l) = 0$ for $l \geq j$ and $\operatorname{dist}_X(f,V_l) \lesssim \|f\|_X$ for all l, so that we also have the inverse estimate

$$\|f\|_{\mathcal{A}^s_\infty(X)} \leq 2^{sj}\|f\|_X, \quad \text{if} \;\; f \in V_j. \tag{3.5.38}$$

By Theorem 3.5.2 we thus have for $\theta \in]0,1[$,

$$\mathcal{A}^{\theta s}_q(X) = [X, \mathcal{A}^s_\infty(X)]_{\theta,q}, \tag{3.5.39}$$

and thus, by reiteration (Theorem 3.5.1), we have for all $q_1, q_2 \in [1,\infty]$,

$$\mathcal{A}^{\theta s}_{q_2}(X) = [X, \mathcal{A}^s_{q_1}(X)]_{\theta,q_2}, \tag{3.5.40}$$

and for $0 < t < s$,

$$[\mathcal{A}^t_{q_1}(X), \mathcal{A}^s_{q_2}(X)]_{\theta,q} = \mathcal{A}^r_q(X) \text{ with } r = (1-\theta)t + \theta s. \tag{3.5.41}$$

Remark 3.5.2 *Theorems 3.5.2 and 3.5.3, as well as the previous remark, easily carry over to the case where* $q \in]0,1[$ *and/or* X *is a quasi-Banach spaces with* $\|.\|_X$ *satisfying the* μ*-triangle inequality of (3.2.22). The proofs are exactly similar, with the exception of the evaluation of* $K(f, 2^{-mj})$ *in terms of the approximation error for the proof of the first theorem: we are forced to use the* μ*-triangle inequality*

$$\|f_j\|_Y \leq [\|f_0\|_Y^\mu + \sum_{l=0}^{j-1} \|f_{l+1} - f_l\|_Y^\mu]^{1/\mu}, \tag{3.5.42}$$

in place of the standard triangle inequality. Similarly, we need to use

$$\|f - P_j f\|_X \leq [\sum_{l \geq j} \|Q_l f\|_Y^\mu]^{1/\mu}, \tag{3.5.43}$$

for the proof of the second theorem. In both cases, a straightforward modification of the Hardy inequalities allows us to reach the same conclusion.

Remark 3.5.3 *The identity (3.5.41) can also be derived in a direct way, using the basic interpolation theory of the sequence spaces*

$$\ell^p_s(\mathbb{N}) := \{(a_n)_{n\geq 0}\ ;\ \|(a_n)_{n\geq 0}\|_{\ell^p_s} := \|(2^{sj}a_n)_{n\geq 0}\|_{\ell^p} < \infty\}, \tag{3.5.44}$$

for $0 < p \leq \infty$ *and* $s \in \mathbb{R}$. *For these weighted spaces, the identities*

$$[\ell^p_s, \ell^p_t]_{\theta,p} = \ell^p_r,\ \ r = (1-\theta)s + \theta t, \tag{3.5.45}$$

can be proved in a straightforward manner, using Hardy's inequalities.

3.6 Characterization of smoothness classes

We now return to the our specific setting of multiresolution spaces and smoothness classes. We shall first give a direct proof of the equivalence between $\mathcal{A}^s_{p,q}$ and $B^s_{p,q}$, that does not require the general mechanism of the previous section.

Theorem 3.6.1 *Under the assumptions of Theorem 3.3.3 (Whitney estimate) and of Theorem 3.4.3 (inverse estimate), we have the norm equivalences*

$$\|f\|_{B^t_{p,q}} \sim \|P_0 f\|_{L^p} + \|(2^{tj}\|Q_j f\|_{L^p})_{j\geq 0}\|_{\ell^q}. \tag{3.6.1}$$

and

$$\|f\|_{\mathcal{A}^t_{p,q}} \sim \|f\|_{B^t_{p,q}}, \tag{3.6.2}$$

for all $t < \min\{n, s\}$, where $n-1$ is the order of polynomial reproduction of the V_j spaces and s is such that $\varphi \in B^s_{p,q_0}$ for some q_0.

Proof Here, we shall directly compare the modulus of smoothness which is involved in the definition of the Besov spaces $B^t_{p,q}$ and the quantities $\|Q_jf\|_{L^p}$.

In one direction, from Theorem 3.3.3, we have

$$\text{dist}_{L^p}(f, V_j) \leq \|f - P_jf\|_{L^p} \lesssim \omega_n(f, 2^{-j})_p, \tag{3.6.3}$$

and thus $\|Q_jf\|_{L^p} \lesssim \omega_n(f, 2^{-j})_p$. It follows that the $\mathcal{A}^s_q(L^p)$ norm and the right hand side of (3.6.1) are both controlled by the $B^s_{p,q}$ norm.

In order to prove the converse result, we remark that the inverse estimate of Theorem 3.4.3 implies the simpler inverse estimate

$$\omega_n(f,t)_p \lesssim [\min\{1, t2^j\}]^s\|f\|_{L^p} \text{ if } f \in V_j. \tag{3.6.4}$$

Indeed this property holds for the values $t = 2^{-l}$, $l \geq j$, by (3.4.4), and the other values of t are treated by the monotonicity of $\omega_n(f,t)_p$.

For $f \in L^p$, we let $f_j \in V_j$ be such that

$$\|f - f_j\|_{L^p} \leq 2\,\text{dist}_{L^p}(f, V_j). \tag{3.6.5}$$

We then have

$$\begin{aligned}
\omega_n(f, 2^{-j}) &\leq \omega_n(f_0, 2^{-j})_p + \textstyle\sum_{l=0}^{j} \omega_n(f_{l+1} - f_l, 2^{-j})_p + \omega_n(f - f_j, 2^{-j})_p \\
&\lesssim 2^{-sj}\|f_0\|_{L^p} + 2^{-sj}[\textstyle\sum_{l=0}^{j-1} 2^{sl}\|f_{l+1} - f_l\|_{L^p}] + \|f - f_j\|_{L^p} \\
&\lesssim 2^{-sj}\|f_0\|_{L^p} + 2^{-sj}[\textstyle\sum_{l=0}^{j} 2^{sl}\|f - f_l\|_{L^p}] \\
&\lesssim 2^{-sj}\|f\|_{L^p} + 2^{-sj}[\textstyle\sum_{l=0}^{j} 2^{sl}\text{dist}_{L^p}(f, V_j)],
\end{aligned}$$

where we have used the inverse estimate (3.6.4). At that point, we find the same estimate as for $K(f, 2^{-mj})$ (with s in place of m) in the second part of the proof of Theorem 3.5.2 and we can thus conclude that the $B^s_{p,q}$ norm is controlled by the $\mathcal{A}^s_{p,q}$ norm, using the Hardy inequality (3.5.27).

We can do the same reasoning with P_jf instead of f_j and replace $\text{dist}_{L^p}(f, V_j)$ by $\|f - P_jf\|_{L^p}$ for the charaterization of $B^s_{p,q}$. Finally, we can use the Hardy inequality (3.5.36), as in the proof of Theorem 3.5.3, to replace $\|f - P_jf\|_{L^p}$ by $\|Q_jf\|_{L^p}$ and conclude the proof. ◇

Another track to obtain the norm equivalence (3.6.1) and (3.6.2) is to use the general mechanism of the previous section. The key observation is that Besov spaces are obtained by the real interpolation applied Sobolev spaces. A more precise result proved in JOHNEN and SCHERRER [1976] is that the terms $\widetilde{K}(f, 2^{-nj}, L^p, W^{n,p})$ and $\omega_n(f, 2^{-j})_p$ are individually equivalent (where $\widetilde{K}$ is the modified K-functional of (3.5.16)). In the above reference, this result is proved to hold on general Lipschitz domains $\Omega \in \mathbb{R}^d$. We give below a simple proof for $\Omega = \mathbb{R}^d$.

Theorem 3.6.2 *One has the equivalence*

$$\omega_n(f,t)_p \sim \widetilde{K}(f, t^n, L^p, W^{n,p}), \tag{3.6.6}$$

for $t \le 1$. *Consequently* $B^s_{p,q} = [L^p, W^{n,p}]_{\theta,q}$, *with* $s = \theta n$, $\theta]0,1[$.

Proof In one direction, we have for all $g \in W^{n,p}$,

$$\begin{aligned}\omega_n(f,t)_p &\le \omega_n(f-g,t)_p + \omega_n(g,t)_p \\ &\lesssim \|f-g\|_{L^p} + t^n |g|_{W^{n,p}},\end{aligned}$$

using (3.3.22). Minimizing over $g \in W^{n,p}$ yields

$$\omega_n(f,t)_p \lesssim K(f, t^n, L^p, W^{n,p}). \tag{3.6.7}$$

In the other direction, for a given f we shall construct a particular function $g = g(f,t)$ such that

$$\|f-g\|_{L^p} + t^n |g|_{W^{n,p}} \lesssim \omega_n(f,t)_p. \tag{3.6.8}$$

To do so, we use a pair of dual scaling functions $(\varphi_{n+1}, \tilde{\varphi}_{n+1,m})$ with compact support, as constructed in §2.11 of Chapter 2: $\varphi_{n+1} = B_n$ is the spline of degree n and m is supposed to be large enough so that $\tilde{\varphi}_{n+1,m}$ is in L^∞. We define their multidimensional analogs $\phi(x) = \varphi_{n+1}(x_1)\cdots\varphi_{n+1}(x_d)$ and $\tilde{\phi}(x) = \tilde{\varphi}_{n+1,m}(x_1)\cdots\tilde{\varphi}_{n+1,m}(x_d)$. For $t \in]2^{-j-1}, 2^{-j}]$, we simply take for g the corresponding projection

$$g := g(f,t) = \sum_{k\in\mathbb{Z}^d} \langle f, \tilde{\phi}_{j,k}\rangle \phi_{j,k}. \tag{3.6.9}$$

From the Whitney estimate in Theorem 3.3.3 and the straightforward property $\omega_n(f, mt, \Omega)_p \le m^n \omega_n(f,t,\Omega)_p$, we are ensured that

$$\|f-g\|_{L^p} \lesssim \omega_n(f, 2^{-j})_p \lesssim \omega_n(f,t)_p \tag{3.6.10}$$

It remains to estimate $|g|_{W^{n,p}}$. To do so, we use the property

$$(\frac{d}{dx})^m B_n = \Delta^m_{-1} B_{n-m}, \quad m \le n. \tag{3.6.11}$$

which comes directly from the definition of splines. This is also a particular instance of the property (2.11.12) (Remark 2.11.3 of §2.11 in Chapter 2) iterated m-times. If $\alpha = (\alpha_1, \cdots, \alpha_d)$ such that $|\alpha| = \alpha_1 + \cdots \alpha_d = n$, we thus have

$$\begin{aligned}\partial^\alpha g &= 2^{nj} \sum_{k\in\mathbb{Z}^d} \langle f, \tilde{\phi}_{j,k}\rangle \tilde{\Delta}^\alpha_{-2^{-j}} \phi^\alpha_{j,k} \\ &= 2^{nj} \sum_{k\in\mathbb{Z}^d} \langle \tilde{\Delta}^\alpha_{2^{-j}} f, \tilde{\phi}_{j,k}\rangle \phi^\alpha_{j,k}.\end{aligned}$$

where $\phi^\alpha(x) := \varphi_{n+1-\alpha_1}(x_1) \cdots \varphi_{n+1-\alpha_d}(x_d) \in L^\infty$ and where $\tilde{\Delta}^\alpha_t$ denotes here the finite difference operator obtained by composing the univariate finite differences operators $\Delta^{\alpha_i}_t$ (in the direction x_i). Since both $\tilde{\phi}$ and ϕ^α are L^∞ and compactly supported, we obtain

$$\|\partial^\alpha g\|_{L^p} \lesssim 2^{nj} \|\tilde{\Delta}^\alpha_{2^{-j}} f\| \lesssim 2^{nj} \omega_n(f, 2^{-j})_p, \tag{3.6.12}$$

(by the L^p-stability result of Theorem 3.3.1) and thus $t^n |g|_{W^{n,p}} \lesssim \omega_n(f,t)_p$ which concludes the proof. ◇

If we combine the above result together with Theorem 3.5.2 (applied with $X = L^p$ and $Y = W^{m,p}$), using the direct estimate (3.3.10) and inverse estimate (3.4.1), we reach almost the same result as in Theorem 3.6.1: the norm equivalence (3.6.1) holds for $s < \min\{n, m\}$ where $n-1$ is the degree of polynomial reproduction in the V_j spaces and m is such that $\varphi \in W^{m,p}$.

Moreover by Theorem 3.5.1, we also obtain $B^s_{p,q_1} = [L^p, B^t_{p,q_2}]_{\theta,q_1}$ if $s = \theta t$, $\theta \in]0,1[$, and $B^r_{p,q} = [B^s_{p,q_1}, B^t_{p,q_2}]_{\theta,q}$ if $r = (1-\theta)s + \theta t$. We can thus reduce the above limitation on s to $s < \min\{n, t\}$, where t is such that $\varphi \in B^s_{p,q}$ for some q, using the inverse estimate of Theorem 3.4.3. Thus, we actually reach exactly the same result as in Theorem 3.6.1, through interpolation theory. Going further (especially in Chapter 4), we shall encounter instances where the interpolation approach has no simple substitute to reach a specific result.

Remark 3.6.1 *One should note the robustness of the norm equivalences (3.6.1) and (3.6.2): many V_j spaces and approximation operators are allowed, provided that they only satisfy the proper direct and inverse theorem. This can be used in particular to prove the equivalence between the definitions of Besov spaces through modulus of smoothness and Littlewood-Paley decompositions, by viewing the $S_j f$ as approximations of f in spaces V_j of band-limited functions: one simply remarks that these approximation spaces satisfy the direct and inverse estimates that are needed to characterize Besov spaces.*

Remark 3.6.2 *From the proof of Theorem 3.6.1, we see that the " $\lesssim$ " part in the equivalence (3.6.1) is obtained from the assumption $t < s$ with*

$\varphi \in B^s_{p,q_0}$, while the converse part is ensured by $t < n$ where $n-1$ is the degree of polynomial reproduction. In practice, the converse part is obtained for a larger range of indices t, due to the properties of refinable functions which imply that $t \leq n$. We have seen such an instance in Theorem 2.8.2 of Chapter 2, in the case where $p = 2$. A similar property is known to hold for $p = 1$ and thus all $p \geq 1$ (see e.g. JIA *[1999]). The limitation of $s < \min\{n,t\}$ in Theorem 3.6.1 can thus be simplified into $s < t$.*

Remark 3.6.3 *The norm equivalence (3.6.1) shows that P_j is a bounded operator in all the intermediate spaces $B^t_{p,q}$, $0 < t < s$, since we have*

$$\begin{aligned} \|P_j f\|_{B^t_{p,q}} &\lesssim \|P_0 f\|_{L^p} + \|(2^{tj}\|Q_l f\|_{L^p})_{0\leq l\leq j-1}\|_{\ell^q} \\ &\leq \|P_0 f\|_{L^p} + \|(2^{tj}\|Q_l f\|_{L^p})_{l\geq 0}\|_{\ell^q} \\ &\lesssim \|f\|_{B^t_{p,q}}. \end{aligned}$$

We end this section with two simple corollaries of our results. The first gives a norm equivalence in terms of the error measured in a Besov norm.

Corollary 3.6.1 *Under the same assumptions as in Theorem 3.6.1, we have the norm equivalences*

$$\|f\|_{B^t_{p,q}} \sim \|P_0 f\|_{B^r_{p,q_1}} + \|(2^{tj}\|Q_j f\|_{B^r_{p,q_1}})_{j\geq 0}\|_{\ell^q}. \tag{3.6.13}$$

and

$$\|f\|_{\mathcal{A}^{t-r}_q(B^r_{p,q_1})} \sim \|f\|_{B^t_{p,q}}, \tag{3.6.14}$$

for all $0 < r < t < \min\{n,s\}$, where $n-1$ is the order of polynomial reproduction of the V_j spaces and s is such that $\varphi \in B^s_{p,q_0}$ for some q_0.

Proof We simply apply Theorem 3.5.2 for (3.6.14) and Theorem 3.5.3 for (3.6.13) with $X = B^r_{p,q_1}$ and $Y = B^s_{p,q_0}$, using the generalized direct and inverse inequalities of Corollaries 3.4.1 and 3.4.2. For the application of Theorem 3.5.3, we also use that P_j is uniformly stable in B^r_{p,q_1} as shown in Remark 3.6.3 above. We then identify $B^t_{p,q}$ with the corresponding interpolation space between X and Y. ◇

The second corollary gives an equivalent norm in terms of the wavelet coefficients. Recall that in the tensor product construction, one needs 2^d-1 wavelets ψ^ε, $\varepsilon \in E := \{0,1\}^d - (0,\cdots,0)$, to characterize the detail spaces, although this is hidden in the notation (3.1.9) that we shall use here.

Corollary 3.6.2 *If $f = \sum_{\lambda\in\nabla} c_\lambda \psi_\lambda$, we have the norm equivalence*

$$\|f\|_{B^t_{p,q}} \sim \left\|\left(2^{tj}2^{d(1/2-1/p)j}\|(c_\lambda)_{\lambda\in\nabla_j}\|_{\ell^p}\right)_{j\geq -1}\right\|_{\ell^q}, \tag{3.6.15}$$

under the assumptions of Theorem 3.6.1.

Proof It suffices to remark that for $j \geq 0$, we have the equivalence

$$\|Q_j f\|_{L^p} \sim 2^{d(1/2-1/p)j}\|(c_\lambda)_{\lambda\in\nabla_j}\|_{\ell^p}, \tag{3.6.16}$$

by the same arguments as for the proof of (3.3.3). ◇

This last result also shows that wavelet bases are unconditional bases for all the Besov spaces in the above range: if $f \in B^t_{p,q}$ the convergence of the series holds in the corresponding norm and is not affected by a rearrangement or a change of sign of the coefficients, since it only depends on the finiteness of the right-hand side of (3.6.15).

Remark 3.6.4 *In the case $p = q$, note that (3.6.15) rewrites in a simpler way*

$$\|f\|_{B^t_{p,p}} \sim \|(\|c_\lambda\psi_\lambda\|_{B^t_{p,p}})_{\lambda\in\nabla}\|_{\ell^p}. \tag{3.6.17}$$

This reveals that the wavelet basis is unconditional in $B^t_{p,p}$. One also says that the multiscale decomposition is stable in $B^t_{p,p}$.

Remark 3.6.5 *The size of a wavelet coefficient $c_\lambda = \langle f, \tilde{\psi}_\lambda\rangle$ can be directly related to the local smoothness of f on the support $\tilde{S}_\lambda$ of $\tilde{\psi}_\lambda$: if the V_j spaces reproduce polynomials of order up to $n-1$, the wavelets $\tilde{\psi}_\lambda$ are orthogonal to Π_{n-1} for $|\lambda| \geq 0$, so that we have*

$$\begin{aligned}\|\langle f, \tilde{\psi}_\lambda\rangle\| &\leq \|\tilde{\psi}_\lambda\|_{L^{p'}} \inf\nolimits_{g\in\Pi_{n-1}} \|f-g\|_{L^p(\tilde{S}_\lambda)}\\ &\lesssim 2^{|\lambda|d(1/2-1/p')}2^{-n|\lambda|/p}|f|_{W^{n,p}(\tilde{S}_\lambda)}\\ &= 2^{|\lambda|[(d-n)/p-d/2]}|f|_{W^{n,p}(\tilde{S}_\lambda)},\end{aligned}$$

where p' is such that $1/p + 1/p' = 1$. This suggests that the size of wavelet coefficients also allows us to characterize the local smoothness of f at some point x by the size of the wavelet coefficients c_λ such that the support of ψ_λ or of $\tilde{\psi}_\lambda$ contains x. Such a program is not exactly possible in these terms and requires a more elaborate notion of local smoothness, related to the concept of microlocal spaces introduced by J.-M. Bony (see JAFFARD *[1991]).*

3.7 L^p-unstable approximation and $0 < p < 1$

So far, we have characterized L^p-Besov spaces by mean of approximation operators P_j that are uniformly bounded in L^p. In particular, interpolation operators - corresponding to the case where φ is interpolatory and $\tilde{\varphi} = \delta_0$ - do not fit in this framework, since they are only bounded in L^∞. This limitation is not clearly justified: we already noticed in §1.6 of Chapter 1 that the univariate Schauder hierarchical basis allows to characterize the Sobolev space $H^1 = B^1_{2,2}$ with a norm equivalence of the type (3.6.15).

Another limitation is that we have only considered values of p in the range $[1, \infty]$, whereas Besov spaces can be defined for $0 < p < 1$. This second restriction is actually related to the first one in the following sense: for $p < 1$, the projectors $P_j f$ are not bounded in the L^p quasi-norm. Moreover, the inner products $\langle f, \tilde{\varphi}_{j,k}\rangle$ are not well defined if we do not make additional assumptions on f.

The treatment of the case $p < 1$ (which turns out to be crucial for the nonlinear approximation results of Chapter 4) necessitates particular attention. A first result is that, although we do not have the L^p boundedness of P_j, we still have some L^p-stability for the scaling function basis. Here, we continue to assume that $(\varphi, \tilde{\varphi})$ are a pair of compactly supported biorthogonal scaling functions, with $\varphi \in L^r$ and $\tilde{\varphi} \in L^{r'}$ for some $r \geq 1$, $1/r' + 1/r = 1$.

Theorem 3.7.1 *Assuming that $\varphi \in L^p$, for $p > 0$, one has the L^p-stability property*

$$\|\sum_{k\in\mathbb{Z}^d} c_k\varphi_{j,k}\|_{L^p} \sim 2^{dj(1/2-1/p)}\|(c_k)_{k\in\mathbb{Z}^d}\|_{\ell^p}, \tag{3.7.1}$$

with constants that do not depend on j.

Proof The upper inequality

$$\|\sum_{k\in\mathbb{Z}^d} c_k\varphi_{j,k}\|_{L^p} \lesssim 2^{dj(1/2-1/p)}\|(c_k)_{k\in\mathbb{Z}^d}\|_{\ell^p}, \tag{3.7.2}$$

can be proved by the same technique as in Theorem 3.3.1: on each dyadic cube $I_{j,l}$ defined by (3.1.7), we still have the local estimate

$$\begin{aligned}\|\textstyle\sum_{k\in\mathbb{Z}^d} c_k\varphi_{j,k}\|_{L^p(I_{j,l})} &\lesssim \sup_{l-k\in \mathrm{Supp}(\varphi)} |c_k|\|\varphi_{j,0}\|_{L^p} \\ &\lesssim 2^{dj(1/2-1/p)}\|(c_k)_{l-k\in\mathrm{Supp}(\varphi)}\|_{\ell^p},\end{aligned}$$

using that all quasi-norms are equivalent in finite dimension.

For the proof of the converse inequality, we cannot use the same argument as in Theorem 3.3.1, since the estimate (3.3.5) is no longer valid for general f.

Here we shall use the *local linear independence* of the scaling functions $\varphi_{j,k}$, $k \in \mathbb{Z}^d$, i.e. the fact that if $\sum_{k\in\mathbb{Z}^d} c_k\varphi_{j,k}$ vanishes on some domain Ω, then $c_k = 0$ whenever $|\mathrm{Supp}(\varphi_{j,k})\cap\Omega| \neq 0$. This property was proved to hold for univariate scaling functions in LEMARIÉ and MALGOUYRES [1991]. Here, we need a simpler result in the sense that we are interested in cubic domains

Ω of sidelength 2^{-j}, but we need this result in several space dimensions, so we prove it below.

For the time being, we assume that there exists a cube $J = [0,m]^d$, where $m > 0$ is an integer, such that the above local linear independence holds on J at scale $j = 0$. By a change of scale, defining $J_{j,k} = 2^{-j}(k+J)$, we thus obtain that

$$\|\sum_{k\in\mathbb{Z}^d} c_k\varphi_{j,k}\|_{L^p(J_{j,l})} \sim 2^{dj(1/2-1/p)}\|(c_k)_{l-k\in \mathrm{Supp}(\varphi)-J}\|_{\ell^p}, \tag{3.7.3}$$

using that all quasi-norms are equivalent in finite dimension. Taking the p-th power and summing up on $l \in \mathbb{Z}$ gives the global stability (3.7.1).

We end by proving the local linear independence property. We first address the univariate case, i.e. $d = 1$. We assume that $\mathrm{Supp}(\varphi) = [a,b]$ and $\mathrm{Supp}(\tilde{\varphi}) = [c,d]$ for some integers $a < b$ and $c < d$ (note that the duality relation implies that these intervals overlap). We set $J = [0,m]$ with $m = \max\{d-c, c+d-2+b-a, b+c-1, d-a-1\}$. With such a choice, we have three properties: (i) since $m \geq d-c$, the functions $\tilde{\varphi}(\cdot - k)$ are supported in J for $k = -c, \cdots, m-d$, (ii) since $m \geq c+d-2+b-a$, the functions $\varphi(x-k)$, $k < c$ have no common support with the functions $\varphi(\cdot - k)$, $k > m-d$ and (iii) $\varphi(x-c+1)$ and $\varphi(\cdot - m+d-1)$ are respectively supported in $]-\infty, m]$ and $[0+\infty[$. Assume now that $\sum_{k\in\mathbb{Z}} c_k\varphi(x-k)$ is identically zero on $[0,m]$. Integrating over $\tilde{\varphi}(\cdot - k)$, we obtain by (i) that $c_k = 0$ for $k = -c, \cdots, m-d$. By (ii), it follows that

$$\sum_{k<c} c_k\varphi(x-k) = \sum_{k>m-d} c_k\varphi(x-k) = 0, \tag{3.7.4}$$

on $[0,m]$. Finally, by (iii), the basis functions overlapping $[0,m]$ in the two above sums are linearly independent in $[0,m]$ by their support properties. It follows that $c_k = 0$ whenever $|\mathrm{Supp}(\varphi(\cdot - k)) \cap [0,m]| \neq 0$.

In the multivariate case with a tensor product scaling function of the form $\phi(x) = \varphi(x_1)\cdots\varphi(x_d)$, we can use induction on the dimension d as follows: assuming that the property holds for the dimension $d-1$, let

$$\sum_{k\in\mathbb{Z}^d} c_k\phi(x-k) = 0, \tag{3.7.5}$$

on $[0,m]^d$. Integrating against a test function $g(x_1, \cdots, x_{d-1})$ supported on $[0,m]^{d-1}$, we thus obtain

$$\sum_{k_d\in\mathbb{Z}} a_{k_d}\varphi(x_d - k_d) = 0, \tag{3.7.6}$$

for all $x_d \in [0, m]$ with

$$a_{k_d} := \int \sum_{l \in \mathbb{Z}^{d-1}} c_{(l,k_d)} g(x_1, \cdots, x_{d-1}) \prod_{i=1}^{d-1} \varphi(x_i - l_i) dx_1 \cdots dx_{d-1}. \quad (3.7.7)$$

From the linear independence in one dimension, it follows that

$$\int \sum_{l \in \mathbb{Z}^{d-1}} c_{(l,k_d)} g(x_1, \cdots, x_{d-1}) \prod_{i=1}^{d-1} \varphi(x_i - l_i) dx_1 \cdots dx_{d-1} = 0, \quad (3.7.8)$$

for all k_d such that $|\mathrm{Supp}(\varphi(\cdot - k_d)) \cap [0, m]| \neq 0$. Since g is an arbitrary test function supported in $[0, m]^{d-1}$, we obtain that

$$\sum_{l \in \mathbb{Z}^{d-1}} c_{(l,k_d)} \varphi(x_1 - l_1) \cdots \varphi(x_{d-1} - l_{d-1}) = 0, \quad (3.7.9)$$

on $[0, m]^{d-1}$ for all k_d such that $|\mathrm{Supp}(\varphi(\cdot - k_d)) \cap [0, m]| \neq 0$. Using the induction hypothesis, we thus obtain that $c_k = 0$ for all k such that $|\phi(\cdot - k)) \cap [0, m]^d| \neq 0$. ◇

Remark 3.7.1 *The local linear independence proved in* LEMARIÉ *and* MALGOUYRES *[1991] is a stronger result in the sense that it actually holds for any non-trivial interval* $[a, b]$. *We could thus have simply used* $J = [0, 1]^d$ *and the cubes* $I_{j,k}$ *in place of* $J_{j,k}$ *in (3.7.3). Here, we have chosen an interval* $[0, m]$ *large enough so that local linear independence is almost straightforward. In particular, we can apply the same reasoning for proving the local linear independence of tensor product wavelets in this weaker sense.*

An immediate consequence of Theorem 3.7.1 is that we can extend the inverse inequalities of Theorems 3.4.2 and 3.4.3 to the case $p < 1$.

Theorem 3.7.2 *Under the assumptions of Theorem 3.7.1, and if* $\varphi \in B^n_{p,q}$ *for some* $q > 0$, *one has*

$$\omega_n(f, t)_p \lesssim [\min\{1, t2^j\}]^n \|f\|_{L^p} \text{ if } f \in V_j. \quad (3.7.10)$$

If $\varphi \in B^s_{p,q}$ *one has*

$$\|f\|_{B^s_{p,q}} \lesssim 2^{sj} \|f\|_{L^p} \text{ if } f \in V_j. \quad (3.7.11)$$

Proof If $f = \sum_{k \in \mathbb{Z}^d} c_k \varphi_{j,k} \in V_j$, we easily obtain

$$\omega_n(f, t)_p \lesssim \omega_n(\varphi_{j,0}, t)_p \|(c_k)_{k \in \mathbb{Z}^d}\|_{\ell^p}, \quad (3.7.12)$$

with constant 1 if $p \leq 1$ (by simply applying the p-triangle inequality). It follows that

$$|f|_{B^s_{p,q}} \lesssim \|(c_k)_{k\in\mathbb{Z}^d}\|_{\ell^p} |\varphi_{j,0}|_{B^s_{p,q}}. \quad (3.7.13)$$

We then observe that we have for $p < 1$ the same estimates for $|\varphi_{j,0}|_{B^s_{p,q}}$ and $\omega_n(\varphi_{j,0}, t)_p$ as in the proof of Theorems 3.4.2 and 3.4.3. Thus, (3.7.10) and (3.7.11) are respectively obtained from (3.7.12) and (3.7.13) combined together with the L^p-stability of Theorem 3.7.1. ◇

If we now want to extend the Whitney estimate of Theorem 3.3.3 to the case $p < 1$, we are facing the problem that L^p functions are not necessarily distributions and that the operator P_j is not a-priori well defined in these spaces (unless we put restrictions on s). One way to circumvent this problem is to consider the error of best approximation $\text{dist}_{L^p}(f, V_j)$ rather than the projection error $\|f - P_j f\|_{L^p}$. In this case, we obtain the following result.

Theorem 3.7.3 *Under the same assumptions as in Theorem 3.7.1, we have*

$$\text{dist}_{L^p}(f, V_j) \lesssim \omega_n(f, 2^{-j})_p, \quad (3.7.14)$$

where $n-1$ is the order of polynomial exactness in V_j.

Proof We consider the cubes $I_{j,k} := 2^j(k + [0,1]^d)$. We denote by E_l (resp. F_k) the set of indices $k \in \mathbb{Z}^d$ (resp. $l \in \mathbb{Z}^d$) such that $|\text{Supp}(\varphi_{j,l}) \cap I_{j,k}| \neq 0$. These sets have the same cardinality N independent of j and k. We then define the larger cubes

$$\tilde{J}_{j,k} = \cup_{l\in F_k} \text{Supp}(\varphi_{j,l}). \quad (3.7.15)$$

For each of these cubes, we choose a polynomial $p_{j,k} \in \Pi_{n-1}$ such that

$$\|f - p_{j,k}\|_{L^p(\tilde{J}_{j,k})} \leq 2 \inf_{g\in\Pi_{n-1}} \|f - g\|_{L^p(\tilde{J}_{j,k})} \lesssim \omega_n(f, 2^{-j}, \tilde{J}_{j,k})_p. \quad (3.7.16)$$

Each of these polynomials admits a (unique) representation

$$p_{j,k} = \sum_{l\in\mathbb{Z}^d} p_{j,k,l}\varphi_{j,l}, \quad (3.7.17)$$

in the scaling function basis. We build an approximation $f_j = \sum_{k\in\mathbb{Z}^d} f_{j,l}\varphi_{j,l}$ of f in V_j, with the choice

$$f_{j,l} := N^{-1}(\sum_{k\in E_l} p_{j,k,l}). \quad (3.7.18)$$

In some sense, f_j is a "blending" of the local polynomial approximations of f.

We now estimate the local approximation error $\|f - f_j\|_{L^p(I_{j,k})}$ by

$$\begin{aligned}\|f - f_j\|_{L^p(I_{j,k})} &\lesssim \|f - p_{j,k}\|_{L^p(I_{j,k})} + \|f_j - p_{j,k}\|_{L^p(I_{j,k})}\\ &\lesssim \omega_n(f, 2^{-j}, I_{j,k})_p + [\sum_{l\in F_k} |p_{j,k,l} - f_{j,l}|^p]^{1/p}\\ &\lesssim \omega_n(f, 2^{-j}, I_{j,k})_p + [\sum_{l\in F_k} \sum_{m\in E_l} |p_{j,k,l} - p_{j,m,l}|^p]^{1/p},\end{aligned}$$

where the first term is obtained by (3.7.16) and the second one by expanding f_j and $p_{j,k}$. In order to estimate the second term, we use the local linear independence of the polynomials on any domain of non-zero measure which gives

$$\begin{aligned}\sum_{m\in E_l} |p_{j,k,l} - p_{j,m,l}|^p &\lesssim \sum_{m\in E_l} \|p_{j,k} - p_{j,m}\|^p_{L^p(\mathrm{Supp}(\varphi_{j,l}))}\\ &\lesssim \sum_{m\in E_l} \|f - p_{j,m}\|^p_{L^p(\mathrm{Supp}(\varphi_{j,l}))}\\ &\lesssim \sum_{m\in E_l} \|f - p_{j,m}\|^p_{L^p(\tilde{J}_{j,m})}\\ &\lesssim \sum_{m\in E_l} \omega_n(f, 2^{-j}, \tilde{J}_{j,m})^p_p,\end{aligned}$$

and therefore

$$\sum_{l\in F_k}\sum_{m\in E_l} |p_{j,k,l} - p_{j,m,l}|^p \lesssim \sum_{l\in F_k}\sum_{m\in E_l} \omega_n(f, 2^{-j}, \tilde{J}_{j,m})^p_p.$$

We conclude as in Theorem 3.3.3: taking the p-th power of the resulting estimate for $\|f - f_j\|_{L^p(I_{j,k})}$ gives

$$\|f - f_j\|^p_{L^p(I_{j,k})} \lesssim \omega_n(f, 2^{-j}, I_{j,k})^p_p + \sum_{l\in F_k}\sum_{m\in E_l} \omega_n(f, 2^{-j}, \tilde{J}_{j,m})^p_p,$$

and we can then sum over k, using the modified modulus of smoothness $\tilde{\omega}_n(f, 2^{-j}, \tilde{J}_{j,m})_p$ of (3.3.15) (the proof of $\tilde{\omega} \sim \omega$ adapts in a straightforward way to the case $p < 1$). ◇

Theorems 3.7.2 and 3.7.3 contain all the ingredients that are needed to extend the identity $\mathcal{A}^s_{p,q} = B^s_{p,q}$ to all possible values of p and q.

Theorem 3.7.4 *Under the assumptions of Theorems 3.7.2 and 3.7.3, we have the norm equivalence*

$$\|f\|_{\mathcal{A}^t_{p,q}} \sim \|f\|_{B^t_{p,q}}, \tag{3.7.19}$$

for all $t < \min\{n, s\}$, where $n-1$ is the order of polynomial reproduction of the V_j spaces and s is such that $\varphi \in B^s_{p,q_0}$ for some $q_0 > 0$.

Proof We proceed exactly as in the proof of (3.6.2) in Theorem 3.6.1, using the p-triangle inequality in the estimate

$$\omega_n(f, 2^{-j})_p^p \leq \omega_n(f_0, 2^{-j})_p^p + \sum_{l=0}^{j} \omega_n(f_{l+1} - f_l, 2^{-j})_p^p + \omega_n(f - f_j, 2^{-j})_p^p.$$

We then conclude by a straightforward adaptation of the Hardy inequality. ◇

Remark 3.7.2 *We also have a direct estimate of the same type as in Corollary 3.3.1. We can thus combine Theorem 3.7.4 (adapted for quasi-normed spaces) and Theorem 3.7.2 to obtain that*

$$B^s_{p,q_1} = [L^p, B^t_{p,q_2}]_{\theta,q_1}, \tag{3.7.20}$$

if $s = \theta t$, $\theta \in]0,1[$. By Remarks 3.5.1 and 3.5.2, we also derive

$$B^r_{p,q} = [B^s_{p,q_1}, B^t_{p,q_2}]_{\theta,q}, \tag{3.7.21}$$

if $r = (1-\theta)s + \theta t$. Here, we have thus used approximation techniques to prove that the Besov spaces $B^s_{p,q}$ defined by moduli of smoothness satisfy the above interpolation identities for all values $0 < p, q \leq \infty$ and $s > 0$.

If we now want to use the specific projector P_j - or the wavelet coefficients - to characterize the $B^s_{p,q}$ norm for $p < 1$, we are obliged to impose conditions on s such that P_j will at least be well defined on the corresponding space. As we already mentioned, this problem is similar to the situation where we want to characterize L^p-Besov spaces for $p < \infty$ with an interpolation operator that is only bounded in L^∞. We shall now see that such characterizations are feasible if s is large enough so that $B^s_{p,q}$ is embedded in some L^r, $r \geq 1$ such that P_j is bounded on L^r.

For this, we need to understand in more detail the embedding properties of Besov spaces. These embeddings rely on the following basic result, which was firstly proved in DEVORE, RIEMENSCHNEIDER and SHARPLEY [1979], for indices r and p in $[1,\infty]$, and that we prove here also for the range $]0,1[$.

Theorem 3.7.5 *Let Ω be either a cube or $\mathbb{R}^d$. If $0 < p < r \leq \infty$, $r \geq 1$, and $f \in L^p$, one has*

$$\omega_n(f, 2^{-j}, \Omega)_r \lesssim \sum_{l \geq j} 2^{dl(1/p-1/r)} \omega_n(f, 2^{-l}, \Omega)_p, \tag{3.7.22}$$

with $f \in L^r(\Omega)$ whenever the right hand side of (3.7.22) is finite. For $r < 1$, we have

$$\omega_n(f, 2^{-j}, \Omega)_r \lesssim \Big(\sum_{l \geq j} [2^{dl(1/p-1/r)} \omega_n(f, 2^{-l}, \Omega)_p]^r\Big)^{1/r}, \tag{3.7.23}$$

in place of (3.7.22).

Proof We start with the domain $\Omega = [0,1]^d$ and consider a multiresolution approximation V_j, $j \geq 0$, adapted to this setting (by tensor product of the construction on the interval, as explained §2.12 of Chapter 2). We assume that it is generated from a scaling function $\varphi \in C^n$ and that polynomials are reproduced up to degree n. Here, we anticipate the results of §3.9, which show that such an adaptation on the cube (as well as on more general domains) allows us to preserve the results of Theorems 3.7.1 to 3.7.4. In particular, we still have the direct estimate

$$\|f - f_j\|_{L^p(\Omega)} \lesssim \omega_n(f, 2^{-j}, \Omega)_p, \tag{3.7.24}$$

where $f_j \in V_j$ is such that $\|f - f_j\|_{L^p(\Omega)} \leq 2\text{dist}_{L^p}(f, V_j)$, as well as the inverse estimate

$$\omega_n(f, t, \Omega)_t \lesssim [\min\{1, t2^j\}]^n \|f\|_{L^p(\Omega)}, \text{ if } f \in V_j. \tag{3.7.25}$$

Note that the stability property of Theorem 3.7.1 shows that any $f_j \in V_j$ also satisfies the inverse estimate

$$\|f_j\|_{L^r(\Omega)} \lesssim 2^{dj(1/p-1/r)} \|f_j\|_{L^p(\Omega)}, \tag{3.7.26}$$

for $r \geq p$ since $\|\cdot\|_{\ell^r} \leq \|\cdot\|_{\ell^p}$. We finally impose that f_0 be a polynomial in Π_{n-1} such that

$$\|f - f_0\|_{L^p(\Omega)} \leq 2 \inf_{g \in \Pi_{n-1}} \|f - g\|_{L^p(\Omega)} \lesssim \omega_n(f, 1, \Omega)_p. \tag{3.7.27}$$

Assuming that $r \geq 1$, we have for $j \geq 0$,

$$\begin{aligned}
\|f\|_{L^r(\Omega)} &\leq \|f_j\|_{L^r(\Omega)} + \textstyle\sum_{l \geq j} \|f_{l+1} - f_l\|_{L^r(\Omega)} \\
&\lesssim 2^{dj(1/p-1/r)} \|f_j\|_{L^p(\Omega)} + \textstyle\sum_{l \geq j} 2^{dl(1/p-1/r)} \|f_{l+1} - f_l\|_{L^p(\Omega)} \\
&\lesssim 2^{dj(1/p-1/r)} \|f\|_{L^p(\Omega)} + \textstyle\sum_{l \geq j} 2^{dl(1/p-1/r)} \|f - f_j\|_{L^p(\Omega)} \\
&\lesssim 2^{dj(1/p-1/r)} \|f\|_{L^p(\Omega)} + \textstyle\sum_{l \geq j} 2^{dl(1/p-1/r)} \omega_n(f, 2^{-l}, \Omega)_p,
\end{aligned}$$

which shows that $f \in L^r(\Omega)$ whenever the right hand side of (3.7.22) is finite. In the case $r < 1$, a straightforward modification, using the r-triangle inequality, gives the same result with the right hand side of (3.7.23).

We clearly have $\omega_n(f_0, 2^{-j}, \Omega)_r = 0$, since the finite difference operator of order n annihilates P_{n-1}. If $r \geq 1$, we thus have for $j \geq 0$,

$$\begin{aligned}\omega_n(f, 2^{-j}, \Omega)_r &= \omega_n(f - f_0, 2^{-j}, \Omega)_r \\ &\leq \sum_{l\geq 0} \omega_n(f_{l+1} - f_l, 2^{-j}, \Omega)_r \\ &\lesssim \sum_{l=0}^{j-1} 2^{n(l-j)} \|f_{l+1} - f_l\|_{L^r(\Omega)} + \sum_{l\geq j} \|f_{l+1} - f_l\|_{L^r(\Omega)},\end{aligned}$$

where we have used the inverse estimate. We now use the local version of (3.7.24) and the direct estimate to treat the above two sums. For the coarse scales, we obtain

$$\begin{aligned}\sum_{l=0}^{j-1} 2^{n(l-j)} \|f_{l+1} - f_l\|_{L^r(\Omega)} &\lesssim \sum_{l=0}^{j-1} 2^{n(l-j)} 2^{dl(1/p-1/r)} \|f_{l+1} - f_l\|_{L^p(\Omega)} \\ &\lesssim \sum_{l=0}^{j-1} 2^{n(l-j)} 2^{dl(1/p-1/r)} \|f - f_l\|_{L^p(\Omega)} \\ &\lesssim \sum_{l=0}^{j-1} 2^{n(l-j)} 2^{dl(1/p-1/r)} \omega_n(f, 2^{-l}, \Omega)_p \\ &\lesssim \sum_{l=0}^{j-1} 2^{(n-1)(l-j)} 2^{dl(1/p-1/r)} \omega_n(f, 2^{-j}, \Omega)_p \\ &\lesssim 2^{dj(1/p-1/r)} \omega_n(f, 2^{-j}, \Omega)_p,\end{aligned}$$

where we have used that n is larger than 1.

For the fine scales, we have

$$\begin{aligned}\sum_{l\geq j} \|f_{l+1} - f_l\|_{L^r(\Omega)} &\lesssim \sum_{l\geq j} 2^{dl(1/p-1/r)} \|f_{l+1} - f_l\|_{L^p(\Omega)} \\ &\lesssim \sum_{l\geq j} 2^{dl(1/p-1/r)} \|f - f_l\|_{L^p(\Omega)} \\ &\lesssim \sum_{l\geq j} 2^{dl(1/p-1/r)} \omega_n(f, 2^{-l}, \Omega)_p.\end{aligned}$$

We have thus proved

$$\omega_n(f, 2^{-j}, \Omega)_r \lesssim \sum_{l\geq j} 2^{dl(1/p-1/r)} \omega_n(f, 2^{-l}, \Omega)_p, \tag{3.7.28}$$

i.e. (3.7.22) on the unit cube and for $j \geq 0$. The same statement for $j \leq 0$ is equivalent to the case $j = 0$ due to the size of Ω.

By a change of scale, it follows that (3.7.28) holds on any cube Ω with a uniform constant and thus also for $\Omega = \mathbb{R}^d$.

In the case $r < 1$, a straightforward modification, using the r-triangle inequality, gives (3.7.23). ◇

Note that the first part of the proof shows that $B^s_{p,q}$, $q = \min\{1, p\}$ is continuously embedded in L^r for $1/p = 1/r + s/d$. In fact we also have a continuous embedding of $B^s_{p,p}$ in L^r, as it will be proved in §4.3 of the next chapter, for all $p > 0$.

An embedding theorem between Besov spaces also appears as an immediate consequence of the above result, combined with the Hardy inequality (3.5.36).

Corollary 3.7.1 *If $0 < p < r \leq \infty$ and $0 < s < t$ are such that $t - s = d(1/p - 1/r)$ we have the embedding relation*

$$\|f\|_{B^s_{r,q}} \lesssim \|f\|_{B^t_{p,q}}, \tag{3.7.29}$$

for all $q > 0$.

Note that since $p < r$, (3.7.29) immediately implies the embedding $\|f\|_{B^s_{r,r}} \lesssim \|f\|_{B^t_{p,p}}$, which was announced in §3.2. Using Theorem 3.7.4, we shall now establish a direct estimate for a possibly L^p-unstable projector.

Theorem 3.7.6 *Assume that $\varphi \in L^r$ and $\tilde{\varphi} \in L^{r'}$ for some $r \in [1, \infty]$, $1/r + 1/r' = 1$, or that $\varphi \in C^0$ and $\tilde{\varphi}$ is a Radon measure in which case we set $r = \infty$. Then, for $0 < p < r$, one has the direct estimate*

$$\|f - P_j f\|_{L^p} \lesssim 2^{-sj} |f|_{B^s_{p,p}}, \tag{3.7.30}$$

for all $s > 0$ such that $d(1/p - 1/r) < s < n$ if $p > 1$ or $d(1/p - 1/r) \leq s < n$ if $p \leq 1$, where $n - 1$ is the order of polynomial reproduction in V_j.

Proof We shall use the cubes $I_{j,k}$ and $\widetilde{I}_{j,k}$ (defined by (3.3.8)) that were already used in the proof of Theorems 3.3.2 and 3.3.3. We then have

$$\begin{aligned}
\|f - P_j f\|_{L^p(I_{j,k})} &\leq 2^{dj(1/r-1/p)} \|f - P_j f\|_{L^r(I_{j,k})} \\
&\lesssim 2^{dj(1/r-1/p)} \omega_n(f, 2^{-j}, \widetilde{I}_{j,k})_r \\
&\lesssim 2^{dj(1/r-1/p)} [\textstyle\sum_{l \geq j} 2^{dl(1/p-1/r)} \omega_n(f, 2^{-l}, \widetilde{I}_{j,k})_p],
\end{aligned}$$

where we have used Hölder's inequality, the local L^r estimate (ensured by the L^r-boundedness of P_j) and (3.7.22) since $r \geq 1$. We now take the p-th power and sum on k. If $p \leq 1$, this gives

$$\begin{aligned}
\|f - P_j f\|^p_{L^p} &\lesssim \textstyle\sum_{k \in \mathbb{Z}^d} \left(2^{dj(1/r-1/p)} [\sum_{l \geq j} 2^{dl(1/p-1/r)} \omega_n(f, 2^{-l}, \widetilde{I}_{j,k})_p]\right)^p \\
&\lesssim \textstyle\sum_{k \in \mathbb{Z}^d} 2^{dj(p/r-1)} \sum_{l \geq j} 2^{dl(1-p/r)} \omega_n(f, 2^{-l}, \widetilde{I}_{j,k})^p_p \\
&\lesssim \textstyle\sum_{l \geq j} 2^{d(l-j)(1-p/r)} \omega_n(f, 2^{-l})^p_p \\
&\lesssim 2^{-spj} \textstyle\sum_{l \geq j} 2^{d(l-j)(1-p/r-sp)} 2^{spl} \omega_n(f, 2^{-l})^p_p \\
&\leq 2^{-spj} \textstyle\sum_{l \geq j} 2^{spl} \omega_n(f, 2^{-l})^p_p \\
&\leq 2^{-spj} |f|^p_{B^s_{p,p}},
\end{aligned}$$

where we have used that $1 - p/r - sp \leq 0$. If $p > 1$, we define p' such that

$1/p + 1/p' = 1$ and we obtain for any $\varepsilon > 0$,

$$\begin{aligned}\|f - P_j f\|_{L^p}^p &\lesssim \sum_{k\in\mathbb{Z}^d}\Big(2^{dj(1/r-1/p)}[\sum_{l\geq j} 2^{dl(1/p-1/r)}\omega_n(f,2^{-l},\tilde{I}_{j,k})_p]\Big)^p\\ &\lesssim \sum_{k\in\mathbb{Z}^d} 2^{dj(p/r-1)}[\sum_{l\geq j} 2^{dl(1-p/r)+lp\varepsilon}\omega_n(f,2^{-l},\tilde{I}_{j,k})_p^p]\\ &\quad\times[\sum_{l\geq j} 2^{-lp'\varepsilon}]^{p/p'}\\ &\lesssim \sum_{k\in\mathbb{Z}^d} 2^{dj(p/r-1)-pj\varepsilon}\sum_{l\geq j} 2^{dl(1-p/r)+lp\varepsilon}\omega_n(f,2^{-l},\tilde{I}_{j,k})_p^p\\ &\lesssim \sum_{l\geq j} 2^{d(l-j)(1-p/r+p\varepsilon/d)}\omega_n(f,2^{-l})_p^p\\ &\lesssim 2^{-spj}\sum_{l\geq j} 2^{d(l-j)(1-p/r-sp+p\varepsilon/d)}2^{spl}\omega_n(f,2^{-l})_p^p\\ &\lesssim 2^{-spj}\sum_{l\geq j} 2^{spl}\omega_n(f,2^{-l})_p^p\\ &\lesssim 2^{-spj}|f|_{B^s_{p,p}}^p,\end{aligned}$$

by choosing ε small enough so that $1-p/r-sp+p\varepsilon/d \leq 0$, which is possible by assumption. ◇

We are now ready to prove the main result of this section.

Theorem 3.7.7 *Assume that $\varphi \in L^r$ and $\tilde{\varphi} \in L^{r'}$ for some $r \in [1,\infty]$, $1/r + 1/r' = 1$, or that $\varphi \in C^0$ and $\tilde{\varphi}$ is a Radon measure in which case we set $r = \infty$. Then, for $0 < p \leq r$, one has the norm equivalence*

$$\|f\|_{B^t_{p,q}} \sim \|P_0 f\|_{L^p} + \|(2^{tj}\|Q_j f\|_{L^p})_{j\geq 0}\|_{\ell^q}, \tag{3.7.31}$$

for all $t > 0$ such that $d(1/p-1/r) < t < \min\{s,n\}$, where $n-1$ is the order of polynomial reproduction in V_j and s is such that $\varphi \in B^s_{p,q_0}$ for some q_0. If $f = \sum_{\lambda\in\nabla} c_\lambda\psi_\lambda$ is the decomposition of f into the corresponding wavelet basis, we also have the norm equivalence

$$\|f\|_{B^t_{p,q}} \sim \|\Big(2^{tj}2^{d(1/2-1/p)j}\|(c_\lambda)_{\lambda\in\nabla_j}\|_{\ell^p}\Big)_{j\geq -1}\|_{\ell^q}, \tag{3.7.32}$$

under the same assumptions.

Proof If we denote by $A^t_{p,q}$ the space of L^r functions such that

$$\|f\|_{A^t_{p,q}} := \|P_0 f\|_{L^p} + \|(2^{tj}\|Q_j f\|_{L^p})_{j\geq 0}\|_{\ell^q}, \tag{3.7.33}$$

is bounded, we clearly get from the direct estimate that

$$B^t_{p,p} \subset A^t_{p,\infty}, \tag{3.7.34}$$

for all $t > 0$ such that $d(1/p - 1/r) < t < n$. By the inverse estimate of Theorem 3.7.2, we also have

$$\begin{aligned}\|f\|_{B^t_{p,p}} &\leq \|P_0 f\|_{B^t_{p,p}} + \sum_{j\geq 0}\|Q_j f\|_{B^t_{p,p}}\\ &\leq \|P_0 f\|_{L^p} + \sum_{j\geq 0} 2^{tj}\|Q_j f\|_{L^p}\\ &\leq \|f\|_{A^t_{p,1}},\end{aligned}$$

if $p \geq 1$ and

$$\|f\|_{B^t_{p,p}} \leq \|f\|_{A^t_{p,p}}, \tag{3.7.35}$$

in the case $p < 1$, for all $t < s$. We thus have

$$A^t_{p,q_p} \subset B^t_{p,p} \subset A^t_{p,\infty}, \tag{3.7.36}$$

with $q_p := \min\{1,p\}$ and $t > 0$ such that $d(1/p - 1/r) < t < \min\{n, s\}$.

We now remark that the $A^t_{p,q}$ are trivially interpolation spaces, in the same sense that they satisfy identities similar to (3.5.41), due to the interpolation properties (3.5.45) of the weighted ℓ^q spaces. We also have a similar structure for Besov spaces, by (3.7.21). If we consider (3.7.36) for t_1 and t_2 such that $d(1/p-1/r) < t_1 < t < t_2 < \min\{n, s\}$, and interpolate with parameters q and θ such that $t = (1-\theta)t_1 + \theta t_2$, we thus obtain, by the reiteration theorem

$$A^t_{p,q} \subset B^t_{p,q} \subset A^t_{p,q}, \tag{3.7.37}$$

i.e. the announced norm equivalence (3.7.31).

The wavelet characterization (3.7.32) is an immediate consequence: we use that

$$\|Q_j f\|_{L^p} \sim 2^{d(1/2-1/p)j} \|(c_\lambda)_{\lambda\in\nabla_j}\|_{\ell^p}, \tag{3.7.38}$$

for $j \geq 0$ which is obtained by a similar reasoning as for the scaling functions in Theorem 3.7.1 (a more general argument, given in §3.9, shows that the L^p-stability of the wavelet basis at level j stems from the L^p-stability of the scaling function basis at level $j+1$). ◇

As an example of application, consider the Schauder hierarchical basis obtained from the univariate version of Chapter 1 by the tensor product method. In this case, polynomials are reproduced up to order 1, i.e. we have $n = 2$. One also easily checks that $\varphi \in B^s_{p,p}$ for $s < 1 + 1/p$, with this upper bound independent of the dimension. Finally, we know that the corresponding projector P_j is only bounded in L^∞. Thus according to Theorem 3.7.6, we can characterize $B^t_{p,q}$ by the Schauder basis for the range

$$d/p < t < \min\{2, 1 + 1/p\}. \tag{3.7.39}$$

If we are interested in the L^2-Sobolev spaces $H^s = B^s_{2,2}$, this range is $]1/2, 3/2[$ if $d = 1$, $]1, 3/2[$ if $d = 2$ and empty for higher dimensions. If we are interested in Hölder spaces $C^s = B^s_{\infty,\infty}$, this range is $]0, 1[$ in all dimensions.

Remark 3.7.3 *One can easily extend the result of Theorem 3.7.6 to the case of operators that are unbounded in all L^p spaces, but bounded in C^m, such as Hermite interpolants. The restriction on t from below is of course more severe ($d/p + m < t$).*

Remark 3.7.4 *We can also use the result of Theorem 3.7.6 to prove the equivalence of the definitions of Besov spaces through modulus of smoothness and Littlewood-Paley decompositions, in the same spirit as explained in Remark 3.6.1. Here, the restriction $s > \min\{0, d(1/p - 1)\}$ is imposed by the fact that $f \mapsto S_j f$ is unbounded in L^p for $p < 1$. Generally speaking, we cannot hope for a characterization of Besov spaces modelled on L^p for $p < 1$ by the size properties of wavelets coefficients for ranges of smoothness below $d(1/p - 1)$.*

Remark 3.7.5 *Another approach to the characterization of Besov spaces for $p < 1$ is proposed in* KYRIASIS *[1996]. It is based on the definition of these spaces using the Hardy spaces H^p instead of L^p in the definition of the modulus of smoothness. The interest of such results is that the Hardy spaces are distribution spaces and that the corresponding Besov spaces can be characterized by inner products with sufficiently smooth wavelets even when $s \leq d(1/p - 1)$. For larger values of s, the L^p-Besov spaces and H^p-Besov spaces coincide, and the results in the above reference are of similar nature to those of this section, although they are obtained by very different techniques. In particular, the approach of Kyriasis considers the homogeneous Besov spaces, for which the scales j in the definition of the Besov norm range from $-\infty$ to $+\infty$. The use of non-homogeneous spaces, as done in the present approach, seems more appropriate for a straightforward subsequent adaptation of our results to bounded domains.*

3.8 Negative smoothness and L^p-spaces

So far, we have only considered Hölder, Sobolev and Besov spaces associated with a smoothness index $s > 0$. For $s < 0$, these spaces are usually defined by duality for $p, q \geq 1$:

$$B^{-s}_{p',q'} := (B^s_{p,q})', \tag{3.8.1}$$

with $1/p + 1/p' = 1$ and $1/q + 1/q' = 1$. The characterization of such dual spaces by means of multiscale decompositions and wavelet coefficients simply relies on the ability of characterizing the corresponding primal space by the dual multiscale decomposition, as shown by the following result.

Theorem 3.8.1 *Assuming that the dual projectors are such that*

$$\|f\|_{B^s_{p,q}} \sim \|P_0^* f\|_{L^p} + \|(2^{sj}\|Q_j^* f\|_{L^p})_{j\geq 0}\|_{\ell^q}, \tag{3.8.2}$$

for some $s > 0$ *and* $p, q \geq 1$, *we then have*

$$\|f\|_{B^{-s}_{p',q'}} \sim \|P_0 f\|_{L^{p'}} + \|(2^{-sj}\|Q_j f\|_{L^{p'}})_{j\geq 0}\|_{\ell^{q'}}, \tag{3.8.3}$$

with $1/p + 1/p' = 1$ *and* $1/q + 1/q' = 1$. *We also have the norm equivalence*

$$\|f\|_{B^{-s}_{p',q'}} \sim \|\left((2^{-sj}2^{d(1/2-1/p')j}\|(c_\lambda)_{\lambda\in\nabla_j}\|_{\ell^{p'}}\right)_{j\geq -1}\|_{\ell^{q'}}, \tag{3.8.4}$$

if $f = \sum_{\lambda\in\nabla} c_\lambda \psi_\lambda$.

Proof By definition, we have

$$\|f\|_{B^{-s}_{p',q'}} := \sup_{g\in\mathcal{D}, \|g\|_{B^s_{p,q}}=1} \langle f, g\rangle, \tag{3.8.5}$$

where $\langle\cdot,\cdot\rangle$ is meant in the sense of the duality $(\mathcal{D}', \mathcal{D})$. If f is also a test function, we have

$$\begin{aligned}
\langle f, g\rangle &= \langle P_0 f + \textstyle\sum_{j\geq 0} Q_j f, P_0^* g + \sum_{j\geq 0} Q_j^* g\rangle \\
&= \langle P_0 f, P_0^* g\rangle + \textstyle\sum_{j\geq 0}\langle Q_j f, Q_j^* g\rangle \\
&\leq \|P_0 f\|_{L^{p'}}\|P_0^* g\|_{L^p} + \textstyle\sum_{j\geq 0}\|Q_j f\|_{L^{p'}}\|Q_j^* g\|_{L^p} \\
&\lesssim \|g\|_{B^s_{p,q}}\left(\|P_0 f\|_{L^{p'}} + \|(2^{-sj}\|Q_j f\|_{L^{p'}})_{j\geq 0}\|_{\ell^{q'}}\right),
\end{aligned}$$

where we have used Hölder's inequality on functions and on sequences, and the fact that $V_0 \perp \widetilde{W}_j$ and $\tilde{V}_0 \perp W_j$ if $j \geq 0$ and that $W_j \perp \widetilde{W}_l = \{0\}$ if $l \neq j$.

We have thus proved that

$$\|f\|_{B^{-s}_{p',q'}} \lesssim \|P_0 f\|_{L^{p'}} + \|(2^{-sj}\|Q_j f\|_{L^{p'}})_{j\geq 0}\|_{\ell^{q'}}. \tag{3.8.6}$$

By density of the test functions, this result holds for all f in $B^{-s}_{p',q'}$.

In the other direction, we shall use the duality between ℓ^r and $\ell^{r'}$ if $1/r + 1/r' = 1$ (which holds when $r = 1$ or $r = \infty$). If f is a test function, there exists a sequence $(a_j)_{j\geq 0}$ with ℓ^q norm equal to 1 such that

$$\|P_0 f\|_{L^{p'}} + \|(2^{-sj}\|Q_j f\|_{L^{p'}})_{j\geq 0}\|_{\ell^{q'}} = \|P_0 f\|_{L^{p'}} + \sum_{j\geq 0} a_j 2^{-sj}\|Q_j f\|_{L^{p'}}.$$

From the norm equivalence

$$\|\sum_{\lambda\in\nabla_j} c_\lambda\psi_\lambda\|_{L^p} \sim \|\sum_{\lambda\in\nabla_j} c_\lambda\tilde{\psi}_\lambda\|_{L^p} \sim 2^{dj(1/2-1/p)}\|(c_\lambda)_{\lambda\in\nabla_j}\|_{\ell^p}, \tag{3.8.7}$$

and the identity

$$\langle \sum_{\lambda\in\nabla_j} c_\lambda\psi_\lambda \, , \, \sum_{\lambda\in\nabla_j} d_\lambda\tilde{\psi}_\lambda\rangle = \sum_{\lambda\in\nabla_j} c_\lambda\overline{d_\lambda}, \tag{3.8.8}$$

we derive that there exist functions $h_j \in \widetilde{W}_j$ such that $\|h_j\|_{L^p} = 1$ and

$$\|Q_j f\|_{L^{p'}} \lesssim \langle Q_j f, h_j\rangle. \tag{3.8.9}$$

Similarly, there exists $g_0 \in \tilde{V}_0$ such that $\|g_0\|_{L^p} = 1$ and

$$\|P_0 f\|_{L^{p'}} \lesssim \langle P_0 f, g_0\rangle. \tag{3.8.10}$$

If we now define

$$g := g_0 + \sum_{j\geq 0} a_j 2^{-sj} h_j, \tag{3.8.11}$$

it follows from (3.8.9) and (3.8.10) that

$$\|P_0 f\|_{L^{p'}} + \|(2^{-sj}\|Q_j f\|_{L^{p'}})_{j\geq 0}\|_{\ell^{q'}} \leq 4\langle f, g\rangle. \tag{3.8.12}$$

Moreover, we have

$$\begin{aligned} \|g\|_{B^s_{p,q}} &\lesssim \|P_0^* g\|_{L^p} + \|(2^{sj}\|Q_j^* g\|_{L^p})_{j\geq 0}\|_{\ell^q} \\ &= \|g_0\|_{L^p} + \|(a_j\|h_j\|_{L^p})_{j\geq 0}\|_{\ell^q} = 2. \end{aligned}$$

Combining this and (3.8.12) with the definition of the $B^{-s}_{p',q'}$ norm, we have thus proved

$$\|P_0 f\|_{L^{p'}} + \|(2^{-sj}\|Q_j f\|_{L^{p'}})_{j\geq 0}\|_{\ell^{q'}} \lesssim \|f\|_{B^{-s}_{p',q'}}. \tag{3.8.13}$$

By density the result follows for all $f \in B^{-s}_{p',q'}$. We have thus proved (3.8.3). The norm equivalence (3.8.4) is an immediate consequence, using the L^p-stability of the wavelet basis at each level, as we already did in the two previous sections. ◇

Remark 3.8.1 *The above proof shows that the projectors P_j and Q_j have uniformly bounded extensions in $B^{-s}_{p',q'}$.*

Remark 3.8.2 *A close result to the above is that the Besov spaces of negative smoothness are well defined by (3.2.19) with the Littlewood-Paley decomposition.*

In the case of Sobolev spaces, we have thus proved the norm equivalence

$$\|f\|_{H^s}^2 \sim \|P_0 f\|_{L^2}^2 + \sum_{j\geq 0} 2^{2js}\|Q_j f\|_{L^2}^2 \sim \sum_{\lambda\in\nabla} 2^{2|\lambda|s}|c_\lambda|^2, \tag{3.8.14}$$

(with $c_\lambda := \langle f, \tilde{\psi}_\lambda\rangle$ the coordinates of f in the multiscale basis) for a regularity index s that is either strictly positive or strictly negative, leaving aside the case $s = 0$ which corresponds to L^2. Note that the above equivalence for $s = 0$ would exactly mean that $\{\psi_\lambda\}_{\lambda\in\nabla}$ constitutes a Riesz Basis for $L^2(\mathbb{R})$. By duality, we would also obtain that $\{\tilde{\psi}_\lambda\}_{\lambda\in\nabla}$ is also a Riesz basis, and thus

$$\|f\|_{L^2}^2 \sim \sum_{\lambda\in\nabla} |c_\lambda|^2 \sim \sum_{\lambda\in\nabla} |\tilde{c}_\lambda|^2, \tag{3.8.15}$$

(with $\tilde{c}_\lambda := \langle f, \psi_\lambda\rangle$).

A possible elementary technique to prove (3.8.15) is to use interpolation theory: L^2 and ℓ^2 can be respectively identified with $[H^{-s}, H^s]_{1/2,2}$ and $[\ell^2_{-s}, \ell^2_s]_{1/2,2}$. Thus (3.8.15) should hold if both primal and dual multiscale decomposition allow us to characterize H^s for some $s > 0$. We shall give here another argument (proposed in DAHMEN [1996]) that will readily adapt to the characterization of $L^2(\Omega)$ if Ω is a bounded domain.

Theorem 3.8.2 *Assume that the projectors P_j are uniformly bounded in L^2 and that there exists a normed space Y such that Y is dense in $L^2(\mathbb{R})$ and such that*

$$\|f - P_j f\|_{L^2} \lesssim 2^{-\varepsilon j}\|f\|_Y \text{ and } \|f - P_j^* f\|_{L^2} \lesssim 2^{-\varepsilon j}\|f\|_Y \tag{3.8.16}$$

with $\varepsilon > 0$ and

$$\|f\|_Y \lesssim 2^{\varepsilon j}\|f\|_X \text{ if } f \in V_j \text{ or } f \in \tilde{V}_j. \tag{3.8.17}$$

Then we have the norm equivalence

$$\|f\|_{L^2}^2 \sim \|P_0 f\|_{L^2}^2 + \sum_{j\geq 0}\|Q_j f\|^2 \sim \|P_0^* f\|_{L^2}^2 + \sum_{j\geq 0}\|Q_j^* f\|^2. \tag{3.8.18}$$

Proof Introducing the dual Y' of Y, we first remark that we can establish from (3.8.16) a similar direct estimate between Y' and L^2: for $f \in L^2$, we have

$$\|f - P_j f\|_{Y'} = \sup_{\|g\|_Y = 1} |\langle f, g - P_j^* g\rangle| \lesssim 2^{-\varepsilon j}\|f\|_{L^2}. \tag{3.8.19}$$

By the symmetry in the assumptions, a similar estimate also holds for P_j^*.

Note that the direct estimate (3.8.16) implies that the union of the V_j spaces is dense in L^2 and thus for any $f \in L^2$, we have

$$\lim_{J\to+\infty} \|P_0 f + \sum_{j=0}^{J} Q_j f - f\|_{L^2} = \lim_{J\to+\infty} \|P_0^* f + \sum_{j=0}^{J} Q_j^* f - f\|_{L^2} = 0. \quad (3.8.20)$$

For $f \in V_{J+1}$, we have

$$\begin{aligned} \|f\|_{L^2}^2 &= \|P_0 f + \textstyle\sum_{j=0}^{J} Q_j f\|_{L^2}^2 \\ &\lesssim \|P_0 f\|_{L^2}^2 + \textstyle\sum_{j\geq 0} \|Q_j f\|_{L^2}^2 + \textstyle\sum_{0\leq j_1<j_2\leq J} \langle Q_{j_1} f, Q_{j_2} f\rangle \\ &\lesssim \|P_0 f\|_{L^2}^2 + \textstyle\sum_{j\geq 0} \|Q_j f\|_{L^2}^2 + \textstyle\sum_{0\leq j_1<j_2\leq J} \|Q_{j_1} f\|_Y \|Q_{j_2} f\|_{Y'}. \end{aligned}$$

The inverse estimate (3.8.17) gives us $\|Q_{j_1} f\|_Y \lesssim 2^{\varepsilon j_1} \|Q_{j_1} f\|_{L^2}$, while the direct estimate (3.8.19) yields

$$\|Q_{j_2} f\|_{Y'} = \|Q_{j_2} f - P_{j_2} f Q_{j_2} f\|_{Y'} \lesssim 2^{-\varepsilon j_2} \|Q_{j_2} f\|_{L^2}. \quad (3.8.21)$$

It follows that we have

$$\begin{aligned} \|f\|_{L^2}^2 &\lesssim \|P_0 f\|_{L^2}^2 + \textstyle\sum_j \|Q_j f\|_{L^2}^2 + \textstyle\sum_{j_1<j_2} 2^{-\varepsilon(j_2-j_1)} \|Q_{j_1} f\|_{L^2} \|Q_{j_2} f\|_{L^2} \\ &\leq \|P_0 f\|_{L^2}^2 + [1 + 2\textstyle\sum_{l>0} 2^{-\varepsilon l}] \textstyle\sum_{j\geq 0} \|Q_j f\|_{L^2}^2 \\ &\lesssim \|P_0 f\|_{L^2}^2 + \textstyle\sum_{j\geq 0} \|Q_j f\|_{L^2}^2. \end{aligned}$$

By density of the spaces V_j, we have thus proved

$$\|f\|_{L^2}^2 \lesssim \|P_0 f\|_{L^2}^2 + \sum_{j\geq 0} \|Q_j f\|_{L^2}^2, \quad (3.8.22)$$

and by symmetry

$$\|f\|_{L^2}^2 \lesssim \|P_0^* f\|_{L^2}^2 + \sum_{j\geq 0} \|Q_j^* f\|_{L^2}^2. \quad (3.8.23)$$

In order to complete the proof, we now remark that

$$\begin{aligned} (\|P_0 f\|_{L^2}^2 + \textstyle\sum_{j\geq 0} \|Q_j f\|_{L^2}^2)^2 &= |\langle f, P_0^* P_0 f + \textstyle\sum_{j\geq 0} Q_j^* Q_j f\rangle|^2 \\ &\leq \|f\|_{L^2}^2 \|P_0^* P_0 f + \textstyle\sum_{j\geq 0} Q_j^* Q_j f\|_{L^2}^2 \\ &\lesssim \|f\|_{L^2}^2 (\|P_0^* P_0 f\|_{L^2}^2 + \textstyle\sum_{j\geq 0} \|Q_j^* Q_j f\|_{L^2}^2) \\ &\lesssim \|f\|_{L^2}^2 (\|P_0 f\|_{L^2}^2 + \textstyle\sum_{j\geq 0} \|Q_j f\|_{L^2}^2), \end{aligned}$$

where we have used (3.8.23) and the uniform boundedness of the projectors P_j^*. Dividing on both sides, we thus obtain

$$\|P_0 f\|_{L^2}^2 + \sum_{j\geq 0} \|Q_j f\|_{L^2}^2 \lesssim \|f\|_{L^2}^2, \quad (3.8.24)$$

and by symmetry

$$\|P_0^* f\|_{L^2}^2 + \sum_{j\geq 0} \|Q_j^* f\|_{L^2}^2 \lesssim \|f\|_{L^2}^2. \tag{3.8.25}$$

We have thus proved the full equivalence (3.8.18). ◇

The above result shows in particular that L^2 is characterized by (3.8.18), if both primal and dual multiresolution approximations satisfy direct and inverse estimates between L^2 and H^ε, for some $\varepsilon > 0$. In practice, this amounts to assuming that φ and $\tilde{\varphi}$ both have H^ε smoothness for some $\varepsilon > 0$, since this in turn implies that both V_j and $\widetilde{V}_j$ reproduce constant functions (from the results in §2.8 of Chapter 2). Form the L^2-stability of the wavelet bases in each subspace W_j and $\widetilde{W}_j$, we also have (3.8.15).

Remark 3.8.3 *Other methods exist for proving (3.8.18) and (3.8.15). The approach proposed in* Cohen, Daubechies *and* Feauveau *[1992] and in* Cohen *and* Daubechies *[1992] gives a direct proof of (3.8.15), making an important use of the Fourier transform and of the translation invariant structure of the bases* $\{\psi_{j,k}\}_{k\in\mathbb{Z}^d}$ *at each level* j. *Such techniques are thus difficult to mimic on a bounded domain* Ω. *A more general argument for the direct proof of (3.8.15) is provided by the vaguelette lemma (see* Meyer *[1990, T.2, Chapter VII]). A vaguelette family is by definition a set of functions* $\{g_{j,k}\}_{j\in\mathbb{Z},k\in\mathbb{Z}^d}$ *such that* $\int g_{j,k} = 0$ *and*

$$|g_{j,k}| + 2^{-j}\sum_{m=1}^{d} |\partial_{x_m} g_{j,k}| \leq 2^{dj/2} g(2^j \cdot -k), \tag{3.8.26}$$

where g *is a positive function such that* $g(x) \leq C(1+|x|)^{-(d+\varepsilon)/2}$ *for some* $\varepsilon > 0$. *Then the vaguelette lemma states that for all* $f \in L^2$, *one has*

$$\sum_{j\in\mathbb{Z}}\sum_{k\in\mathbb{Z}^d} |\langle f, g_{j,k}\rangle|^2 \lesssim \|f\|_{L^2}^2. \tag{3.8.27}$$

Such a result immediately yields one side of the equivalence (3.8.15) for both primal and dual bases, and the other side is then easily proved by duality arguments. One disadvantage of the vaguelette approach is that it requires a bit more than the minimal smoothness assumptions $\varphi, \tilde{\varphi} \in H^\varepsilon$ *(the differentiability of* $g_{j,k}$ *can still be lowered down to a Hölder regularity assumption).*

From the definition of Besov spaces by (3.2.19) using the Littlewood-Paley decomposition, we see that L^2 identifies with $B^0_{2,2}$. The norm equivalences (3.8.15) and (3.8.18) are thus the natural extension to the case $s = 0$

of the characterization of Besov spaces that we have obtained so far. More generally, for $1 \leq p \leq \infty$, we can use the elementary interpolation properties (3.5.45) of the weighted sequence spaces ℓ^p_s, to obtain the norm equivalence

$$\|f\|_{B^0_{p,q}} \sim \|\left(2^{d(1/2-1/p)j}\|(c_\lambda)_{\lambda\in\nabla_j}\|_{\ell^p}\right)_{j\geq -1}\|_{\ell^q}, \tag{3.8.28}$$

provided that the wavelet basis allows us to characterize B^{ε}_{p,q_1} and $B^{-\varepsilon}_{p,q_2}$ for some $\varepsilon > 0$ and $q_1, q_2 > 0$. However, we cannot identify L^p with $B^0_{p,q}$ for any $q > 0$ if $p \neq 2$. This reflects the well known fact that the L^p spaces (and more generally the $H^{s,p}$ spaces) do not belong to the scale of Besov spaces when $p \neq 2$.

Concerning such L^p spaces, we can formulate two basic questions: does the wavelet expansion of an arbitrary function $f \in L^p$ converge unconditionally in L^p ? Is there a simple characterization of L^p by the size of the wavelets coefficients ?

The answer to these questions is clearly negative for L^∞ since this space is non-separable. It is also negative for L^1 which is known to possess no unconditional basis.

For the case $1 < p < \infty$, a positive answer to the first question is provided by the "real value" theory, developed since the 1950's by Calderon and Zygmund in order to study the continuity properties of operators. By definition, a Calderon-Zygmund operator is an L^2-bounded operator T with the integral form

$$Tf(x) = \int_{\mathbb{R}^d} K(x,y)dxdy, \tag{3.8.29}$$

where the kernel function K satisfies decay estimates of the type

$$|K(x,y)| \lesssim |x-y|^{-d}, \tag{3.8.30}$$

and

$$\sum_{m=1}^{d} [|\partial_{x_m} K(x,y)| + |\partial_{y_m} K(x,y)|] \lesssim |x-y|^{-d-1}, \tag{3.8.31}$$

away from the diagonal, i.e. for $x \neq y$.

One of the most famous results in the study of such operators states that T is also bounded in L^p for $1 < p < \infty$, with a norm that only depends (for fixed p) on its L^2 norm, p and the constants in the estimates (3.8.30) and (3.8.31). A proof of this result can be found in MEYER [1990, chapter VII] and DAUBECHIES [1992, chapter 9] (see also STEIN [1970] for a general introduction on singular integral operators).

Returning to the L^p-unconditionality of wavelet bases, we can formulate this property in terms of operators: it holds if and only if the projectors

$$T_\Gamma f = \sum_{\lambda\in\Gamma} \langle f, \tilde{\psi}_\lambda\rangle \psi_\lambda, \tag{3.8.32}$$

are uniformly bounded in L^p independently of the finite or infinite subset $\Gamma \subset \nabla$. More generally, with any sequence $(a_\lambda)_{\lambda\in\nabla}$ such that $|a_\lambda| \le 1$, we associate an operator T_a defined by

$$T_a(\sum_{\lambda\in\nabla} c_\lambda\psi_\lambda) = \sum_{\lambda\in\nabla} a_\lambda c_\lambda \psi_\lambda. \tag{3.8.33}$$

Clearly T_Γ corresponds to the choice $a_\lambda = 1$ if $\lambda \in \Gamma$, 0 otherwise. Assuming that φ and $\tilde{\varphi}$ have C^1 smoothness, we then remark that the associated kernels

$$K_a(x,y) = \sum_{\lambda\in\nabla} a_\lambda \psi_\lambda(x)\tilde{\psi}_\lambda(y), \tag{3.8.34}$$

satisfy the estimates (3.8.30) and (3.8.31) with constants that are independent of the eigenvalue sequence a_λ. Indeed, we can estimate

$$\begin{aligned} |K_a(x,y)| &\le \textstyle\sum_{\lambda\in\nabla} |\psi_\lambda(x)\tilde{\psi}_\lambda(y)| \\ &= \textstyle\sum\sum_{j\ge -1}\sum_{\lambda\in\nabla_j} |\psi_\lambda(x)\tilde{\psi}_\lambda(y)|, \end{aligned}$$

by remarking that due to the support size of the basis function, the non-trivial contributions in the last sum are limited to $j \le J$ if $|x-y| \ge C2^{-J}$ for some fixed constant C. Due to the controlled overlapping property (3.1.15) at each level, we also have

$$\sum_{\lambda\in\nabla_j} |\psi_\lambda(x)\tilde{\psi}_\lambda(y)| \lesssim 2^{d|\lambda|}, \tag{3.8.35}$$

and the estimate (3.8.30) follows. A similar technique is used to proved (3.8.31), remarking that

$$\sum_{m=1}^{d}\sum_{\lambda\in\nabla_j} [|\partial_{x_m}\psi_\lambda(x)\tilde{\psi}_\lambda(y)| + |\psi_\lambda(x)\partial_{y_m}\tilde{\psi}_\lambda(y)|] \lesssim 2^{(d+1)|\lambda|}. \tag{3.8.36}$$

Assuming that $(\psi_\lambda)_{\lambda\in\nabla}$ and $(\tilde{\psi}_\lambda)_{\lambda\in\nabla}$ are biorthogonal Riesz bases for L^2, we also clearly have that the T_a are uniformly bounded in L^2. We can thus derive from the Calderon-Zygmund theory that these operators are uniformly L^p bounded for $1 < p < \infty$, i.e. the wavelet bases are also L^p-unconditional.

It should be noted that the assumption (3.8.31) can be weakened in such a way that the same conclusion holds with φ and $\tilde{\varphi}$ only in C^ε for some $\varepsilon > 0$ (see e.g. MEYER [1990, Chapter VII]). We summarize these results below.

Theorem 3.8.3 *If the assumptions of Theorem 3.8.2 are satisfied and if the generators of the primal and dual wavelet bases are both in C^ε for some $\varepsilon > 0$, then these bases are also unconditional for L^p, $1 < p < \infty$.*

Remark 3.8.4 *In our case of interest, the L^2 boundedness of the operators T_a is a straightforward consequence of the L^2-stability of the multiscale bases. More generally, if $K(x,y)$ only satifies (3.8.30) and (3.8.31), the possible non-integrability in each variable means that the L^2-boundedness of T should stem from certain cancellation properties in the Kernel K. Such requirements were made more precise by the celebrated $T(1)$ theorem of* DAVID *and* JOURNÉ *[1984], which states that the image of the constant function by T and T^* should belong to the space BMO.*

The characterization of L^p norms from the size properties of the wavelet coefficients is also possible for $1 < p < \infty$, by means of a *square function*, defined for $f = \sum_{\lambda\in\nabla} c_\lambda\psi_\lambda$ by

$$Sf(x) = [\sum_{\lambda\in\nabla} |c_\lambda|^2 |\psi_\lambda(x)|^2]^{1/2}. \tag{3.8.37}$$

Clearly, we have $\|f\|_{L^2} \sim \|Sf\|_{L^2}$. A remarkable result is that we also have

$$\|f\|_{L^p} \sim \|Sf\|_{L^p}, \tag{3.8.38}$$

for $1 < p < \infty$. The proof of such a result relies on the classical Khinchine inequality (see e.g. STEIN [1970, p.277]): if $(\varepsilon_\lambda)_{\lambda\in\nabla}$ is a sequence of independent random variables, with law $(\delta_{-1} + \delta_1)/2$ (i.e. 1 or -1 with equal probability), then one has the equivalence between the expectations

$$\Big(E(|\sum_{\lambda\in\nabla} \varepsilon_\lambda x_\lambda|^p)\Big)^{1/p} \sim \Big(E(|\sum_{\lambda\in\nabla} \varepsilon_\lambda x_\lambda|^q)\Big)^{1/q}, \tag{3.8.39}$$

for any fixed $p, q \in]0, \infty[$. For $q = 2$, the orthogonality of the ε_λ yields

$$\Big(E(|\sum_{\lambda\in\nabla} \varepsilon_\lambda x_\lambda|^p)\Big)^{1/p} \sim \Big(\sum_{\lambda\in\nabla} |x_\lambda|^2\Big)^{1/2}. \tag{3.8.40}$$

Applying (3.8.40) to $x_\lambda = c_\lambda\psi_\lambda(x)$ and integrating in x after taking the p-th power gives us

$$\|Sf\|_{L^p} \lesssim E(\|T_\varepsilon f\|_{L^p}), \tag{3.8.41}$$

where T_ε is defined by (3.8.33). Since $\|T_\varepsilon f\|_{L^p} \lesssim \|f\|_{L^p}$, we immediately obtain one side of the equivalence (3.8.37). The other side follows by remarking that since $T_\varepsilon^2 = I$, we also have $\|f\|_{L^p} \lesssim \|T_\varepsilon f\|_{L^p}$.

Remark 3.8.5 *Similar characterizations are available for general potential spaces $H^{t,p}$ and in particular for $W^{m,p} = H^{m,p}$ if m is an integer, under additional smoothness assumptions on φ. They rely on the modified square function*

$$S_t f(x) = [\sum_{\lambda\in\nabla} 4^{t|\lambda|}|c_\lambda|^2|\psi_\lambda(x)|^2]^{1/2}. \quad (3.8.42)$$

and have the form

$$\|f\|_{H^{t,p}} \lesssim \|S_t f\|_{L^p}. \quad (3.8.43)$$

The proof of such a result can be found in MEYER *[1990, p.165]. The $H^{s,p}$ spaces are members of the general family of Triebel-Lizorkin spaces $F^s_{p,q}$ which are also defined by Littlewood-Paley decompositions: $f \in F^s_{p,q}$ if and only if $S^q_t f(x) = (|S_0(x)|^q + \sum_{j\geq 0}|2^{sj}\Delta_j f(x)|^q)^{1/q}$ belongs to L^p. One has in particular $H^{s,p} = F^s_{p,2}$ (see* TRIEBEL *[1983]).*

3.9 Bounded domains

The goal of this section is to explain how the results obtained so far for function spaces on $\mathbb{R}^r$ in the previous sections can be adapted to the framework of bounded domains: we are interested in characterizing function spaces related to a bounded domain $\Omega \subset \mathbb{R}^d$ in terms of their multiscale decomposition.

There are two different approaches to the definition of function spaces $X(\Omega)$ on bounded domains: one can either view them as the *restrictions* of the corresponding function spaces $X(\mathbb{R}^d)$ defined on the whole of $\mathbb{R}^d$, or search for an appropriate *inner description* of $X(\Omega)$. In §3.2 we have given inner descriptions for Hölder, Sobolev and Besov spaces on domains, and we shall continue here with this approach.

Remark 3.9.1 *For fairly general domains, it is well known that the two approaches yield the same spaces for a wide of smoothness classes. A classical instance is the case of the Sobolev spaces $W^{m,p}$, for $1 \leq p \leq \infty$ and $m \in \mathbb{N}$. Clearly, the restriction to Ω of $f \in W^{m,p}(\mathbb{R}^d)$ belongs to $W^{m,p}(\Omega)$ defined as in §3.2. Conversly, if Ω is a Lipschitz domain, it is known from* STEIN *[1970, p.181], that any function $f \in W^{m,p}(\Omega)$ can be extended into $\mathcal{E}f \in W^{m,p}(\mathbb{R}^d)$ by a bounded linear extension operator $\mathcal{E}$. As a consequence, we find an analogous result for Besov spaces: since $B^s_{p,q}(\Omega)$ can* *be identified with $[L^p, W^{m,p}]_{\theta,q}$, $\theta = s/m$ for any Lipschitz domain Ω as for $\mathbb{R}^d$ (by the result of* JOHNEN *and* SCHERRER *[1976] that we proved in Theorem 3.6.2 in the special case $\Omega = \mathbb{R}^d$), the extension operator E is also bounded from $B^s_{p,q}(\Omega)$ to $B^s_{p,q}(\mathbb{R}^d)$ by interpolation. Extension results for Besov spaces with $p < 1$ on general Lipschitz domains have also been obtained by* DEVORE *and* SHARPLEY *[1993]. As it will be remarked later, the subsequent results will provide ad-hoc concrete extension operators for such spaces.*

Of interest to us is also the case of function spaces $X_b(\Omega)$ with prescribed boundary conditions, which are usually defined by the closure in $X(\Omega)$ of a set of smooth functions which vanish up to some order at $\Gamma := \partial\Omega$. Their introduction is natural for the study of boundary value problems, as well as for the proper definition of negative smoothness by duality. The adaptation of our results to such spaces will be the object of the next section.

In §2.12 and §2.13 of the previous chapter, we have described the two main approaches to multiscale decompositions and wavelet bases on bounded domains: a domain decomposition strategy, allowing us to exploit the simpler construction on the interval, and a strategy based on the multiscale decomposition of finite element spaces.

Both approaches have similar features to the tensor product construction on $\mathbb{R}^d$: local bases, polynomial reproduction up to some prescribed degree, prescribed smoothness properties for the basis functions. This will often allow us to adapt in a straightforward way most of the results that we have stated for $\Omega = \mathbb{R}^d$, at the cost of heavier notations. For this reason, we shall only elaborate below when the adaptation is less trivial.

At this stage, we fix some general assumptions on our domain: Ω should have a simple geometry, expressed by a conformal partition

$$\Omega = \cup_{i=1}^n S_i, \tag{3.9.1}$$

into simplicial subdomains: each S_i is the image of the unit simplex

$$S := \{0 \le x_1 + \cdots + x_d \le 1\}, \tag{3.9.2}$$

by an affine transformation. By "conformal" we mean that a face of an S_i is either part of the boundary Γ or coincides with a face of another S_j. We also assume that Ω is connected in the sense that for all $j, l \in \{1, \cdots, n\}$, there exists a sequence $i_0, \cdots, i_m$ such that $i_0 = j$ and $i_m = l$ and such that S_{i_k} and $S_{i_{k+1}}$ have a common face. Clearly, polygons and polyhedrons fall into this category. More general curved domains or manifolds are also included here, provided that they can be smoothly parametrized by such simple reference domains, as explained in §2.12 and §2.13 of the previous chapter.

For the characterization of Besov spaces $B^s_{p,q}(\Omega)$ with $s > 0$, we directly place ourselves in the general setting of §3.7 where we allow p to be less than 1. Our goal is to show here that all the statements in that section stay valid in the present setting of bounded domains.

Recall that the basic ingredient for proving the results of §3.7 was the covering of $\mathbb{R}^d$ by the dyadic cubes $I_{j,k}$, $\tilde{I}_{j,k}$, $J_{j,k}$ and $\tilde{J}_{j,k}$ for $k \in \mathbb{Z}^d$ (the cubes $I_{j,k}$, $\tilde{I}_{j,k}$ were also used for the direct estimates of §3.3). Also recall (Remark 3.7.1) that we could actually have taken $J_{j,k} = I_{j,k}$.

The adaptation of these results to Ω thus relies on the ability to cover Ω in an analogous manner. More precisely, at each scale j, we need a covering of the type

$$\Omega = \cup_{\mu \in \Theta_j} I_\mu, \tag{3.9.3}$$

with $\operatorname{diam}(I_\mu) \lesssim 2^{-|\mu|}$ (here we also use the convention $|\mu| = j$ if $\mu \in \Theta_j$). The domains I_μ will play the same role as the $I_{j,k}$ and $J_{j,k}$ cubes in the $\mathbb{R}^d$ case. With this covering, we associate $\tilde{I}_\mu$ and $\tilde{J}_\mu$ defined if $|\mu| = j$ by

$$\tilde{I}_\mu = \cup_{\lambda \in F_\mu} \operatorname{Supp}(\tilde{\varphi}_\lambda), \tag{3.9.4}$$

and

$$\tilde{J}_\mu = \cup_{\lambda \in F_\mu} \operatorname{Supp}(\varphi_\lambda), \tag{3.9.5}$$

with

$$F_\mu = \{\lambda \in \Gamma_j \ ; \ |\operatorname{Supp}(\varphi_\lambda) \cap I_\mu| \neq 0\}. \tag{3.9.6}$$

These larger subdomains also have diameter of order $2^{-|\mu|}$. They are intended to play the same role as the cubes $\tilde{I}_{j,k}$ and $\tilde{J}_{j,k}$ which were used in the proof of Theorems 3.7.3 and 3.7.6. In the following, we shall analyze the properties that need to be fulfilled by these coverings in order to adapt the proofs of the Theorems 3.7.1 to 3.7.7 to Ω, and we shall explain how to design them.

In order to turn local estimates into global estimates, these domains should not overlap too much in the sense that there exists a constant C such that for all $x \in \Omega$,

$$\#\{\mu \in \Theta_j \ ; \ x \in I_\mu\} \leq C. \tag{3.9.7}$$

Clearly, (3.9.7) implies the same property for the families $\tilde{I}_\mu$ and $\tilde{J}_\mu$, $\mu \in \Theta_j$, due to the local supports of the basis functions.

In order to prove Theorems 3.7.1 and 3.7.2 the stability estimate

$$\|\sum_{\lambda \in \Gamma_j} c_\lambda \varphi_\lambda\|_{L^p(I_\mu)} \sim \|(c_\lambda)_{\lambda \in F_\mu}\|_{\ell^p}, \quad \mu \in \Theta_j, \tag{3.9.8}$$

should hold with a constant that does not depend on μ.

The main headache is to select a covering $(I_\mu)_{\mu\in\Theta_j}$ that satisfies this last prescription, i.e. uniform stability (3.9.8). This is feasible in a way that depends on the type of multiscale decomposition which is used for Ω.

Domain decomposition approach: remark first that for $\Omega = [0,1]^d$, we obtain a covering that satisfies the third prescription by simply taking those $I_{j,k} = 2^{-j}(k + [0,1]^d)$, for all $k \in \{0, \cdots, 2^j - 1\}^d$, i.e. such that $I_{j,k} \subset [0,1]^d$. Indeed, we have already proved by (3.7.3) the uniform stability (3.7.5) for the $I_{j,k}$ in the interior of $[0,1]^d$ which only meet the supports of unmodified interior scaling functions. For $I_{j,k}$ near the boundary meeting the support of a modified scaling function, we still have local linear independance of these basis functions on $I_{j,k}$. To check this, we can reduce to the univariate case, i.e. $\Omega = [0,1]$, by the same reasoning as in the proof of Theorem 3.7.1 (induction on the dimension d of the cube). Then, local linear independence near the edges is easily checked, recalling that the edge functions $\varphi^e_{j,q}$ are obtained as linear combinations of the locally independent scaling functions $\varphi_{j,k}$ with Vandermonde type coefficients. The uniformity of the constant in (3.9.8) is then ensured by change of scale. For a more general Ω, recall (see §2.12 of Chapter 2) that Ω is patched into subdomains Ω_i, $i = 0, \cdots, n$ which are the image of the reference cube $[0,1]^d$ by smooth isoparametric maps T_i. The basis functions of $V_j^{\Omega_i}$ are the images of the basis functions of $V_j^{[0,1]^d}$ by the map T_i and V_j^Ω is defined by the direct sum of $V_j^{\Omega_i}$ intersected with $C^0(\Omega)$. A natural choice for the covering of Ω is then provided by

$$\{I_\mu\}_{\mu\in\Theta_j} = \bigcup_{i=1}^{n} \{T_i I_{j,k} \; ; \; k \in \{0, \cdots, 2^j - m\}^d\}. \tag{3.9.9}$$

With such a choice, the uniform stability (3.9.8) appears as a simple consequence of the same property on the unit cube.

Finite element approach: in this setting, as explained in §2.13 of Chapter 2, one starts form a coarse decomposition $\mathcal{T}_0$ of the type (3.9.1) and obtains a hierarchy of decompositions $\mathcal{T}_j$, $j \geq 0$, by iterative uniform subdivision of the simplices S_j using the mid-point rule. The V_j spaces are then simplicial finite element spaces associated with the decomposition $\mathcal{T}_j$. In this setting, local linear independence of the nodal basis functions is well known to hold on each simplex of $\mathcal{T}_j$, and the stability property (3.9.8) holds uniformly by rescaling. We can thus simply take for $\{I_\mu\}_{\mu\in\Theta_j}$ the set of simplices $\mathcal{T}_j$ (thus a covering by disjoint sets).

With such a covering, we can thus adapt the proof of Theorem 3.7.1 and obtain a global stability estimate

$$\|\sum_{\lambda\in\Gamma_j} c_\lambda\varphi_\lambda\|_{L^p} \sim 2^{jd(1/2-1/p)}\|(c_\lambda)_{\lambda\in\Gamma_j}\|_{\ell^p}. \tag{3.9.10}$$

Theorem 3.7.2 is then immediately adapted with an unchanged statement. However, it should be remarked that in the case of parametrized curved domains, the smoothness of the map enters in as a limitation of the range of s in the inverse estimate. For example, in the domain decomposition approach, if we only impose C^0 continuity on the interface, we find the limitation $s < 1 + 1/p$: even if the initial generator φ has high smoothness, the functions φ_λ sitting on the interfaces between the subdomains are not in $B^s_{p,q}$ for $s > 1 + 1/p$.

Remark 3.9.2 *In the case where $p \geq 1$, one does not need to work so hard on the design of the I_μ: using a dual scaling function basis $\tilde{\varphi}_\lambda$, $\lambda \in \nabla_j$ in L^∞, one can directly prove the stability property, as well as the direct estimate, in the same way as it is done in §3.3.*

In order to prove Theorem 3.7.3, we need additional properties for the domains $\widetilde{J}_\mu$: on the one hand, a local Whitney estimate

$$\inf_{g\in\Pi_n} \|f-g\|_{L^p(\widetilde{J}_\mu)} \lesssim \omega_{n+1}(f, 2^{-|\mu|}, \widetilde{J}_\mu)_p, \tag{3.9.11}$$

should hold with a constant that does not depend on μ. On the other hand, in order to turn the local error estimates into global estimates, we need to use the modified modulus of smoothness $\tilde{\omega}(f, 2^{-|\mu|}, \widetilde{J}_\mu)_p$ defined by (3.3.15), and thus we should have

$$\tilde{\omega}(f, 2^{-|\mu|}, \widetilde{J}_\mu)_p \lesssim \omega(f, 2^{-|\mu|}, \widetilde{J}_\mu)_p, \tag{3.9.12}$$

with a constant that does not depend on μ.

One of the problems here is that the domains $\widetilde{J}_\mu$ are not cubes, for which these estimates are known to hold according to (3.2.26) and (3.3.17). To our knowledge, there is no proof of such results for domains of general shape. Note however that in both domain decomposition and finite element approaches, the domains $\widetilde{J}_\mu$ are the images of a reference domain within a finite set $\{\widetilde{J}_1, \cdots, \widetilde{J}_N\}$ by affine transformations T_μ which are the product of an orthogonal transform with an isotropic scaling by $2^{-|\mu|}$. Such transformations leave invariant the constants in (3.9.11) and (3.9.12), which only depend on the shape of the domain, so that we are reduced to proving (3.2.26) and (3.3.17) for the reference domains $\{\widetilde{J}_1, \cdots, \widetilde{J}_N\}$.

Once again, these reference domains are not cubes, but have a simple geometry as Ω, i.e. expressed by a partition of the type (3.9.1). Our work can thus be reduced to proving (3.2.26) and (3.2.17) for a domain Ω of the type (3.9.1).

A simple method of proof is the following: we first remark that since (3.2.26) and (3.2.17) hold for cubes, by a change of variable it follows that these estimates also hold on any domain of the form $T[0,1]^d$ where T is an invertible affine transformation. The constants are of course affected by the properties of T. We next remark that for a domain Ω of the type (3.9.1), we can easily find invertible affine transformations $T_1, \cdots, T_m$ such that

$$\Omega = \cup_{j=1}^m R_j, \quad R_j := T_j[0,1]^d, \tag{3.9.13}$$

and such that $|R_j \cap R_{j+1}| \neq 0$. An example of such an overlapping decomposition is shown in Figure 3.9.1 in the case of a polygon.

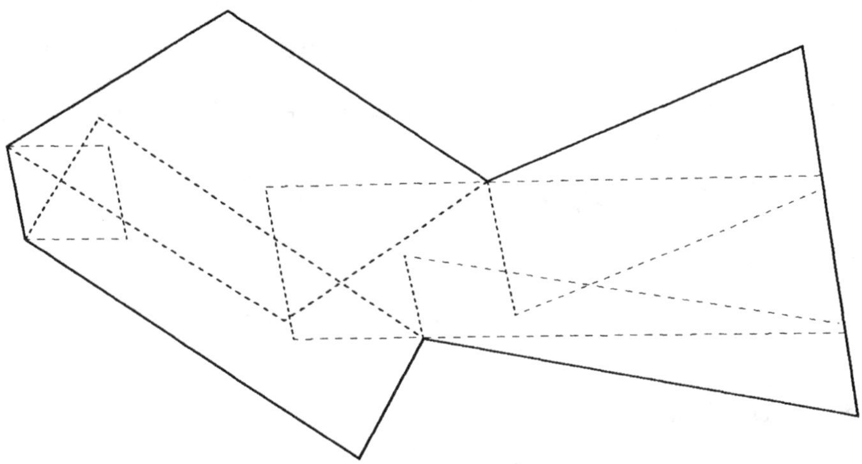

Figure 3.9.1: Overlapping partition of Ω by the simpler domains R_i

In this case, the overlapping is crucial in order to turn the estimates (3.2.26) and (3.2.17) which hold on each R_j into similar estimates for the whole of Ω.

Theorem 3.9.1 *If Ω is of the type (3.9.13), we have*

$$\inf_{g \in \Pi_{n-1}} \|f - g\|_{L^p(\Omega)} \lesssim \omega_n(f, 1, \Omega)_p, \tag{3.9.14}$$

and

$$\omega_n(f,1,\Omega)_p \lesssim \tilde{\omega}_n(f,1,\Omega)_p. \tag{3.9.15}$$

Proof On each R_i, there exists a polynomial $p_i \in \Pi_{n-1}$ such that

$$\|f - p_i\|_{L^p(R_i)} \lesssim \omega_n(f,1,R_i)_p. \tag{3.9.16}$$

Considering first $i = 1, 2$, we then have

$$\|p_2 - p_1\|_{L^p(R_1 \cap R_2)} \lesssim \omega_n(f,1,R_1)_p + \omega_n(f,1,R_2)_p \lesssim \omega_n(f,1,R_1 \cup R_2)_p,$$

and thus

$$\|p_2 - p_1\|_{L^p(R_1 \cup R_2)} \lesssim \omega_n(f,1,R_1 \cup R_2)_p, \tag{3.9.17}$$

from the equivalence of all quasi-norms in finite dimensions. It follows that

$$\|f - p_2\|_{L^p(R_1 \cup R_2)} \lesssim \omega_n(f,1,R_1 \cup R_2)_p. \tag{3.9.18}$$

By iteration of this reasoning, we end up with

$$\|f - p_m\|_{L^p(\Omega)} \lesssim \omega_n(f,1,\Omega)_p, \tag{3.9.19}$$

i.e. the desired estimate (3.9.14). In order to prove (3.9.15), we remark that if $|A \cap B| \neq 0$, we have

$$\omega_n(f,1,A \cup B)_p \lesssim \omega_n(f,1,A)_p + \omega_n(f,1,B)_p, \tag{3.9.20}$$

so that

$$\omega_n(f,1,\Omega)_p \lesssim \sum_{i=1}^{m} \omega_n(f,1,R_i)_p \lesssim \sum_{i=1}^{m} \tilde{\omega}_n(f,1,R_i)_p. \tag{3.9.21}$$

Since we clearly have

$$\tilde{\omega}_n(f,1,A)_p + \tilde{\omega}_n(f,1,B)_p \lesssim \tilde{\omega}_n(f,1,A \cup B)_p, \tag{3.9.22}$$

this concludes the proof. ◇

From this result, we conclude that the statements of Theorems 3.7.3 and 3.7.4 also hold for multiscale decompositions on Ω. Note that this also implies the interpolation results stated in Remark 3.7.1, and that the argument that we have used in the proof of Theorem 3.7.5(based on multiresolution approximation on the cube $[0,1]^d$) is now justified.

The next step is to observe that the estimates (3.7.22) and (3.7.23) of Theorem 3.7.5 also hold for the domains Ω of the type (3.9.1) in place of a simple cube, with a constant that only depends on the shape of Ω. This allows us to mimic the proof of Theorem 3.7.6, with the I_μ in place of the $I_{j,k}$ and $\widetilde{I}_m u$ in place of $\widetilde{I}_{j,k}$. Here again, we use that the $\widetilde{I}_\mu$ are rescaled versions of simple reference domains $\widetilde{I}_1, \cdots, \widetilde{I}_N$ in order to have a local estimate of $\|f - P_j f\|_{L^r(\widetilde{I}_\mu)}$ and to apply (3.7.22) on $\widetilde{I}_\mu$.

The proof of the first statement (3.7.31) in Theorem 3.7.7 is then a straightforward adaptation. For the second statement, i.e. the wavelet characterization by (3.7.32), we need an analog of the uniform local stability property (3.9.10) with wavelets in place of scaling functions.

Theorem 3.9.2 *If (3.9.10) holds, then we also have*

$$\|\sum_{\lambda\in\nabla_j} c_\lambda \psi_\lambda\|_{L^p} \sim 2^{jd(1/2-1/p)} \|(c_\lambda)_{\lambda\in\nabla_j}\|_{\ell^p}, \tag{3.9.23}$$

i.e. the L^p-stability of the wavelet basis $\{\psi_\lambda\}_{\lambda\in\nabla_j}$ independently of the level j.

Proof We remark that we can express a combination of wavelet coefficients at scale j as a combination of scaling functions at scale $j+1$, i.e.

$$\sum_{\lambda\in\nabla_j} c_\lambda \psi_\lambda = \sum_{\gamma\in\Gamma_{j+1}} d_\gamma \varphi_\gamma. \tag{3.9.24}$$

We thus have, according to (3.9.10),

$$\|\sum_{\lambda\in\nabla_j} c_\lambda \psi_\lambda\|_{L^p} \sim 2^{jd(1/2-1/p)} \|(d_\gamma)_{\gamma\in\Gamma_{j+1}}\|_{\ell^p}. \tag{3.9.25}$$

The c_λ and d_γ are linked by local discrete two scale relations. From the fine to coarse decomposition

$$c_\lambda = \sum_{\gamma\in\Gamma_{j+1}} \langle \varphi_\gamma, \tilde{\psi}_\lambda\rangle d_\gamma, \tag{3.9.26}$$

it follows that $\|(c_\lambda)_{\lambda\in\nabla_j}\|_{\ell^p} \lesssim \|(d_\gamma)_{\gamma\in\Gamma_{j+1}}\|_{\ell^p}$, while from the coarse to fine reconstruction

$$d_\gamma = \sum_{\lambda\in\nabla_j} \langle \psi_\lambda, \tilde{\varphi}_\gamma\rangle c_\lambda, \tag{3.9.27}$$

it follows that $\|(d_\gamma)_{\gamma\in\Gamma_{j+1}}\|_{\ell^p} \lesssim \|(c_\lambda)_{\lambda\in\nabla_j}\|_{\ell^p}$. We thus obtain here the stability of the wavelet basis as a consequence of the stability of the scaling function basis. ◇

Remark 3.9.3 *In both domain decomposition and finite element approaches, the scaling functions that are not supported in the interior of Ω have natural extensions to $\mathbb{R}^d$. The simplest example is the case of $I = [0,1]$, for which the extension is obtained by removing from the sums (2.12.17) and (2.12.18) the scaling functions such that $|\mathrm{Supp}(\varphi_{j,k}) \cap \Omega| = 0$ and viewing the resulting combination as a function defined on $\mathbb{R}$. This allows us to build operators $\mathcal{E}_j$ that extend the functions $f \in V_j$ up to a distance of order 2^{-j} from the boundary, by simply keeping the same coefficients with the extended basis. Such an extension at level j satisfies two useful properties: firstly, if the scaling function basis is L^p-stable in the sense of Theorem 3.7.1, then $\|\mathcal{E}_j f\|_{L^p} \lesssim \|f\|_{L^p}$. Secondly, under the assumptions of Theorem 3.7.2, we also have the inverse estimate (3.7.10) and (3.7.11) with $\mathcal{E}_j f$ on the left and f on the right. From this, one can easily check that*

$$\mathcal{E} f = \mathcal{E}_0 P_0 f + \sum_{j\geq 0} \mathcal{E}_{j+1} Q_j f, \tag{3.9.28}$$

defines a bounded extension operator from $B^s_{p,q}(\Omega)$ to $B^s_{p,q}(\mathbb{R}^d)$, provided that the first space is characterized by the wavelet decomposition according to (3.7.31). In the case where $0 < s \leq d/p - d$ (which is not covered by Theorem 3.7.7) it suffices to replace $P_0 f$ by f_0 and $Q_j f$ by $f_{j+1} - f_j$ in (3.9.28), where f_j is a near-best approximation to f in V_j (we then use the characterization of $B^s_{p,q}$ by Theorem 3.7.4).

3.10 Boundary conditions

Before addressing the adaptation of the results in §3.8 (negative smoothness and L^p-spaces) to bounded domains, we need to consider the case of Sobolev and Besov spaces with prescribed homogeneous boundary conditions on $\Gamma := \partial\Omega$. Here, we maintain the same geometrical assumptions on Ω as in the previous section. For the sake of simplicity, we shall only consider $B^s_{p,q}$ and $W^{s,p}$ with $p \geq 1$. The case $p < 1$ can be treated by a similar approach, up to some technical modifications, as explained in Remark 3.10.6.

We denote by $C^\infty(\Gamma, m)$ the space of smooth functions defined on Ω which vanish at order m on Γ. This means that $f \in C^\infty(\Gamma, m)$ if and only if $f \in C^\infty(\Omega)$ and $|f(x)| \leq C[\mathrm{dist}(x, \Gamma)]^{m+1}$. We then define the spaces $W^{s,p}(\Gamma, m)$ (resp. $B^s_{p,q}(\Gamma, m)$) as the closure of $C^\infty(\Gamma, m)$ in $W^{s,p}$ (resp. $B^s_{p,q}$). We recall below some basic facts which are are well-known for Sobolev spaces (see e.g. GRISVARD [1983, chapter 1]), and are also easy to prove for Besov spaces.

By truncation and regularization near the boundary, one can prove that

$C^\infty(\Gamma, m)$ is dense in $B^s_{p,q}$ and $W^{s,p}$ for all $m \geq 0$ if $s < 1/p$, i.e.

$$W^{s,p}(\Gamma, m) = W^{s,p} \text{ and } B^s_{p,q}(\Gamma, m) = B^s_{p,q}, \quad s < 1/p. \tag{3.10.1}$$

For values of s above this range, it should also be noticed that for m_0 and m_1 strictly larger than $s - 1/p$, we have

$$W^{s,p}(\Gamma, m_0) = W^{s,p}(\Gamma, m_1) \text{ and } B^s_{p,q}(\Gamma, m_0) = B^s_{p,q}(\Gamma, m_1). \tag{3.10.2}$$

For such values of the flatness parameter m, these spaces can also be obtained as the closure of the space of test functions $\mathcal{D}(\Omega)$. These spaces are classically denoted by $W^{s,p}_0$ or H^s_0. Thus, if $n + 1/p < s \leq n + 1/p + 1$, we should only make the distinction between the spaces $W^{s,p}(\Gamma, m)$ for $m = 0, \cdots, n$ and similarly for Besov spaces.

There is an alternative definition of such spaces based on traces. If s is strictly larger than $n + 1/p$ and $f \in W^{s,p}$, the traces $\gamma_m f$ of the normal derivatives of f up to order $m = n$ are $W^{s-1/p-m,p}$ functions on the smooth regions of Γ (i.e. on the faces of the S_i in (3.9.1) that are contained in Γ). Then, for $m \leq n$, we have

$$f \in W^{s,p}(\Gamma, m) \Leftrightarrow f \in W^{s,p} \text{ and } \gamma_k f = 0, \quad k = 0, \cdots, m. \tag{3.10.3}$$

The same holds with $B^s_{p,q}$ in place of $W^{s,p}$. From this, it follows that

$$W^{s,p}(\Gamma, m) = W^{s,p} \cap X(\Gamma, m) \text{ and } B^s_{p,q}(\Gamma, m) = B^s_{p,q} \cap X(\Gamma, m), \tag{3.10.4}$$

for $X = W^{t,p}$ or $B^t_{p,r}$, with t such that $m + 1/p < t < s$. For example, we have $H^2(\Gamma, 0) = H^2 \cap H^1(\Gamma, 0) = H^2 \cap H^1_0$, the space of H^2 functions with homogeneous Dirichlet boundary conditions.

Remark 3.10.1 *More generally, if B is a closed subset of Γ, we can in a similar way define the spaces $B^s_{p,q}(B, m)$ and $W^{s,p}(B, m)$ as the closure of smooth functions with flatness m near B. We obtain similar properties as the above when B has a non-zero $d - 1$ dimensional measure. One can also define spaces with mixed boundary conditions involving a linear combination of several normal derivatives, or with different boundary conditions on different regions of Γ. For the sake of simplicity, our subsequent development only deals with the characterization of the spaces $B^s_{p,q}(\Gamma, m)$ with the help of multiresolution spaces that are adapted to such boundary conditions. It can be extended to more general boundary conditions provided that one can build the corresponding multiresolution spaces.*

In order to characterize the $W^{s,p}(\Omega, m)$ and $B^s_{p,q}(\Omega, m)$ in the same way as we did for function spaces without boundary conditions, we need that the functions in V_j vanish with the proper order near the boundary, while preserving some notion of polynomial exactness. This polynomial exactness can no longer be global, since for general domains, only the trivial polynomial vanishes on $\partial\Omega$. A natural intuition is that the characterization is feasible if all polynomials are locally reproduced up to some degree $n-1$ such that $n > s$ by the scaling functions that sit in the interior of the domain, while next to the boundary are locally reproduced only those polynomials of degree $n-1$ which vanish up to order m near $\partial\Omega$. We shall provide here a precise result which confirms this intuition.

The construction of multiscale approximations with such modified polynomial reproduction properties near the boundary is a classical task in the finite element setting (for example $m = 0$ with the $\mathbb{P}_n$ Lagrange finite elements, one simply imposes zero values at the nodes situated on $\partial\Omega$). In the domain decomposition setting, it amounts to imposing such modifications at certain faces of the cube $[0,1]^d$ (which are related to the boundary $\partial\Omega$ by the isoparametric maps). By tensor product, this reduces to a similar construction at the edge 0 and/or 1 of the interval $[0,1]$, which is simply obtained by removing the functions $\varphi^{\varepsilon}_{j,q}$, $q = 0,\cdots,m$, $\varepsilon = 0,1$ from the generators of $V_j^{[0,1]}$, i.e. retaining only those which vanish at the edge with order m flatness.

In both approaches, we can build (in a similar way as in the beginning of this section) a covering $(I_\mu)_{\mu\in\Theta_j}$ which satisfies the same basic prescriptions (in particular (3.9.7) and (3.9.8)). We can then summarize as follows the modified polynomial reproduction properties of the V_j spaces.

Interior behaviour: there exists a constant C and an integer j_0 such that for all $j \geq j_0$, if $\text{dist}(I_\mu, \partial\Omega) \geq C2^{-|\mu|}$, then the restrictions of V_j to $\tilde{I}_\mu$ and to $\tilde{J}_\mu$ contain the polynomials of degree n (with $\tilde{I}_\mu$ and $\tilde{J}_\mu$ defined by (3.9.4) and (3.9.5)).

Boundary behaviour: we denote by $\tilde{\Theta}_j$ the subset of Θ_j consisting of those μ such that $\text{dist}(I_\mu, \partial\Omega) \leq C2^{-|\mu|}$. From the modifications near the boundary, there exists a family K_μ, $\mu \in \tilde{\Theta}_j$ such that (i) each K_μ is obtained from a reference domain of the type (3.9.1) among a finite collection $K_1,\cdots,K_N$ by the product of an orthogonal affine transformation with a scaling of $2^{-|\mu|}$ (similarly to the $\tilde{I}_\mu$ and $\tilde{J}_\mu$), (ii) $\tilde{I}_\mu$ and $\tilde{J}_\mu$ are contained in K_μ and $\partial K_\mu \cap \partial\Omega$ has non-zero $d-1$ dimensional measure, and (iii) the restriction of V_j to K_μ contains those polynomials of degree n that vanish with flatness m at $\partial K_\mu \cap \partial\Omega$.

We say that such multiresolution approximation spaces reproduce polynomials of degree n with flatness m at the boundary. Note that for $m \geq n$ no polynomial is reproduced near the boundary, i.e. we can always assume that $m \leq n$. Our goal is now to show that such spaces can be used to characterize the spaces $B^s_{p,q}(\Omega, m)$ by norm equivalences of the type (3.6.1) and (3.6.15). Our first step is to establish direct and inverse estimates between L^p and $W^{s,p}(\Omega, m)$.

Inverse estimates operate with exactly the same arguments as when no boundary conditions were imposed on the V_j spaces. We thus obtain a straightforward adaptation of the results in §3.4 and of Theorems 3.7.1 and 3.7.2.

In order to prove a direct estimate, we cannot proceed exactly as in the previous section. In particular, the local Whitney estimate of Theorem 3.2.2 cannot be exploited directly, and we shall instead make use of the following generalization of the Deny-Lions theorem.

Theorem 3.10.1 *Let Ω be a connected bounded domain and let B be a closed subset of $\partial\Omega$ of non-zero $d-1$ dimensional measure. We denote by $\Pi_{n,m}$ the set of polynomials $g \in \Pi_n$ such that $|g(x)| \lesssim [\mathrm{dist}(x, B)]^{m+1}$, and we then have*

$$\inf_{g \in \Pi_{n,m}} \|f - g\|_{L^p} \lesssim |f|_{W^{s,p}}, \tag{3.10.5}$$

for all $f \in W^{s,p}(B, m)$ with $m + 1/p \leq s \leq n + 1$.

Proof We shall apply a general abstract result of functional analysis: if X, Y and Z are Banach spaces and if

$$\|f\|_X \sim \|Af\|_Y + \|Kf\|_Z, \tag{3.10.6}$$

where $A : X \to Y$ and $K : X \to Z$ are respectively bounded and compact operators, then $\mathrm{Ker}(A)$ is finite dimensional and $\mathrm{Ran}(A)$ is closed. It follows that A is an isomorphism between the quotient space $X/\mathrm{Ker}(A)$ and $\mathrm{Ran}(A)$, and that

$$\inf_{g \in \mathrm{Ker}(A)} \|f - g\|_X \lesssim \|Af\|_Y. \tag{3.10.7}$$

Before proving this result, let us see how it applies to our specific context. Here, we take $X = W^{s,p}(B, m)$, $Z = L^p$ and K the (compact) injection of $W^{s,p}(B, m)$ in L^p. Note that for $m = s = 0$ and $p = \infty$ we have $X = Z$ and no compact injection, but (3.10.5) is trivial in this case.

In the case where $s = l \in \mathbb{N} - \{0\}$, we simply take $A : f \mapsto (\partial^\alpha f)_{|\alpha|=l}$ and $Y = (L^p)^{n(l)}$ with $n(l) = \#\{\alpha \in \mathbb{N}^d \ ; \ |\alpha| = l\}$.

In the case where $l - 1 < s < l$ is fractional, we take $A : f \mapsto (g_\alpha)_{|\alpha|=l}$ with $g_\alpha(x, y) = (\partial^\alpha f(x) - \partial^\alpha f(y))/|x-y|^{s-l+1+d/p}$ and $Y = (L^p(\Omega \times \Omega))^{n(l)}$.

In both cases, (3.10.5) holds by definition of the Sobolev norm, and the kernel Ker(A) consists of the polynomials of degree q that belong to X. Since $l \geq m + 1/p$, these are exactly the polynomials $\Pi_{l,m}$. We thus obtain

$$\inf_{g \in \Pi_{l,m}} \|f - g\|_{L^p} \lesssim |f|_{W^{s,p}(B,m)}, \quad (3.10.8)$$

and thus (3.10.5) since $\Pi_{l,m} \subset \Pi_{n,m}$.

We end by proving the abstract result. If $(f_n)_{n\geq 0}$ is a sequence in Ker(A) such that $\|f_n\|_X = 1$, then since $\|f_n - f_m\|_X \sim \|Bf_n - Bf_m\|_Z$ and B is compact, there exists a subsequence of $(f_n)_{n\geq 0}$ that converges in X. Thus, the unit ball of Ker(A) is compact, i.e. Ker(A) is finite dimensional (note that in our specific situation this statement does not need to be proved since Ker(A) consists of polynomials of degree less than q).

Now let $(g_n)_{n\geq 0}$ be a sequence in Ran(A) which converges to some $g \in Y$. Since Ker(A) is finite dimensional, we can associate $f_n \in X$ such that $g_n = Af_n$ and

$$\|f_n\|_X = \min_{g \in \mathrm{Ker}(A)} \|f_n - g\|_X = \|f_n\|_{X/\mathrm{Ker}(A)}. \quad (3.10.9)$$

We want to prove that $g \in$ Ran(A). If $g = 0$ this is trivial. If not, we can assume (up to a subsequence) that g_n does not vanish and define $h_n = f_n/\|f_n\|_X$. In particular, we have $\|h_n\|_X = \|h_n\|_{X/\mathrm{Ker}(A)} = 1$ so that (up to a subsequence) we can assume that Kh_n converges. We can also assume (up to a subsequence) that $\|f_n\|_X$ converges to a finite or infinite limit. If this limit is infinite, then $Ah_n = g_n/\|f_n\|_X$ goes to zero in Y, and thus from (3.10.6), h_n converges to some limit h in X which should both satisfy $\|Ah\|_Y = 0$ and $\|h\|_{X/\mathrm{Ker}(A)} = 1$, i.e. a contradiction. If $\|f_n\|_X$ goes to some finite limit, then (up a subsequence) Kf_n converges, and thus from (3.10.6) f_n converges to some f in X. Thus $g = Af \in$ Ran(A), which concludes the proof. ◇

We are now ready to prove the direct estimate for Sobolev spaces with boundary conditions.

Theorem 3.10.2 *Assume that the V_j spaces reproduce polynomials of degree $n-1$ with flatness $m \leq n-1$ at the boundary of Ω. Then, for $s \leq n$ we have*

$$\mathrm{dist}_{L^p}(f, V_j) \lesssim 2^{-sj}\|f\|_{W^{s,p}}, \quad (3.10.10)$$

for all $f \in W^{s,p}(\Gamma, m)$.

Proof Let us first assume that $s \geq m + 1/p$. In the interior of Ω, for $\mu \in \Theta_j \setminus \widetilde{\Theta}_j$, we use the classical Deny-Lions theorem to build polynomials

$p_\mu \in \Pi_{n-1}$ such that

$$\|f - p_\mu\|_{L^p(\tilde{J}_\mu)} \lesssim 2^{-sj}|f|_{W^{s,p}(\tilde{J}_\mu)}. \qquad (3.10.11)$$

Near the boundary, for $\mu \in \tilde{\Theta}_j$, we use Theorem 3.10.1 to build polynomials $p_\mu \in \Pi_{n-1}$ which vanish with flatness m on $\partial K_\mu \cap \Gamma$ and such that

$$\|f - p_\mu\|_{L^p(K_\mu)} \lesssim 2^{-sj}|f|_{W^{s,p}(K_\mu)}. \qquad (3.10.12)$$

We then proceed as in the proof of Theorem 3.7.3 to build the approximant $f_j \in V_j$, using the expression of the polynomials p_μ in the local basis on $\tilde{J}_\mu$ and K_μ. By a similar reasoning, we end up with

$$\|f - f_j\|_{L^p(I_\mu)} \lesssim 2^{-sj}[\sum_{\nu\in A_\mu} \|f|^p_{W^{s,p}(K_\nu)} + \sum_{\nu\in B_\mu} \|f|^p_{W^{s,p}(\tilde{J}_\nu)}]^{1/p}, \qquad (3.10.13)$$

(replaced by a supremum in ν if $p = \infty$). The sets A_μ and B_μ are of the type

$$A_\mu = \{\nu\ ;\ \mathrm{dist}(I_\mu, K_\nu) \le C2^{-j}\} \text{ and } B_\mu = \{\nu\ ;\ \mathrm{dist}(I_\mu, \tilde{J}_\nu) \le C2^{-j}\},$$

with C a fixed constant independent of μ and j.

From the properties of the Sobolev norms with respect to the union of sets with controlled overlapping (observed in Remark 3.3.5, and which can be generalized to covering with noncubic domains by (3.9.15) of Theorem 3.9.1), we obtain the direct estimate (3.10.10) with the $W^{s,p}$ semi-norm in place of the full norm (i.e. a stronger result).

Assume now that we have $s < m + 1/p$. We claim that the direct estimate (3.10.10) holds if V_j reproduces polynomials of some degree q with $s \le q$ in the interior, without any requirement on the boundary, and thus under our assumptions. To see this, assume that $l - 1 + 1/p \le s < l + 1/p$ for an integer l such that $0 \le l \le m$ (if $l = 0$, we only consider the range $0 < s \le 1/p$). We thus cannot directly apply Theorem 3.10.1, but we remark that $W^{s,p}(\Gamma, m) = W^{s,l}(\Gamma, l)$.

In the case where $l - 1 + 1/p \le s \le l$, according to the discussion for $s \ge m + 1/p$, we thus have the direct estimate (3.10.10) (again with the semi-norm $W^{s,p}$) if V_j reproduces polynomials of degree q with flatness q at the boundary, i.e. no requirement on the boundary. This clearly holds with our assumptions.

In the case where $l < s < l + 1/p$, if we again build the approximation f_j in a similar way as in the proof of Theorem 3.7.3, we still obtain an interior estimate of the type

$$\|f - f_j\|_{L^p(\Omega_j)} \lesssim 2^{-sj}|f|_{W^{s,p}}, \qquad (3.10.14)$$

where Ω_j consists of the union of the I_μ for all μ such that $|A_\mu| = 0$.

We are left with evaluating $\|f\|^p_{L^p(M_j)}$ where $M_j := \Omega \setminus \Omega_j$ is a margin of width 2^{-j} next to the boundary, since we do not assume any polynomial reproduction in this area of Ω. Since for $s < m + 1/p$, $W^{s,p}(\Gamma, m)$ is also obtained as the closure of test functions, we shall assume that f is in $\mathcal{D}(\Omega)$. By local maps (using the simple geometry of Ω), it is sufficient to consider only the following simpler situation: f is a C^∞ function on $[0,1]^d$ which vanishes near the face $x_1 = 0$ and

$$M_j = [0, 2^{-j}] \times [0,1]^{d-1}. \tag{3.10.15}$$

We consider r such that $s - l = 1/p - 1/r$. We thus have $p < r < \infty$, and by the Hölder inequality in the x_1 variable,

$$\|f\|^p_{L^p(M_j)} \le 2^{-j(1-p/r)} \int_{[0,1]^{d-1}} [\int_0^{2^{-j}} |f(x_1, \cdots, x_d)|^r dx_1]^{p/r} dx_2 \cdots dx_d.$$

By integration from $x_1 = 0$, we thus obtain

$$\|f\|^p_{L^p(M_j)} \lesssim 2^{-spj} \int_{[0,1]^{d-1}} [\int_0^{2^{-j}} |(\partial/\partial x_1)^l f(x_1, \cdots, x_d)|^r dx_1]^{p/r} dx_2 \cdots dx_d.$$

We now apply the Sobolev embedding of $W^{s,p}$ in $W^{r,l}$ (or L^r if $l = 0$) in one variable which gives

$$\begin{aligned} \|f\|^p_{L^p(M_j)} &\lesssim 2^{-spj} \textstyle\int_{[0,1]^{d-1}} \|f(\cdot, x_2, \cdots, x_d)\|^p_{W^{s,p}([0,1])} dx_2 \cdots dx_d \\ &\le 2^{-sj} \|f\|^p_{W^{s,p}([0,1]^d)}, \end{aligned}$$

and we thus finally obtain (3.10.10) by combining this with the interior estimate (3.10.14). ◇

Remark 3.10.2 *As it can be seen in the proof, the full $W^{s,p}$ norm instead of the semi-norm is needed in (3.10.10) only for certain values of s, namely $l < s < l + 1/p$ for $0 \le l \le m$. The following simple example (corresponding to $l = 0$) shows that the full norm is indeed necessary: if $0 < s < 1/p$, the constant $f = 1$ belongs to $W^{s,p}_0(\Omega)$ with null semi-norm. Since the V_j spaces reproduce constants only in the interior of Ω, there is no hope of a direct estimate with the semi-norm.*

Remark 3.10.3 *In the case where the generators of V_j are such that P_j is uniformly bounded in L^p we can use $P_j f$ in place of f_j in the direct estimate.*

In order to proceed further, we need a better understanding of the interpolation properties of Besov and Sobolev spaces with boundary conditions. We first derive a partial statement.

Theorem 3.10.3 *If $s_0 < s_1$ are both in one of the intervals $]m+1/p,+\infty[$, $]0,1/p[$ and $]l-1+1/p,l+1/p[$, $l=1,\cdots,m$, and if $s=(1-\theta)s_0+\theta s_1$, $\theta \in]0,1[$, then*

$$B^s_{p,q}(\Gamma,m) = [W^{s_0,p}(\Gamma,m), W^{s_1,p}(\Gamma,m)]_{\theta,q}. \tag{3.10.16}$$

Proof The proof of this result is derived from the basic properties that we gave on Sobolev and Besov spaces with boundary conditions (in particular (3.10.4)), which show that, under our assumptions, we have

$$B^s_{p,q}(\Gamma,m) = W^{s_0,p}(\Gamma,m)\cap B^s_{p,q}, \text{ and } W^{s_1,p}(\Gamma,m) = W^{s_0,p}(\Gamma,m)\cap W^{s_1,p}.$$

One concludes with the following observation which is an easy consequence of the definition of real interpolation spaces: if A, $C \subset B$ are Banach spaces and if $A\cap B$ and $A\cap C$ are respectively closed in B and C, then $A\cap[B,C]_{\theta,q}$ is also closed in $[B,C]_{\theta,q}$ and we have

$$[A\cap B, A\cap C]_{\theta,q} = A\cap [B,C]_{\theta,q}. \tag{3.10.17}$$

Applying (3.10.17) to $A=W^{s_0,p}(\Gamma,m)$, $B=W^{s_0,p}$ and $C=W^{s_1,p}$ yields (3.10.16). ◇

Our next result, which should be viewed as the adaptation of Theorem 3.7.4 to bounded domains with boundary conditions, will also provide additional information on the interpolation of the spaces $B^s_{p,q}(\Gamma,m)$ and $W^{s,p}(\Gamma,m)$.

Theorem 3.10.4 *Assume that V_j reproduces polynomials of degree $n-1$ with flatness m near the boundary and that the scaling functions which generate V_j belong to $B^s_{p,q_0}(\Gamma,m)$ for some $q_0>0$. Then, for all t in $]0,\min\{n,s\}[$ such that $t-1/p$ is not an integer among $0,\cdots,m$, we have $\mathcal{A}^t_{p,q} = B^t_{p,q}(\Gamma,m)$ with equivalent norms, i.e.*

$$\|f\|_{B^t_{p,q}} \sim \|(2^{tj}\mathrm{dist}_{L^p}(f,V_j))_{j\geq 0}\|_{\ell^q}, \tag{3.10.18}$$

for all $f\in B^t_{p,q}(\Gamma,m)$.

Proof By assumption, t sits in one of the intervals $]0,1/p[$, $]m+1/p,+\infty[$ or $]l-1+1/p,l+1/p[$, $l=1,\cdots,m$. Let s_0 and s_1 be in the same interval as t and such that $s_0<t<s_1<\min\{n,s\}$.

From the inverse estimate generalizing Theorem 3.7.2 and the direct estimate established in Theorem 3.10.2, we can use the same simple arguments as in Corollaries 3.4.1 and 3.4.2 to derive direct and inverse estimates between $B^{s_0}_{p,p}(\Gamma,m)$ and $B^{s_1}_{p,p}(\Gamma,m)$, namely

$$\inf_{f_j\in V_j}\|f-f_j\|_{B^{s_0}_{p,p}} \lesssim 2^{-j(s_1-s_0}\|f\|_{B^{s_1}_{p,p}}, \tag{3.10.19}$$

for all $f\in B^{s_1}_{p,p}(\Gamma,m)$ and

$$\|f\|_{B^{s_1}_{p,p}} \lesssim 2^{j(s_1-s_0}\|f\|_{B^{s_0}_{p,p}}, \quad \text{if } f\in V_j. \tag{3.10.20}$$

We can thus combine Theorem 3.5.2 with Theorem 3.10.3 to derive that $B^t_{p,q}(\Gamma,m)=\mathcal{A}^{t-s_0}_q(B^{s_0}_{p,p})$, i.e. (using Hardy's inequality),

$$\|f\|_{B^t_{p,q}} \lesssim \|(2^{(t-s_0)j}\|f_j-f_{j-1}\|_{B^{s_0}_{p,p}})_{j\geq 0}\|_{\ell^q}, \tag{3.10.21}$$

for any $f\in B^t_{p,q}(\Gamma,m)$ where f_j is a near-best approximation of f in the $B^{s_0}_{p,p}$ norm and $f_{-1}:=0$.

Since by the inverse estimate

$$\|f_j-f_{j-1}\|_{B^{s_0}_{p,p}} \lesssim 2^{s_0 j}\|f_j-f_{j-1}\|_{L^p}, \tag{3.10.22}$$

we derive the " $\lesssim$ " part of the equivalence (3.10.18) from (3.10.21). The converse part is obtained by a direct argument using the inverse estimate as in the proof of Theorem 3.6.1. ◇

Remark 3.10.4 *We can combine the above result together with the known interpolation properties of approximation spaces in order to obtain more information on the interpolation of Sobolev and Besov spaces with boundary conditions: for $t-1/p$ not an integer among $0,\cdots,m$, we have*

$$B^t_{p,q}(\Gamma,m)=[L^p,W^{s,p}(\Gamma,m)]_{\theta,q}, \quad t=\theta s, \tag{3.10.23}$$

as well as

$$B^t_{p,q}(\Gamma,m)=[B^{s_0}_{p,q_0}(\Gamma,m),B^{s_1}_{p,q_1}(\Gamma,m)]_{\theta,q}, \quad t=(1-\theta)s_0+\theta s_1. \tag{3.10.24}$$

Among others, we see that this implies $[L^p,W^{s,p}_0]_{\theta,p}=W^{t,p}_0$, $t=\theta s$ for $t\notin\mathbb{N}$ (except if $p=2$) and $t-1/p$ not an integer.

Remark 3.10.5 *If $t-1/p$ is an integer out of $0,\cdots,m$, if and $t=\theta s$, one still has the identity of the space $\mathcal{A}^t_{p,q}$ and of $[L^p,W^{s,p}(\Gamma,m)]_{\theta,q}$ but this last space differs from $B^s_{p,q}(\Gamma,m)$. For example, if $p=q=2$, $m\geq 0$ and $t=1/2$, we obtain the space $H^{1/2}_{0,0}$ which is strictly smaller than $H^{1/2}_0$.*

The adaptation of Theorems 3.7.6 and 3.7.7 (still in the case where $p \geq 1$) are easy tasks, with the help of the previous results. For Theorem 3.7.6, we have the following modifications: (3.7.30) holds with the full $B^s_{p,p}$ norm, for all $f \in B^s_{p,p}(\Omega, m)$, adding the assumption that V_j reproduces polynomials of degree $n-1$ and flatness m at the boundary. Here we proceed in a slightly different way in evaluating the projection error: for $\mu \in \Theta_j \setminus \widetilde{\Theta}_j$, i.e. in the interior of the domain, we define t such that $d/p - d/r = s - t$, and we write

$$\begin{aligned} \|f - P_j f\|_{L^p(I_\mu)} &\lesssim 2^{dj(1/r-1/p)}\|f - P_j f\|_{L^r(I_\mu)} \\ &\lesssim 2^{-sj} \inf_{p\in\Pi_{n-1}} |f - p|_{W^{t,r}(\widetilde{I}_\mu)} \\ &\lesssim 2^{-sj}|f|_{W^{s,p}(\widetilde{I}_\mu)}. \end{aligned}$$

Here, we have first used that P_j reproduces polynomials of degree $n-1$, and then applied the Sobolev embedding together with Deny-Lions theorem. This yields an interior estimate of the same type as (3.10.14) for $P_j f$. The estimate on the margin M_j (with possibly the full $W^{s,p}$ norm) is derived by similar modifications as in the proof of Theorem 3.10.2.

The adaptation of Theorem 3.7.7 is then straightforward and we simply give its statement below.

Theorem 3.10.5 *Assume that the primal scaling functions are in L^r and the dual scaling functions in $L^{r'}$ for some $r \in [1,\infty]$, $1/r + 1/r' = 1$, or that the primal scaling functions are continuous and the dual scaling functions are Radon measures in which case we set $r = \infty$. Also assume that V_j reproduces polynomials of degree $n-1$ with flatness m at the boundary and that $\varphi \in B^s_{p,q_0}$ for some q_0. Then, for $1 \leq p \leq r$, one has the norm equivalence*

$$\|f\|_{B^t_{p,q}} \sim \|P_0 f\|_{L^p} + \|(2^{tj}\|Q_j f\|_{L^p})_{j\geq 0}\|_{\ell^q}, \tag{3.10.25}$$

for all $f \in B^t_{p,q}(\Gamma, m)$ with $t > 0$ such that $d(1/p - 1/r) < t < \min\{s, n\}$ and $t - 1/p$ not an integer out of $0, \cdots, m$. If $f = \sum_{\lambda\in\nabla} c_\lambda \psi_\lambda$ is the decomposition of f into the corresponding wavelet basis, we also have the norm equivalence

$$\|f\|_{B^t_{p,q}} \sim \|\left((2^{tj}2^{d(1/2-1/p)j}\|(c_\lambda)_{\lambda\in\nabla_j}\|_{\ell^p}\right)_{j\geq -1}\|_{\ell^q}, \tag{3.10.26}$$

under the same assumptions.

Remark 3.10.6 *The case $p < 1$, which was left aside here for the sake of simplicity, can be treated by a similar approach. The spaces $B^s_{p,q}(\Gamma, m)$ are defined in a similar manner, and have the same basic properties. In* *order to obtain the same results (in particular Theorem 3.10.5 above with $p < 1$), one simply needs to prove Theorem 3.10.1 with the seminorm $|f|_{B^s_{p,p}}$ in place of $|f|_{W^{s,p}}$ on the right side of (3.10.5) and $p < 1$. This is done by similar functional analytic arguments, conveniently adapted to the setting of quasi-Banach spaces: (i) from the equivalence $\omega \sim \tilde{\omega}$ given by Theorem 3.9.1, we can again write $\|f\|_{B^s_{p,p}} \sim \|f\|_{L^p} + \|Af\|_{L^p}$ where A is an operator involving the finite differences of f and (ii) by the approximation results of the previous section, we still have compactness of the unit ball of $B^s_{p,p}$ in L^p on a bounded domain if $s > 0$ (in the sense of the metric topology induced by the distance $d(f,g) = \|f - g\|^p_{L^p}$). Here again, the characterization of $B^s_{p,q}(\Gamma, m)$ by multiscale decomposition or wavelet coefficients is only feasible if $s - 1/p$ is not an integer out of $0, \cdots, m$.*

We end this section by explaining the adaptation of the results of §3.8, which is much simpler. For $s < 0$, $p, q \geq 1$, the space $B^s_{p,q}$ is properly defined as a space of distibution on Ω by

$$B^s_{p,q}(\Omega) := (B^{-s}_{p',q'}(\Gamma, m))', \tag{3.10.27}$$

for $1/p' + 1/p = 1/q' + 1/q = 1$ and any m such that $m > -s - 1/p$ (recall that, for such values of m, $B^{-s}_{p',q'}(\Gamma, m)$ is also the closure of $\mathcal{D}(\Omega)$ in $B^{-s}_{p',q'}(\Omega)$)

With such a definition, the adaptation of Theorem 3.8.1 is straighforward: $B^s_{p,q}$ is characterized by approximation in the primal spaces V_j if $B^{-s}_{p',q'}(\Gamma, m)$ is characterized by approximation in the dual spaces $\widetilde{V}_j$, and we have the same norm equivalences (3.8.3) and (3.8.4).

Theorem 3.8.2 is independent of the domain Ω. In particular, $L^2(\Omega)$ is characterized by (3.8.15) if $H^\varepsilon(\Omega)$ is characterized by both approximations in V_j and $\widetilde{V}_j$. Note that since $H^\varepsilon(\Omega) = H^\varepsilon_0(\Omega) = H^\varepsilon(\Gamma, m)$ for $\varepsilon < 1/2$, the characterization of L^2 is feasible with V_j and $\widetilde{V}_j$ constituted of functions that vanish near the boundary.

Finally, the characterization of L^p spaces easily extends to the type of domains that we are considering here, since the Calderon-Zygmund theory (which allows us to establish the uniform L^p-boundedness of the operators $T_E f := \sum_{\lambda \in E} \langle f, \tilde{\psi}_\lambda \rangle \psi_\lambda$ independently of the subset $E \subset \nabla$) also operates in such domains.

3.11 Multilevel preconditioning

One of the interests of multiscale discretizations is their ability to precondition large systems arising from elliptic operator equations. A general setting for such equations is the following: H is a Hilbert space embedded in $L^2(\Omega)$ and $a(\cdot,\cdot)$ is a bilinear form on $H \times H$ such that

$$a(u,u) \sim \|u\|_H^2. \tag{3.11.1}$$

Given $f \in H'$, we search for $u \in H$ such that

$$a(u,v) = \langle f, v\rangle, \text{ for all } v \in H. \tag{3.11.2}$$

From the Lax-Milgram lemma, this problem has a unique solution u. If A is defined by $\langle Au, v\rangle = a(u,v)$ for all $v \in H$, (3.11.1) implies that A is an isomorphism from H to H', so that u is also the unique solution in H of

$$Au = f. \tag{3.11.3}$$

If V_h is a subspace of H, the Galerkin approximation of u in V_h is classically defined by $u_h \in V_h$ such that

$$a(u_h, v_h) = \langle f, v_h\rangle, \text{ for all } v_h \in V_h. \tag{3.11.4}$$

If V_h is finite dimensional, the approximated problem (3.11.4) amounts to solving a linear system. By its definition, u_h is also the projection of u onto V_h in the sense of the energy norm $\|u\|_a := \sqrt{a(u,u)}$, so that we have the error estimate

$$\|u - u_h\|_H \lesssim \|u - u_h\|_a = \inf_{v_h \in V_h} \|u - v_h\|_a \lesssim \inf_{v_h \in V_h} \|u - v_h\|_H. \tag{3.11.5}$$

Of interest to us is the practical situation where H is an L^2-Sobolev space. Classical instances are given by the *Poisson* equation $-\Delta u = f$ with Dirichlet boundary condition $u = 0$ on $\Gamma := \partial\Omega$, for which $H = H_0^1$, the *Helmholtz* equation $u - \Delta u = f$ with Neumann boundary condition $\partial u/\partial n = 0$ on Γ, for which $H = H^1$, and the *Bilaplacian* equation $\Delta^2 u = f$ with condition $u = \partial u/\partial n = 0$ on Γ, for which $H = H_0^2$.

For such equations, it is well known that the matrices resulting from Galerkin discretizations in finite element spaces are ill-conditioned in the sense that their condition number grows like h^{-2s} where h is the mesh size and s is the order of the corresponding Sobolev space H (i.e. $2s$ is the order of the operator). A particular instance was evoked in §1.6 of Chapter 1 for the Poisson equation.

Remark 3.11.1 *Elliptic equations involving integral operators of negative order also enter the above class of problems with the modification that H' is now embedded in $L^2(\Omega)$. Instances are the single (resp. double) layer* *potential operators for which $H = H^{-1/2}$ (resp. L^2). This unified point of view was proposed in* DAHMEN, PRÖSSDORF *and* SCHNEIDER *[1993,1994], together with matrix compression techniques that will be evoked in §4.6 of the next chapter.*

The use of multilevel methods for preconditioning such matrices is tied to the possibility of characterizing the L^2-Sobolev space H by means of multiscale decompositions. In this particular setting, the norm equivalences that were established in the previous sections have the form

$$\|f\|_{H^s}^2 \sim \|P_0 f\|_{L^2}^2 + \sum_{j\geq 0} 2^{2sj}\|Q_j f\|_{L^2}^2, \tag{3.11.6}$$

or in terms of wavelet coefficients

$$\|f\|_{H^s}^2 \sim \sum_{\lambda\in\nabla} 2^{2s|\lambda|}|c_\lambda|^2. \tag{3.11.7}$$

In Theorem 3.7.6, the range of s for which such equivalences hold is described in terms of three basic properties of the multicale decomposition: smoothness, polynomial reproduction and L^p-stability of the associated projector. We have seen that these equivalences also hold for the spaces $H^s(\Omega)$ and $H_0^s(\Omega)$ with appropriate adaptations of the multiscale decomposition near the boundary, as well as for negative values of s. Such norm equivalences will allow us to analyze the performance of several multilevel preconditioning schemes.

Example 1. Wavelet diagonal preconditioning

A particular instance of wavelet diagonal preconditioning was already discussed in §1.6 of Chapter 1. We also refer to this section for the basic concept of preconditioner for an iterative scheme. Let us consider the Galerkin discretization (3.11.4) on a multiresolution approximation space $V_J \subset H$ corresponding to a mesh size 2^{-J} and we denote by u_J the corresponding solution. We assume that $\cup_{j\geq 0} V_j$ is dense in H (which ensures in particular that $\|u_J - u\|_H$ goes to zero as $J \to +\infty$). For the practical computation of u_J, we can use either the nodal basis $\Phi_J = \{\varphi_\gamma\ ;\ |\gamma| = J\}$ or the multiscale basis $\Psi_J = \{\psi_\lambda\ ;\ |\lambda| < J\}$.

With the nodal basis, we obtain a system

$$A_J U_J = F_J, \tag{3.11.8}$$

where U_J is the coordinate vector of u_J in the basis Φ_J, $F_J = (\langle f, \varphi_\gamma\rangle)_{|\gamma|=J}$, and $A_J = (\langle A\varphi_\gamma, \varphi_\beta\rangle)_{|\gamma|=|\beta|=J}$ is the correponding stiffness matrix.

Similarly, with the multiscale basis, we obtain a system

$$\widetilde{A}_J \widetilde{U}_J = \widetilde{F}_J, \tag{3.11.9}$$

where $\widetilde{U}_J$ is the coordinate vector of u_J in the basis Ψ_J, $F_J = (\langle f, \psi_\lambda \rangle)_{|\lambda|<J}$, and $A_J = (\langle A\psi_\lambda, \psi_\lambda \rangle)_{|\lambda|,|\mu|<J}$ is the stiffness matrix.

The link between these two systems is provided by the linear transformation R_J representing the fast wavelet reconstruction algorithm: we clearly have

$$U_J = R_J \widetilde{U}_J, \quad \widetilde{F}_J = R_J^* F_J \text{ and } \widetilde{A}_J = R_J^* A_J R_J. \tag{3.11.10}$$

Recall that R_J is implemented by a coarse to fine $\mathcal{O}(N)$ algorithm. Note that R_J^* has the same fine to coarse structure as the decomposition algorithm, but uses the same low-pass and high-pass filters as the reconstruction R_J: it differs from R_J^{-1}, except if orthonormal wavelets are used.

In the following, we shall denote by $\langle \cdot, \cdot \rangle_d$ and $\|\cdot\|_d$ the discrete inner product and hilbertian norm.

A first result relates the norm equivalences of the type (3.11.2) to diagonal preconditioning of A in the multiscale basis.

Theorem 3.11.1 *Consider the the diagonal matrix $D_J = (2^{2s|\lambda|}\delta_{\lambda,\mu})_{|\lambda|,|\mu|<J}$, indexed by the multiscale discretization. The following two statements are equivalent:*

1. *H is characterized by a norm equivalence*

$$\|f\|_H^2 \sim \sum_{\lambda \in \nabla} 2^{2s|\lambda|} |c_\lambda|^2, \tag{3.11.11}$$

2. *The condition number $\mathcal{K}(D_J^{-1}\widetilde{A}_J) = \mathcal{K}(D_J^{-1/2}\widetilde{A}_J D_J^{-1/2})$ is bounded independently of J.*

Proof The first property is equivalent to

$$\langle D_J U, U \rangle_d \sim \langle A_J U, U \rangle_d, \tag{3.11.12}$$

with constants independent of the vector U and the scale level J. From the definition of A_J, this can also be expressed by

$$\|v_J\|_a^2 \sim \sum_{|\lambda|<J} 2^{2s|\lambda|} |c_\lambda|^2, \tag{3.11.13}$$

for all $v_J = \sum_{|\lambda|<J} c_\lambda \psi_\lambda$ in V_J. Since $\|\cdot\|_a \sim \|\cdot\|_H$, (3.11.13) is equivalent to (3.11.11) for all $f \in V_J$. By density of the V_J spaces, this is equivalent

to (3.11.11) for all $f \in H$. ◇.

From this result, we see that for a suitable choice of τ, the preconditioned iteration

$$\widetilde{U}_J^{n+1} = \widetilde{U}_J^n + \tau D_J^{-1}(\widetilde{F}_J - \widetilde{A}_J\widetilde{U}_J^n), \quad (3.11.14)$$

decreases the error $E_J^n = U_J^n - U_J$ with a convergence rate

$$\rho = [1 - \mathcal{K}(D_J^{-1}\widetilde{A}_J)]/[1 + \mathcal{K}(D_J^{-1}\widetilde{A}_J)], \quad (3.11.15)$$

which is independent of the number of levels J.

A particular interest of the nodal basis discretization is the sparse structure of A_J in the case where A is a partial differential operator which has a local action: the cost of one application of A_J is then in $\mathcal{O}(N_J)$ where $N_J = \dim(V_J) \sim 2^{dJ}$. Since $\widetilde{A}_J = R_J^* A_J R_J$ and R_J can also be implemented in $\mathcal{O}(N_J)$ operations, this optimal complexity is preserved in the cost of the above iteration.

Since $D_J^{-1/2}\widetilde{A}_J D_J^{-1/2} = D_J^{-1/2} R_J^* A_J R_J D_J^{-1/2}$, we can also view the matrix $B_J := R_J D_J^{-1} R_J^*$ as a preconditioner for A_J. The corresponding preconditioned iteration

$$U_J^{n+1} = U_J^n + \tau B_J(F_J - A_J U_J^n), \quad (3.11.16)$$

which directly acts in the nodal discretization, can be directly obtained by application of R_J on (3.11.14) and produces the same rate of decay for the error.

The norm equivalence (3.11.11) is also useful in controlling the progress of these algorithms from the evaluation of their residuals $G_J^n := F_J - A_J U_J^n$ and $\widetilde{G}_J^n := \widetilde{F}_J - \widetilde{A}_J \widetilde{U}_J^n$. These quantities satisfy the equations

$$A_J E_J^n = G_J^n \text{ and } \widetilde{A}_J \widetilde{E}_J^n = \widetilde{G}_J^n, \quad (3.11.17)$$

so that we can evaluate the norm of the corresponding error $e_J \in V_J$ as

$$\|e_J^n\|_H^2 \sim \langle \widetilde{A}_J \widetilde{E}_J^n, \widetilde{E}_J^n \rangle_d \sim \langle \widetilde{A}_J^{-1} \widetilde{G}_J^n, \widetilde{G}_J^n \rangle_d \sim \langle D_J^{-1} \widetilde{G}_J^n, \widetilde{G}_J^n \rangle_d \quad (3.11.18)$$

for the iteration (3.7.15) and similarly by $\|e_J^n\|_H^2 \sim \langle B_J G_J^n, G_J^n \rangle_d$ for the iteration (3.7.16).

Remark 3.11.2 *Both iterations (3.11.14) and (3.11.16) only involve the filters in the primal refinement equations*

$$\varphi_\gamma = \sum_{\beta \in \Gamma_{j+1}} h_{\beta,\gamma} \varphi_\beta \text{ and } \psi_\lambda = \sum_{\mu \in \Gamma_{j+1}} g_{\beta,\lambda} \varphi_\beta, \quad (3.11.19)$$

(for $\gamma \in \Gamma_j$ and $\lambda \in \nabla_j$), which are used in the application of R_J and R_J^*. These filters are finitely supported if the primal scaling functions and wavelets also have compact support, in which case R_J and R_J^* apply in $\mathcal{O}(N_J)$ operations. We can thus allow the dual scaling functions and wavelets to have infinite support as in several examples of splines and finite element wavelets presented in §2.11 and §2.13 of Chapter 2.

Remark 3.11.3 *Further improvement is achieved by using the preconditioned conjugate gradient algorithm, which simply amounts to the application of a conjugate gradient algorithm to the preconditioned equation*

$$D_J^{-1/2} R_J^* A_J R_J D_J^{-1/2} X_J = Y_J, \tag{3.11.20}$$

where $Y_J := D_J^{-1/2} R_J^* F_J$ and $U_J = R_J D_J^{-1/2} X_J$. It is well known that the convergence rate for such an algorithm is given by

$$\tilde{\rho}_J = (\sqrt{\mathcal{K}(B_J A_J)} - 1)/(\sqrt{\mathcal{K}(B_J A_J)} + 1), \tag{3.11.21}$$

improving on the rate $\rho_J = (\mathcal{K}(B_J A_J) - 1)/(\mathcal{K}(B_J A_J) + 1)$ of the iteration (3.11.16) (obtained with an optimal choice of τ), although this improvement is less significant when $\mathcal{K}(B_J A_J)$ stays close to 1.

Remark 3.11.4 *Diagonal wavelet preconditioning was also proposed in the more general setting Petrov-Galerkin discretization in* DAHMEN, PRÖSSDORF *and* SCHNEIDER *[1994]: the problem is now to find $u_J \in V_J$ such that*

$$a(u_J, v_J) = \langle f, v_J \rangle, \quad v_J \in X_J, \tag{3.11.22}$$

where X_J has the same dimension as V_J. Such schemes include in particular collocation for which the basis functions of X_J are Dirac masses on the grid indexing the nodal basis of V_J. The above reference considers the setting of periodic boundary conditions for which the well-posedness of (3.11.22), the resulting error estimates and the possibility of multiscale preconditioning can be completely analyzed. Such an analysis appears to be more difficult for other types of boundary conditions.

Remark 3.11.5 *In this example, we are using the multiresolution spaces V_j for two distinct purposes: discretization at the finest level and preconditioning by B_J. In practice, one could consider the idea of decoupling those two tasks, i.e. use a certain space V_J for the Galerkin discretization and precondition with a discrete multiscale transform R'_J which relates to different spaces V'_j. Such a strategy is more difficult to analyze in its full generality and might even fail in some cases. This idea is still particularly attractive* *in the case where V_J has high approximation order: the filters involved in the application of R_J might then have large supports, so that the computational cost of one iteration will be substantially reduced by using a discrete multiscale transform with shorter filters.*

Example 2. BPX preconditioning
The wavelet diagonal preconditioning in the iteration (3.11.16) is by essence a parallel technique: the residual is decomposed into multiscale blocks which are treated independently by the scaling D_J^{-1} and then reassembled.

Similar parallel multilevel preconditioners were introduced in BRAMBLE, PASCIAK and XU [1990], which do not require the existence of a wavelet basis for characterizing the details, but only involve the nodal basis for the mutliscale approximation spaces. The construction and analysis of these preconditioners are again related to the norm equivalence

$$\|f\|_H^2 \sim \|P_0 f\|_{L^2}^2 + \sum_{j\geq 0} 2^{2js}\|Q_j f\|_{L^2}^2. \quad (3.11.23)$$

We assume here that $s > 0$ and we recall the variant

$$\|f\|_H^2 \sim \|f\|_{L^2}^2 + \sum_{j\geq 0} 2^{2js}\|f - P_j f\|_{L^2}^2. \quad (3.11.24)$$

If the spaces V_j have enough smoothness and polynomial reproduction properties, we know from the previous sections that (3.11.23) and (3.2.22) are always satisfied with P_j being the L^2-orthogonal projector onto V_j. We also assume that we have an L^2-stable scaling (or nodal) function basis φ_γ, $\gamma \in \Gamma_j$ to characterize the V_j spaces.

For $j \leq J$, we also define the discrete spaces $V_{J,j}$ corresponding to the elements of V_j expressed in the nodal basis of V_J, and $P_{J,j}$ the orthogonal projector from $V_{J,J}$ (the full discrete space) onto $V_{j,J}$ with respect to the discrete inner product. We also define $Q_{J,j} := P_{J,j+1} - P_{J,j}$ the projector on the orthogonal complement $W_{J,j}$ of $V_{J,j}$ into $V_{J,j+1}$ with respect to the discrete inner product. If G is the coordinate vector of $g \in V_J$ in the nodal basis, we thus have

$$\begin{aligned}\langle A_J G, G\rangle_d &\sim \|g\|_H^2 \\ &\sim \|g\|_{L^2}^2 + \textstyle\sum_{j=0}^{J-1} 2^{2js}\|g - P_j g\|_{L^2}^2 \\ &\sim \|G\|_d^2 + \textstyle\sum_{j=0}^{J-1} 2^{2js}\|G - P_{J,j}G\|_d^2 \\ &\sim \|P_{J,0}G\|_d^2 + \textstyle\sum_{j=0}^{J-1} 2^{2js}\|Q_{J,j}G\|_d^2,\end{aligned}$$

where we have used the ellipticity property and the L^2-stability of the nodal basis of V_J.

It follows that A_J is spectrally equivalent to the self-adjoint discrete operator $Z_J := P_{J,0} + \sum_{j\geq 0} 2^{2js} Q_{J,j}$. As we noticed in §1.6 of Chapter 1, this also means that A_J^{-1} is spectrally equivalent to Z_J^{-1} which is given by $Z_J^{-1} = P_{J,0} + \sum_{j\geq 0} 2^{-2js} Q_{J,j}$ using the discrete othogonality of the spaces $V_{J,0}$ and $W_{J,j}$, $j = 0, \cdots, J-1$. We thus have

$$\begin{aligned}\langle A_J^{-1}G, G\rangle_d &\sim \|P_{J,0}G\|_d^2 + \textstyle\sum_{j=0}^{J-1} 2^{-2js}\|Q_{J,j}G\|_d^2 \\ &\sim \textstyle\sum_{j=0}^{J} 2^{-2js}\|P_{J,j}G\|_d^2\end{aligned}$$

where we have used that $\|Q_{J,j}G\|_d^2 = \|P_{J,j+1}G\|_d^2 - \|P_{J,j}G\|_d^2$.

An $\mathcal{O}(1)$ preconditioner for A_J is thus given by

$$B_J = \sum_{j=0}^{J} 2^{-2js} P_{J,j}. \tag{3.11.25}$$

The main drawback of this choice is that, except in the case of orthonormal scaling functions, the computation of $P_{J,j}$ is not local so that B_J cannot exactly be applied in $\mathcal{O}(N_J)$ operations: one needs to invert the mass matrix $(\langle \varphi_{J,\gamma}, \varphi_{J,\beta}\rangle_d)_{\gamma,\beta\in\Gamma_j}$ where $\varphi_{J,\gamma}$ is the coordinate vector of φ_γ in the nodal basis of V_J.

A possibility to circumvent this drawback is to make the following observation: one has the uniform stability property

$$\|\sum_{\gamma\in\Gamma_j} c_\gamma \varphi_{J,\gamma}\|_d^2 \sim \sum_{\gamma\in\Gamma_j} |c_\gamma|^2, \tag{3.11.26}$$

from the L^2-stability of the nodal bases. As was observed in §2.2 of Chapter 2, this implies the *frame property*

$$\|G\|_d^2 \sim \sum_{\gamma\in\Gamma_j} |\langle G, \varphi_{J,\gamma}\rangle_d|^2, \tag{3.11.27}$$

for $G \in V_{J,j}$, with constants also independent of j and J. It follows that

$$\langle A_J^{-1}G, G\rangle_d \sim \sum_{j=0}^{J} 2^{-2js} \sum_{\gamma\in\Gamma_j} |\langle G, \varphi_{J,\gamma}\rangle_d|^2, \tag{3.11.28}$$

which shows that an alternate choice for a preconditioner is given by

$$\widetilde{B}_J = \sum_{j=0}^{J} 2^{-2js} I_j, \tag{3.11.29}$$

where $I_j := \sum_{\gamma\in\Gamma_j}\langle G,\varphi_{J,\gamma}\rangle_d\varphi_{J,\gamma}$ is a discrete quasi-interpolant operator. In comparison to (3.11.25), this choice is better in the sense that I_j is a local operator. Note that the wavelet preconditioner $R_J D_J^{-1} R_J^*$ of the previous example can also be expressed as

$$R_J D_J^{-1} R_J^* G = A_0 G + \sum_{j=0}^{J-1} 2^{-2js} \sum_{\lambda\in\nabla_j} \langle G,\psi_{J,\lambda}\rangle_d \psi_{J,\lambda}, \qquad (3.11.30)$$

with $\psi_{J,\gamma}$ being the coordinates of ψ_λ in the nodal basis of V_J. The so-called BPX preconditioner $\widetilde{B}_J$ is thus implemented in a similar manner to the wavelet diagonal preconditioner, with the distinction that the multiscale transforms only involve the low pass filters and the approximation coefficients.

Remark 3.11.6 *Note that I_j can also be interpreted as a first order approximation to $P_{J,j}$ in the sense that I_jG is obtained after one iteration in the iterative inversion of the mass matrix $(\langle\varphi_{J,\gamma},\varphi_{J,\beta}\rangle_d)_{\gamma,\beta\in\Gamma_j}$ for the computation of $P_{J,j}G$. Another point of view on the BPX preconditioner is given in* OSWALD *[1993], in terms of the frame properties of the family $\cup_{j\geq 0}\{\varphi_\gamma\}_{\gamma\in\Gamma_j}$ in the space H.*

Remark 3.11.7 *In contrast to the wavelet diagonal preconditioning, the constants in the equivalence $\langle A_J^{-1}G,G\rangle_d \sim \langle\widetilde{B}_J^{-1}G,G\rangle_d$ deteriorate as s goes to zero and the equivalence does no longer hold for the value $s=0$ (one can still prove an $\mathcal{O}(\log(J)^2)$ condition number). This limitation of the BPX preconditioner to positive order operators makes it inadequate for problems of the type $p(x)u(x)-\nabla[q(x)\nabla u(x)]=f$, if the first coefficient $p(x)$ is numerically dominant, i.e. the L^2 term dominates in the energy norm. A more general remark is that the condition numbers resulting from multiscale preconditioning can be strongly affected by the variations in the coefficients of the operator, even if the space H does not change. We shall discuss in the next example possible solutions to circumvent this difficulty.*

Remark 3.11.8 *Both BPX and wavelet preconditioners are additive: one iteration treats the different levels of scale in parallel and adds up the result, according to (3.11.29) or (3.11.30). In comparison, most algorithms previously developed in multigrid theory are multiplicative, i.e. treat the scales in a sequential way. These techniques also amount to $\mathcal{O}(1)$ preconditioning, although their analysis is less simple than in the additive case. We shall not discuss them here, although it should be mentionned that in several test problems, multiplicative multilevel techniques yield slightly better results in terms of condition numbers than parallel preconditioners. There* *is now a huge amount of published litterature on multigrid methods in various contexts. General treatments can be found in* HACKBUSCH *[1985] and* BRAMBLE *[1993]. The reader will also find in* OSWALD *[1993] an analysis of both multilevel preconditioning and domain decomposition techniques from a subspace splitting point of view.*

Example 3. Adaptive preconditioning

Wavelet preconditioning provides uniform bounds for the condition number of $D_J^{-1/2}\widetilde{A}_J D_J^{-1/2}$. However, these bounds might be large in practice, in relation to both the multiscale decomposition and the operator.

Indeed, in most constructions of wavelets on bounded domains, the multiscale decomposition cannot be carried out down to the coarsest level of discretization: even in the simple case of the inteval, treated in §2.12 of Chapter 2, we have seen that the adaptation near the boundary often imposes a minimal level $j_0 \geq 0$ allowing us to decouple the construction of the modified scaling functions and wavelets at the edges. This affects in practice the ratio between the constants in the norm equivalence (3.11.11) and thus the resulting condition number of $D_J^{-1}\widetilde{A}_J$. Recall also that there is a certain flexibility in the construction of the modified basis functions at the edges (especially in the biorthogonal case) and certain choices might result in better constants than others.

In addition, the ratio between the constants of ellipticity and continuity of the bilinear form in (3.11.1) also affects the condition number of $D_J^{-1}\widetilde{A}_J$. This ratio deteriorates in particular in the case of non-constant coefficient operators: if for instance $Au = -\nabla[q(x)\nabla u]$, we find that the quantity $\sup|q(x)|/\inf|q(x)|$ enters in the resulting condition number.

In order to remedy these problems, a good preconditioner should somehow take into account such finer information on the multilevel decomposition and on the operators.

A first basic idea is to eliminate the problem of the coarse scale level by a full inversion of the stiffness matrix. More precisely, in the preconditioning of $\widetilde{A}_J$, we replace the purely diagonal matrix D_J by a block diagonal matrix X_J which consists of a purely diagonal part identical to D_J for the detail scales $|\lambda| \geq j_0$, and of the coarse scale stiffness matrix A_{j_0} for the basic scale $|\lambda| = j_0 - 1$, for which we have by convention $\{\psi_\lambda\}_{\lambda\in\nabla_{j_0-1}} = \{\varphi_\gamma\}_{\gamma\in\Gamma_{j_0}}$. The entries of the new matrix X_J are thus given by

$$X_J(\lambda,\mu) = 2^{2s|\lambda|}\delta_{\lambda,\mu} \text{ if } |\lambda| \geq j_0 \text{ or } |\mu| \geq j_0, \tag{3.11.31}$$

and

$$X_J(\lambda,\mu) = a(\psi_\lambda,\psi_\mu) \text{ if } |\lambda| = |\mu| = j_0. \tag{3.11.32}$$

Note that this matrix is easily inverted since the non-diagonal part is of small size.

Another simple idea is to replace the simple scaling by $2^{2s|\lambda|}$ in the diagonal part (3.11.31) by the corresponding diagonal elements of the stiffness matrix $\widetilde{A}_J$. This defines a new block diagonal matrix Y_J which is identical to X_J for the coarse components (3.11.32) and such that

$$Y_J(\lambda,\mu) = a(\psi_\lambda,\psi_\mu)\delta_{\lambda,\mu} \text{ if } |\lambda| \geq j_0 \text{ or } |\mu| \geq j_0. \tag{3.11.33}$$

By such a modification, we expect to be more adaptive with respect to the variation of the coefficients of the operator, as well as to the modifications of the scaling functions and wavelets near the edges.

A simple example which reveals the effectiveness of this approach is the singular perturbation problem

$$u - \varepsilon\Delta u = f \text{ in } \Omega, \quad u = 0 \text{ on } \partial\Omega. \tag{3.11.34}$$

The corresponding bilinear form is given by

$$a_\varepsilon(u,u) = \int_\Omega |u(x)|^2 dx + \varepsilon \int_\Omega |\nabla u(x)|^2 dx. \tag{3.11.35}$$

On $H = H_0^1$, we still have $a_\varepsilon(u,u) \sim \|u\|_{H_0^1}^2$ but we see that the ratio between the constants in this equivalence is of the order $1/\varepsilon$. As a consequence, we can only expect a condition number $\mathcal{K}(D_J^{-1}A_J)$ of the order $1/\varepsilon$.

We now remark that if $g \in V_J$ has coordinate vector $G = (c_\lambda)_{\lambda\in\nabla}$ in the wavelet basis, we have

$$\begin{aligned}
\langle \widetilde{A}_J G, G\rangle &= a_\varepsilon(g,g)\\
&\sim \textstyle\sum_{|\lambda|<J}(1+\varepsilon 2^{2s|\lambda|})|c_\lambda|^2\\
&\sim \textstyle\sum_{|\lambda|<J} a_\varepsilon(\psi_\lambda,\psi_\lambda)|c_\lambda|^2\\
&\sim \langle Y_J G, G\rangle,
\end{aligned}$$

with constants that are independent of both J and ε, provided that the wavelet basis characterizes both L^2 and H_0^1. Thus $R_J Y_J^{-1} R_J^*$ is more adapted to precondition A_J than the simpler B_J of Example 1.

In the case of varying coefficients, the effectiveness of such an adaptive method is also observed in practice. However it is not proved in full theoretical generality. Consider an operator of the type $Au := -\nabla[p(x)\nabla u]$ with scalar coefficients $0 < p_{\min} \leq p(x) \leq p_{\max}$. Ideally, one would like to prove that $\mathcal{K}(Y_J^{-1}\widetilde{A}_J)$ is independent of the ratio $p_{\max}/p_{\min}$, i.e. that the equivalence

$$\int_{\Omega} p(x)|\nabla u(x)|^2 \sim \sum_{|\lambda|<J} p_\lambda |c_\lambda|^2, \;\; p_\lambda := \int_{\Omega} p(x)|\nabla \psi_\lambda|^2 dx, \tag{3.11.36}$$

holds without dependence on $p_{\max}/p_{\min}$. Such a result is certainly false without any additional assumptions, for instance some bounds on $p(x)$ in a smoother norm than L^∞.

If one wants to improve further on the resulting condition number, a possibility is to also drop the diagonal structure of Y_J^{-1} in the high scales $|\lambda| \geq j_0$ and look for a preconditioner which is still "almost diagonal". This approach is proposed in COHEN and MASSON [1999], in which the simple inversion of $a(\psi_\lambda, \psi_\lambda)$ that yields Y_J^{-1} is replaced by a local Petrov-Galerkin problem which amounts to the inversion of a small matrix. A close strategy is proposed in LAZAAR, LIANDRAT and TCHAMITCHIAN [1994], using "vaguelettes" and freezing coefficients techniques in the periodic setting. Generally speaking, such preconditioners search for a compromise between the simple wavelet diagonal preconditioner $R_J D_J^{-1} R_J^*$ which has very low cost but might still yield large (although bounded) condition numbers, and the inverse A_J^{-1} of the operator which is too heavy to apply. The best compromize B_J should optimize the overall computational cost to reach an error ε in some prescribed norm, i.e. $\mathcal{C}(\varepsilon) = n(\varepsilon)C_0$ where $n(\varepsilon)$ is the number of needed iterations (which depends on the $\mathcal{K}(B_J A_J)$) and C_0 is the cost of one iteration (which depends on the number of unknowns at level J and on the sparsity of B_J). The position of such a compromise clearly depends on the type of problem at hand.

Example 4. An optimal complexity multiscale solver
If we want to evaluate the computational cost of the global resolution of (3.11.4) using the preconditioned iteration (3.11.16), we need to fix a certain target accuracy ε, that we shall allow in the $\|\cdot\|_H$ norm related to our problem. We then choose a resolution level J such that

$$\|u_J - u\|_H \leq \varepsilon. \tag{3.11.37}$$

where u_J is the Galerkin solution of (3.11.4) in V_J. Such a scale can be crudely obtained using a-priori information on the smoothness of the solution: for example, in the case of the Laplace equation $-\Delta u = f$ on a smooth domain with Dirichlet boundary conditions, we know that $u \in H^{r+2} \cap H_0^1$ if $f \in H^r$, $r \geq -1$, so the estimate

$$\|u - u_J\|_{H_0^1} \lesssim \operatorname{dist}_{H_0^1}(u, V_J) \lesssim 2^{-(r+1)J} \|u\|_{H^{r+2}}, \tag{3.11.38}$$

is ensured by the results of this chapter, provided that V_J has the appropriate polynomial reproduction properties. More generally, we assume here that the smoothness of u ensures that

$$\|u_j - u\|_H \le C(u)2^{-tj}, \ \ j \ge 0, \tag{3.11.39}$$

and we denote by $N_J := \dim(V_J) \sim 2^{dJ}$ the dimension of the space V_J which is selected to reach the accuracy $\varepsilon \sim C(u)2^{-tJ}$. Since $C(u)$ is in practice a smoother norm than the H norm, it is reasonable to assume in addition that

$$\|u\|_H \le C(u). \tag{3.11.40}$$

We are thus allowed to accept a similar error between u_J and the solution u_J^n obtained after n iterations of (3.11.16). Recall that this error e_J^n can be controlled by the residual $G_J^n = F_J - A_J U_J^n$, according to

$$\|e_J^n\|_H \sim \langle B_J G_J^n, G_J^n\rangle_d. \tag{3.11.41}$$

The preconditioned iteration decreases the error by a factor $\rho < 1$ which is independent of J. Therefore, if the algorithm is initialized from $U_J^0 = 0$, we can attain the error $\varepsilon \sim C(u)2^{-tJ}$ in a number n_J of iteration such that $\rho^{n_J} \lesssim 2^{-tJ}$ (since $\|e_J^0\|_H = \|u_J\|_H \lesssim C(u)$). Since one iteration has $\mathcal{O}(N_J)$ complexity, the overall computation cost is given by

$$\mathcal{C}(\varepsilon) \sim n_J N_J \lesssim N_J \log(N_J). \tag{3.11.42}$$

One can actually remove the $\log(N_J)$ factor with a simple elaboration of this multiscale solver, which simply consists in applying the first iterations only on the coarse scale and progressively refining the discretization. More precisely, we proceed as follows:

1. Start from a Galerkin discretization at the coarsest scale (here assumed to be $j = 0$ and apply the iteration (3.11.16) with $J = 0$ and initial condition $u_0^0 = 0$.

2. Stop after a number n of iterations such that $\|u_0^n - u_0\|_H \lesssim C(u)$ (using the residual evaluation).

3. Discretize the problem at the scale $j = 1$, and apply the iteration (3.11.16) with $J = 1$ and initial condition $u_1^0 = u_0^n \in V_0 \subset V_1$.

4. Stop after a number n of iterations such that $\|u_1^n - u_1\|_H \lesssim C(u)2^{-t}$ (using the residual evaluation).

5. Iterate this procedure up from $j = 2$ to $j = J$, requiring at the scale j the error $\|u_j^n - u_j\|_H \lesssim C(u)2^{-tj}$.

Clearly the result of this algorithm satisfies the required error estimate. Let us now compute the computational cost at each scale level: the initial error is estimated by

$$\begin{aligned}\|u_j - u_j^0\|_H &= \|u_j - u_{j-1}^n\|_H \\ &\le \|u_{j-1} - u_{j-1}^n\|_H + \|u_j - u_{j-1}\|_H \\ &\lesssim C(u)2^{-tj} + \|u - u_j\|_H + \|u - u_{j-1}\|_H \\ &\lesssim C(u)2^{-tj},\end{aligned}$$

where we have used the assumption (3.11.39). Therefore, a fixed number n_0 of iterations is required at each scale level. Each iteration having complexity $\mathcal{O}(N_j)$, we find that the overall computational cost is estimated by

$$\mathcal{C}(\varepsilon) \lesssim n_0 \sum_{j=0}^{J} 2^{dj} \lesssim 2^{dJ} \lesssim N_J. \tag{3.11.43}$$

We have thus obtained an algorithm of *optimal complexity*: the amount of computation is directly proportional to the number of parameters required to achieve the prescribed accuracy.

Remark 3.11.9 *This last optimal algorithm should be viewed as a reformulation of the nested iteration in the full multigrid algorithm introduced in* BANK *and* DUPONT *[1981] which also has optimal complexity.*

Remark 3.11.10 *Throughout this section, we only made use of the characterization of Sobolev spaces of positive order. The characterization of negative order Sobolev spaces can be required for the preconditioning of negative order operators. In several instances, these operators appear as a compact perturbation of the identity, and do not directly require a preconditioner. We shall see in §4.9 of the next chapter that the norm equivalences for negative order Sobolev spaces are needed to develop an a-posteriori analysis in the context of adaptive methods for solving elliptic equations. They can also be used to derive stabilization techniques for non-coercive problems, as shown in* BERTOLUZZA *[1998].*

3.12 Conclusions

The general multiscale approximation tools that we have developed in the previous chapter allow us to characterize a large number of smoothness classes, on $\mathbb{R}^d$ or on a bounded domain, with or without prescribed boundary conditions. The basic requirement is that these spaces satisfy proper

Jackson and Bernstein inequalities. Such estimates respectively rely on the polynomial reproduction and smoothness properties of the basis functions for the V_j spaces.

The resulting characterizations can be expressed in three different ways: in terms of the size properties (decay and summability) of the approximation error $\text{dist}_X(f, V_j)$ or $\|f - P_j f\|_X$ in a certain metric X, in terms of the size properties of the detail terms $\|P_{j+1} - P_j f\|_X$ or $\|f_{j+1} - f_j\|_X$, where f_j is a near-best approximation in V_j, or in terms of the size properties of the wavelet coefficients $d_\lambda := \langle f, \tilde{\psi}_\lambda \rangle$.

Such results are interesting from a theoretical point of view since they somehow reduce the description of smoothness classes to simple sequence spaces. The multiscale decomposition into wavelet bases can thus be viewed here as a tool of harmonic analysis which is much more flexible than the Fourier series and the Littlewood-Paley decomposition, since it can easily be adapted to general bounded domains.

From an application point of view, such characterizations (in the case of L^2-Sobolev spaces) can be used to design and analyze multilevel preconditioners for elliptic operator equations. Other types of applications involving more general Besov spaces for $p \neq 2$ will appear in our next chapter.

3.13 Historical notes

The possibility of characterizing the smoothness of a function by means of approximation properties became clear following the thesis of Jackson in 1911. At that time, the zoology of function spaces was limited to a few basic classes, and so were the existing approximation tools (algebraic and trigonometric polynomials).

The theory of function spaces came to flourish in the 1950's and 1960's (a nice historical survey can be found in TRIEBEL [1983]). The $B^s_{p,q}$ spaces (denoted by $\Lambda^s_{p,q}$ in several references) were introduced by Besov in BESOV [1959], who immediately pointed out the characterization of such spaces by approximation procedures based on trigonometric polynomials.

This period also corresponded to the development of spline functions and the systematic study of their approximation properties. A characterization of Besov spaces from the rate of approximation by "dyadic splines", i.e. the V_j spaces of spline functions with equally spaced knots at dyadic points $2^{-j}k$, was completed in CIESIELSKI [1973]. The analogous result using wavelet bases was proved in MEYER [1990].

Another vision of Besov spaces came through the real interpolation method introduced in LIONS and PEETRE [1964]. One of the main motivations for the introduction of this method was the theoretical study of

PDE's. In this setting, interpolation results for spaces with boundary conditions were proved in LIONS and MAGENES [1972] (for smooth domains and L^2-Sobolev spaces).

Approximation spaces were studied in relation to interpolation theory in PEETRE and SPAAR [1972] and BUTZER and SCHERER [1972]. Theorem 3.5.2 connecting these two concepts is given the same form in chapter VII of DEVORE and LORENTZ [1993], which also provides an overview of the identified relations between the analytical properties of functions and their rate of approximation by specific procedures.

The first ideas on multilevel iterative solvers for elliptic PDE's came out in the 1960's, in the work of FEODORENKO [1964] and BAKHVALOV [1966]. Multigrid theory was then developed in the 1970's in the pioneering work of Brandt, in the context of finite difference discretizations (see BRANDT [1977]). Important contributions to the theoretical understanding of multigrid methods were made in BANK and DUPONT [1981], which introduces and analyzes the full-multigrid technique, and in BRAESS and HACKBUSCH [1983] where a proof of the $\mathcal{O}(1)$ preconditioning property for the so-called V-cycle multigrid algorithm was given. There is now a huge amount of litterature on multigrid methods in various contexts. We refer to HACKBUSCH [1985] and BRAMBLE [1993] for a general treatment, and to OSWALD [1993] for an analysis of such methods from an approximation theoretical point of view. The use of multilevel preconditioners based on wavelets was proposed in JAFFARD [1992], while their connexions with the BPX-type preconditioners was analyzed in DAHMEN and KUNOTH [1992].

Chapter 4

Adaptivity

4.1 Introduction

Among those relevant phenomena which are modelled by partial differential or integral equations, countless are the instances where the mathematical solutions exhibit ***singularities***. Perhaps the most classic examples are elliptic equations on domains with re-entrant corners, or nonlinear hyperbolic systems of conservation laws. While such singularities are sources of obvious theoretical difficulties - classical solutions should be abandoned in favour of weak solutions - they are also an obstacle to the convergence of numerical approximation methods, in the sense that they deteriorate the rate of decay of the error with respect to the size of the discrete problem: achieving a prescribed accuracy will typically require finer resolution and therefore heavier computational cost and memory storage, in comparison to the approximation of smooth solutions.

More precisely, the results in the previous chapter have shown that a variety of smoothness classes can be characterized by the rate of decay of the multiscale approximation error. Roughly speaking, we have seen that

$$\mathrm{dist}(f, V_j)_{L^p} \lesssim 2^{-sj}, \tag{4.1.1}$$

if and only if f has s derivative in L^p. Such results express that the convergence of approximations to a function f at some prescribed rate, as the discretization step or mesh size tends ***uniformly*** to zero, always implies a certain amount of smoothness for this function. In turn, we cannot hope for a high rate of approximation in L^p (more generally in $W^{t,p}$ or $B^t_{p,q}$) if f has poor smoothness in L^p. If we work on a bounded domain $\Omega \subset \mathbb{R}^d$,

(4.1.1) also writes

$$\sigma_N(f) \lesssim N^{-s/d} \tag{4.1.2}$$

where $\sigma_N(f)$ is the best approximation error in $S_N = V_j$ defined by (3.1.1) with $X = L^p$ and $N := \dim(V_j) \sim 2^{dj}$ represents the number of parameters that describe the approximation. Therefore if s/d is small, N should be large in order to achieve a prescribed approximation error ε.

Let us remark that singularities often have a physical relevance: they represent the concentration of stress in elasticity, boundary layers in viscous fluid flows, shock waves in gas dynamics...It is therefore a legitimate requirement that they should be accurately resolved by the numerical method. In this context, the use of adaptive methods appears as a natural solution to improve the approximation at a reasonable computational cost. Here, the word *adaptivity* has a twofold meaning: (i) the discretization is allowed to be refined only locally, in particular near the singularities of the solution, and (ii) the resolution algorithm uses information gained during a given stage of the computation in order to derive a new refined discretization for the next stage. The most typical example is *adaptive mesh refinement* based on *a-posteriori* error estimates in the finite element context. In particular, the *rate of convergence* of the adaptive algorithm, which describes the tradeoff between the accuracy and the complexity of the approximation, is not always clearly understood.

From the point of view of approximation theory introduced in §3.1, adaptive finite element spaces can be described as follows: the number of parameters N is proportional to the number of elements, but for a given budget N the partition $\mathcal{T}$ and the finite element space $V_{\mathcal{T}}$ associated with this partition are allowed to be locally refined in a way which depends on the function f to be approximated. It is therefore natural to define the approximation spaces S_N as

$$S_N := \cup_{|\mathcal{T}| \le N} V_{\mathcal{T}}, \tag{4.1.3}$$

i.e. the union of all finite element spaces of a fixed type associated with a partition of at most N elements. It should be well understood that the S_N are not linear vector spaces (the sum of two elements does not in general fall in S_N when their triangulations do not match) but any $g \in S_N$ is still described by $\mathcal{O}(N)$ parameters, which encode both its triangulation $\mathcal{T}$ and its coordinates in $V_{\mathcal{T}}$. The requirement of adaptivity has thus led us to the concept of *nonlinear approximation*.

In the context of wavelet discretizations, adaptive approximations are obtained by using only a limited set of indices λ as the scale $|\lambda|$ grows, which depends on the function to be approximated and typically corresponds to those wavelets whose supports are close to its singularities. It is therefore natural to define the approximation spaces S_N as the set of all N-term combinations

$$S_N := \{\sum_{\lambda\in\Lambda} d_\lambda\psi_\lambda \ ; \ |\Lambda| \leq N\}. \tag{4.1.4}$$

(in this chapter, we shall systematically use the general notations that were proposed in the introduction of the previous chapter: $\{\psi_\lambda\}_{\lambda\in\nabla}$ will denote a wavelet basis allowing us to characterize function spaces defined on Ω, where Ω is either $\mathbb{R}^d$ or a bounded domain of the type addressed in §3.9 of the previous chapter). Again this is obviously not a linear space, since we allow ourself to approximate a function by choosing the best N terms which correspond with different sets of indices from one function to another. Note that we still do have $S_N + S_N = S_{2N}$.

Both adaptive finite element and wavelet frameworks have obvious similarities. However, the answer to the two basic questions raised in §3.1 - what are the properties of f which govern the decay of the best approximation error $\sigma_N(f)$ in some norm X and how to compute in a simple way a near-optimal approximation of f in S_N - is only fully understood in the wavelet framework.

Concerning the first question, a striking result due to DeVore and his collaborators is the following: roughly speaking with $X := W^{t,p}$, best N-term wavelet approximations satisfy

$$f \in W^{t+r,q} \Rightarrow \sigma_N(f) \leq CN^{-r/d}, \tag{4.1.5}$$

with q and r connected by the relation $1/q = 1/p + r/d$, assuming that the multiresolution approximation spaces associated with the wavelet basis are in $W^{t,p}$ and contain the polynomials of degree strictly less than $t+r$. Such an estimate should be compared with the linear estimate (3.1.6): the same convergence rate is governed by a much weaker smoothness condition on f, especially when s becomes large (in which case q might be less than 1). This smoothness condition essentially measures the ***sparsity*** of the wavelet decomposition of f, i.e. the fact that the $W^{t,p}$ norm is concentrated on a small number of coefficients. In particular, a function f which has isolated singularities - and therefore a sparse wavelet decomposition - usually has a smaller amount of smoothness $s+t$ when measured in L^p than when measured in L^q with $1/q = 1/p + t/d$, and therefore an adaptive discretization will be substantially more efficient than a uniform discretization.

The answer to the second question is relatively simple when one has access to the wavelet expansion $f = \sum_{\lambda\in\nabla} d_\lambda\psi_\lambda$ of the function f to be approximated. In that case, a natural N-term approximation in X is provided by the choice

$$f_N = \sum_{\lambda \in \Lambda_N} c_\lambda \psi_\lambda, \tag{4.1.6}$$

where $\Lambda_N = \Lambda_N(f, X)$ is the set of indices corresponding to the N largest contributions $\|c_\lambda \psi_\lambda\|_X$. Then, we shall see that for several interesting choices of X, we have

$$\|f - f_N\|_X \lesssim \sigma_N(f), \tag{4.1.7}$$

i.e. a simple thresholding of the largest contributions in the wavelet decomposition provides a near-optimal N-term approximation. This fact is obvious when $X = L^2$ using the orthonormal basis property. It is a remarkable property of wavelet bases that it also holds for more general function spaces. In summary, thresholding plays for best N-term wavelet approximation the same role as projection for linear approximation.

All these results will be proved in §4.2 for $X = L^2$ or a more general Besov space, and in §4.3 for $X = L^p$ or $W^{m,p}$, $p \neq 2$. As in the linear case, "if and only if" statements will be obtained up to slight technical modifications in the statement of (4.1.5).

We devote §4.4 to a review of the adaptive finite element counterpart of these results. Nonlinear approximation theory is a rich subject that also includes approximation by rational functions and dictionaries of bases. We review in §4.5 some important examples in this general theory and compare them to N-term wavelet approximation, in particular from the perspective of the *curse of dimensionality* when dealing with anisotropic singularities. We also evoke here the celebrated application of nonlinear wavelet approximation to image data compression.

These results suggest reconsidering the notion of smoothness for solutions of PDE's that develop singularities: such solutions might possess significantly higher smoothness in the scale of function spaces which governs the rate of nonlinear approximation in a given norm (or equivalently the rate of sparsity in the wavelet basis) than in the scale which governs the rate of linear approximation in the same norm. In order to understand the potential of adaptive discretizations, we should thus turn to an analysis of PDE's in terms of sparsity rather than of classical smoothness. Such an analysis is, at the present time, still in its infancy, and we can only provide here an uncomplete picture. At the start, one needs to understand how the action of the operators involved in partial differential as well as integral equations affects the sparsity of a function. An important observation, that we discuss in §4.6, is that for a large class of such operators the matrices resulting from wavelet discretizations are sparse and have local approximations. In turn, their action on a function preserves the sparsity of its wavelet coefficients. We then review in §4.7 several instances of PDE's for which a theoretical analysis shows that, despite the presence of singularities in the solution, high order nonlinear approximation can be achieved. These results show that if u is the solution of such equations, the rate of decay of $\sigma_N(u)$ is significantly higher for best N-term approximation than for the projection on uniform finite element spaces, hence advocating the use of adaptive discretizations of such PDE's.

On the other hand these results also provide an ideal *benchmark* for adaptive discretizations of the equation, since $\sigma_N(u)$ represents the best accuracy which can be achieved by N parameters. In the wavelet case these parameters are typically the N largest coefficients of the exact solution u. However, in the practice of solving a PDE, these coefficients are not known, and neither is the set Λ_N of the indices of the N largest contributions $\|d_\lambda \psi_\lambda\|$. It is therefore necessary to develop appropriate *adaptive resolution strategies* as a substitute for the thresholding procedure. Such strategies aim at detecting the indices of the largest coefficients of the solutions and to compute them accurately, in a similar way that adaptive mesh refinement strategies aim at constructing the optimal mesh for finite element approximation. We end this chapter with a review of the two main lines of thought which have been followed in the development of adaptive wavelet strategies.

The first approach, described in §4.8, applies to a general class of *evolution problems*

$$\partial_t u = \mathcal{E}(u) \tag{4.1.8}$$

for which we have at our disposal a reliable numerical scheme which operates on a fine uniform grid. Wavelet decompositions are then used as an adaptive post-processing device, which aims to reduce the computational cost and memory storage required by this scheme, while preserving its initial accuracy.

The second approach, described in §4.9, applies to a general class of *stationary problems*

$$\mathcal{F}(u) = 0, \tag{4.1.9}$$

expressed by a variational formulation and should be thought of as an adaptive *space refinement* strategy in which the set of wavelet coefficients Λ that describes the approximate solution is iteratively updated. This approach has led to a new paradigm which contrasts significantly with the standard adaptive finite element approach. In several cases, one can prove that it is optimal with respect to the above benchmark in the sense that the adaptive solution $u_N \in S_N$ converges to u at the same rate as the best N term approximation and can be computed at $\mathcal{O}(N)$ cost.

4.2 Nonlinear approximation in Besov spaces

As a starter, we consider N-term approximation in the L^2-norm: we are interested in the behaviour of $\text{dist}_{L^2}(f, S_N)$ as N goes to infinity, where S_N is defined by (4.1.4). For the sake of simplicity in this first example, let us assume here that $\{\psi_\lambda\}_{\lambda\in\nabla}$ is an orthonormal basis for L^2. Thus any $f \in L^2(\Omega)$ can be decomposed into

$$f = \sum_{\lambda\in\nabla} c_\lambda \psi_\lambda, \quad c_\lambda = \langle f, \psi_\lambda \rangle, \tag{4.2.1}$$

and we can define the set $\Lambda_N = \Lambda_N(f) \subset \nabla$ of the N largest coefficients of f, i.e. such that $|\Lambda_N| = N$ and

$$\lambda \in \Lambda_N, \lambda' \notin \Lambda_N \Rightarrow |c_{\lambda'}| \leq |c_\lambda|. \tag{4.2.2}$$

Note that there might be several possible choices for Λ_N if the moduli of several coefficients of f take the same value. In such a case, we simply take for Λ_N any of the sets of cardinality N that satisfy (4.2.2). Since $\{\psi_\lambda\}_{\lambda\in\nabla}$ is an orthonormal basis, we clearly have

$$\text{dist}_{L^2}(f, S_N) = \|f - \sum_{\lambda\in\Lambda_N} c_\lambda \psi_\lambda\|_{L^2} = \Big(\sum_{\lambda\notin\Lambda_N} |c_\lambda|^2\Big)^{1/2}. \tag{4.2.3}$$

If we define by $(c_n)_{n>0}$ any rearrangement of the coefficients $(c_\lambda)_{\lambda\in\nabla}$ with decreasing moduli, i.e. such that $|c_{n+1}| \leq |c_n|$, we can express (4.2.3) as follows

$$\text{dist}_{L^2}(f, S_N) = \Big(\sum_{n>N} |c_n|^2\Big)^{1/2}. \tag{4.2.4}$$

Let us now consider the spaces $B^s_{q,q}$ where $s > 0$ and q is such that $1/q = 1/2 + s/d$. We assume here that the conditions of Theorem 3.7.6 (or of its generalization to a bounded domain Ω), are satisfied so that $B^s_{q,q}$ is characterized by (3.7.32). For such indices, we note that this equivalence can be simplified into

$$\|f\|_{B^s_{q,q}} \sim \|(c_\lambda)_{\lambda\in\nabla}\|_{\ell^q}. \tag{4.2.5}$$

A first immediate consequence of (4.2.5) is the embedding of $B^s_{q,q}$ in L^2 (announced in §3.7 of the previous chapter), since ℓ^q is trivially embedded in ℓ^2. Note that such an embedding is not compact: the canonical sequence $(s_n)_{n\geq 0} = ((\delta_{n,k})_{k\geq 0})_{n\geq 0}$ is uniformly bounded in ℓ^q but does not contain any subsequence that converges in ℓ^2.

Since $(c_n)_{n>0}$ is a decreasing rearrangement of the coefficients $(c_\lambda)_{\lambda\in\nabla}$, we have

$$n|c_n|^q \leq \sum_{k>0} |c_k|^q \sim \|f\|^q_{B^s_{q,q}}, \tag{4.2.6}$$

which yields

$$|c_n| \lesssim \|f\|_{B^s_{q,q}} n^{-1/q}. \tag{4.2.7}$$

Combining (4.2.7) with (4.2.4) yields

$$\begin{aligned} \mathrm{dist}_{L^2}(f, S_N) &= \left(\sum_{n>N} |c_n|^2\right)^{1/2} \\ &\lesssim \left(\sum_{n>N} n^{-2/q}\right)^{1/2} \|f\|_{B^s_{q,q}} \\ &\lesssim N^{-1/q+1/2} \|f\|_{B^s_{q,q}} \\ &= N^{-s/d} \|f\|^q_{B^s_{q,q}}. \end{aligned}$$

We have thus obtained a Jackson-type estimate

$$\mathrm{dist}_{L^2}(f, S_N) \lesssim N^{-s/d} \|f\|^q_{B^s_{q,q}}, \tag{4.2.8}$$

with respect to the nonlinear spaces S_N.

Remark 4.2.1 *While we have used that for $q < 2$*

$$|c_n| \lesssim n^{-1/q}, \tag{4.2.9}$$

implies

$$\left(\sum_{n>N} |c_n|^2\right)^{1/2} \lesssim N^{1/2-1/q}, \tag{4.2.10}$$

it is readily seen that these two properties are actually equivalent: if (4.2.10) holds, it follows that

$$n^{1/2}|c_{2n}| \leq [\sum_{k=n}^{2n-1} |c_n|^2]^{1/2} \lesssim n^{1/2-1/q} \tag{4.2.11}$$

and therefore (4.2.9) also holds. Another equivalent property is that for all $\varepsilon > 0$

$$\#\{\lambda \in \nabla\ ;\ |c_\lambda| \geq \eta\} \lesssim \eta^{-q}, \tag{4.2.12}$$

i.e. $(c_\lambda)_{\lambda\in\nabla}$ belongs to the weak space ℓ^q_w. Intuitively, the parameter q measures the level of sparsity of the sequence $(c_\lambda)_{\lambda\in\nabla}$ since it controls the growth of the number of terms above a threshold η as it tends to 0: the sequence is sparser if q is smaller.

Let us now observe that if $f \in S_N$, we also have by Hölder's inequality

$$\|f\|_{B^s_{q,q}} \lesssim \|(c_\lambda)_{\lambda\in\nabla}\|_{\ell^q} \leq N^{1/q-1/2}\|(c_\lambda)_{\lambda\in\nabla}\|_{\ell^2} = N^{s/d}\|f\|_{L^2}, \quad (4.2.13)$$

i.e. a Bernstein-type estimate.

The equivalence (4.2.5) also shows that for $0 < t < s$ and $1/r = 1/2+t/d$, we have the interpolation identity

$$B^t_{r,r} = [L^2, B^s_{q,q}]_{\theta,r}, \quad \theta = t/s, \quad (4.2.14)$$

which is a simple re-expression of $[\ell^2, \ell^q]_{\theta,r} = \ell^r$.

We thus have in our hand the three ingredients - direct estimates, inverse estimates and interpolation theory - that were used in the previous chapter to characterize function spaces by the error of linear approximation. At this point, we need to take into account the fact that the spaces S_N are not linear spaces: if these spaces were linear, defining for $j \geq 0$,

$$\Sigma_j := S_{2^{dj}}, \quad (4.2.15)$$

a straightforward application of Theorem 3.5.2 with $V_j = \Sigma_j$ would give

$$\|f\|_{B^t_{r,r}} \lesssim \|f\|_{L^2} + \|(2^{jt}\mathrm{dist}_{L^2}(f,\Sigma_j))_{j\geq 0}\|_{\ell^r}. \quad (4.2.16)$$

It turns out that such a result also holds in the present nonlinear context, using the following adaptation of Theorem 3.5.2. Here, as well as in all the following, the approximation spaces $\mathcal{A}^s_q(X)$ and $\mathcal{A}^s_{p,q}$ have the same definition as in §3.5 of Chapter 3 with the spaces Σ_j in place of V_j.

Theorem 4.2.1 *Assume that X and Y are quasi-normed spaces (satisfying the μ-triangle inequality (3.2.22)) and that Σ_j, $j \geq 0$, is a sequence of nonlinear approximation spaces*

$$\Sigma_j \subset \Sigma_{j+1} \subset \cdots \subset Y \subset X, \quad (4.2.17)$$

such that for some $m > 0$, one has a Jackson-type estimate

$$\mathrm{dist}_X(f,\Sigma_j) = \inf_{g\in\Sigma_j} \|f-g\|_X \lesssim 2^{-mj}\|f\|_Y, \quad (4.2.18)$$

and a Bernstein-type estimate

$$\|f\|_Y \lesssim 2^{mj}\|f\|_X \text{ if } f \in \Sigma_j. \quad (4.2.19)$$

Moreover assume that there exists a fixed integer a such that

$$\Sigma_j + \Sigma_j \subset \Sigma_{j+a}, \quad j \geq 0. \quad (4.2.20)$$

Then, for $s \in]0,m[$, *one has the norm equivalence*

$$\|(2^{js}K(f,2^{-mj}))_{j\geq 0}\|_{\ell^q} \sim \|f\|_X + \|(2^{js}\text{dist}_X(f,\Sigma_j))_{j\geq 0}\|_{\ell^q}, \quad (4.2.21)$$

and thus $[X,Y]_{\theta,q} = \mathcal{A}^s_q(X)$ *for* $s = \theta m$.

Proof We mimic the proof of Theorem 3.5.2 (and its adaptation to quasi-norms as explained in Remark 3.5.2). In this proof, the linearity of the approximation spaces is only involved in the estimate

$$\|f_j\|_Y \lesssim [\|f_0\|_X^\mu + \sum_{l=0}^{j-1} 2^{lm}\|f_{l+1} - f_l\|_X^\mu]^{1/\mu}, \quad (4.2.22)$$

where we use that $f_{l+1} - f_l \in V_{l+1}$ in order to apply the inverse estimate. Here, we can still obtain this result up to a change in the constant, since $f_{l+1} - f_l \in V_{l+1+a}$ by (4.2.20). $\diamond$.

In the above result, the assumption (4.2.20) should be viewed as a limitation on the nonlinearity of the spaces Σ_j. In the present context of N-term approximation, we already noted that $S_n + S_m \subset S_{n+m}$, so that this assumption holds with $a = 1$. We thus conclude that the norm equivalence (4.2.16) indeed holds.

Remark 4.2.2 *From the monotonicity of the sequence* $\text{dist}_{L^2}(f,S_N)$, *we have the equivalence*

$$\sum_{j\geq 0}[2^{jt}\text{dist}_{L^2}(f,\Sigma_j)]^r \sim \sum_{N\geq 1} N^{-1}[N^{t/d}\text{dist}_{L^2}(f,S_N)]^r. \quad (4.2.23)$$

The finiteness of the above quantities for $r < \infty$ *is a slightly stronger property than* $\text{dist}_{L^2}(f,S_N) \lesssim N^{-t/d}$ *which was initially obtained in the direct estimate (4.2.8). According to the above theorem, this last property characterizes the intermediate space*

$$[L^2, B^s_{q,q}]_{\theta,\infty} = \mathcal{A}^t_{2,\infty}, \quad (4.2.24)$$

with $t = \theta s$, *which cannot be thought as of a Besov space. According to Remark 4.2.1, this space is also characterized by the property that* $(c_\lambda)_{\lambda\in\nabla}$ *belongs to the weak space* ℓ^r_w. *More generally,* $\mathcal{A}^t_{2,r'}$ *identifies with the Lorentz space* $\ell^{r,r'} = [\ell^r,\ell^q]_{\theta,r'}$ *(one has in particular* $\ell^{r,r} = \ell^r$ *and* $\ell^{r,\infty} = \ell^r_w$*).*

It is not difficult to extend the above results to the case where the error is measured in more general Besov spaces of the type $B^s_{p,p}$. Recall that such spaces include Hölder spaces ($p = \infty$), and Sobolev spaces (of non-integer smoothness if $p \neq 2$). Here, we leave aside the orthonormality assumption on the basis $\{\psi_\lambda\}_{\lambda\in\nabla}$ which is irrelevant at this stage of generality. A first result is that a near-best N-term approximation in $B^s_{p,p}$ is still provided by a simple thresholding procedure.

Lemma 4.2.1 *Assuming that $B^s_{p,p}$ admits a wavelet characterization of the type (3.7.32), and if $f = \sum_{\lambda\in\nabla} c_\lambda\psi_\lambda$, then*

$$\mathrm{dist}_{B^s_{p,p}}(f, S_N) \lesssim \|f - \sum_{\lambda\in\Lambda_N} c_\lambda\psi_\lambda\|_{B^s_{p,p}}, \tag{4.2.25}$$

where $\Lambda_N = \Lambda_N(f, B^s_{p,p})$ is the set of indices corresponding to the N largest contributions $\|c_\lambda\psi_\lambda\|_{B^s_{p,p}}$ or equivalently the N largest $2^{(s+d/2-d/p)|\lambda|}|c_\lambda|$.

Proof The norm equivalence (3.7.32) shows that $\|\psi_\lambda\|_{B^s_{p,p}} \sim 2^{(s+d/2-d/p)|\lambda|}$. It can thus be reformulated as

$$\|f\|_{B^s_{p,p}} \sim \|(\|c_\lambda\psi_\lambda\|_{B^s_{p,p}})_{\lambda\in\nabla}\|_{\ell^p}. \tag{4.2.26}$$

Clearly, the N-term approximation $\sum_{\lambda\in\Lambda_N} c_\lambda\psi_\lambda$ minimizes the distance between f and S_N when measured in this equivalent norm for $B^s_{p,p}$. It is thus a near minimizer for the $B^s_{p,p}$ norm in the sense of (4.2.25). $\diamond$.

We are now ready to prove the general result of nonlinear wavelet approximation in Besov spaces.

Theorem 4.2.2 *Assume that the spaces $B^t_{q,q}$, $t - s = d/q - d/p$ admit a wavelet characterization of the type (3.7.32) for $t \in [s, s']$, $s' > s$. Then, for $t \in]s, s'[$, $t - s = d/q - d/p$, we have the norm equivalence*

$$\|f\|_{B^t_{q,q}} \sim \|f\|_{B^s_{p,p}} + \|(2^{j(t-s)}\mathrm{dist}_{B^s_{p,p}}(f, \Sigma_j))_{j\geq 0}\|_{\ell^q}. \tag{4.2.27}$$

Proof If $t > s$ and $t - s = d/q - d/p$, the norm equivalence (3.7.32) can be rewritten as

$$\|f\|_{B^t_{q,q}} \sim \|(\|c_\lambda\psi_\lambda\|_{B^s_{p,p}})_{\lambda\in\nabla}\|_{\ell^q}, \tag{4.2.28}$$

where $c_\lambda = \langle f, \tilde\psi_\lambda\rangle$ are again the coefficients of f. We can then proceed in a similar way as in the particular case of approximation in L^2. Denoting by $(c_n)_{n>0}$ the decreasing rearrangement of the sequence $(\|c_\lambda\psi_\lambda\|_{B^s_{p,p}})_{\lambda\in\nabla}$, we remark that since

$$nc_n^q \leq \sum_{k>0} c_k^q \lesssim \|f\|^q_{B^t_{q,q}}, \tag{4.2.29}$$

we have the estimate

$$c_n \lesssim n^{-1/q} \|f\|_{B^t_{q,q}}. \quad (4.2.30)$$

Denoting by Λ_N a set of indices defined as in Lemma 4.2.1, we obtain

$$\begin{aligned} \mathrm{dist}_{B^s_{p,p}}(f, S_N) &\leq \|f - \textstyle\sum_{\lambda\in\Lambda_N} c_\lambda \psi_\lambda\|_{B^s_{p,p}} \\ &\lesssim (\textstyle\sum_{n\geq N} c_n^p)^{1/p} \\ &\lesssim N^{1/p-1/q} \|f\|_{B^t_{q,q}}. \end{aligned}$$

We have thus established the Jackson-type estimate

$$\mathrm{dist}_{B^s_{p,p}}(f, S_N) \leq N^{-(t-s)/d} \|f\|_{B^t_{q,q}}. \quad (4.2.31)$$

If $f \in S_N$, we also have by (4.2.28) and Hölder's inequality

$$\|f\|_{B^t_{q,q}} \lesssim N^{1/q-1/p} \|(\|c_\lambda\psi_\lambda\|_{B^s_{p,p}})_{\lambda\in\nabla}\|_{\ell^p} = N^{(t-s)/d} \|f\|_{B^s_{p,p}}, \quad (4.2.32)$$

i.e. the corresponding Bernstein-type estimate. The equivalence (4.2.28) also shows that the spaces $B^t_{q,q}$, $t - s = d/q - d/p$ are interpolation spaces: for $s < t < s'$, $t - s = d/q - d/p$, and $s' - s = d/p' - d/p$, we have the interpolation identity

$$B^t_{q,q} = [B^s_{p,p}, B^{s'}_{p',p'}]_{\theta,q}, \quad \theta = (t-s)/(s'-s), \quad (4.2.33)$$

which is a simple re-expression of $[\ell^p, \ell^{p'}]_{\theta,q} = \ell^q$. By Theorem 4.2.1, we thus obtain the norm equivalence (4.2.27). ◇

Remark 4.2.3 *Note that we did not assume here that p or q are larger than 1. This is particularly important for the index q: if we fix the space $B^s_{p,p}$ in which the error is measured, q always become smaller than 1 for large values of t, due to the relation $1/q - 1/p = t/d - s/d$. Thus, the correct description of those functions that can be approximated at a high rate by an adapted combination of N wavelets necessarily involves Besov spaces with $q < 1$.*

Remark 4.2.4 *Similarly to Remark 4.2.1 in the case of L^2 approximation, for all $q < p$ there is an equivalence between the properties*

$$c_n \lesssim n^{-1/q}, \quad (4.2.34)$$

and

$$\Big(\sum_{n\geq N} c_n^p\Big)^{1/p} \lesssim N^{1/p-1/q}, \quad (4.2.35)$$

with the usual modification if $p = \infty$. Another equivalent statement is that $(\|c_\lambda\psi_\lambda\|_{B^s_{p,p}})_{\lambda\in\nabla}$ belongs to the weak space ℓ^q_w which does not identify with a classical Besov space.

Remark 4.2.5 *The linear approximation result (Corollary 3.6.1) tells us that the same error rate* $\mathcal{O}(N^{-(t-s)/d})$ *in* $B^s_{p,p}$ *is achieved by a linear method for functions in* $B^t_{p,p}$*, i.e. a smaller space than* $B^t_{q,q}$*. It should be noted that, as* t *becomes large, the functions in the space* $B^t_{p,p}$ *become smooth in the classical sense, while* $B^t_{q,q}$ *might still contain discontinuous functions. As an example, consider a function* f *defined on a bounded interval which coincides with* C^∞ *functions except on a finite set of discontinuities. One easily checks that* $f \in B^s_{p,p}$ *if and only if* $s < 1/p$*, so that the* L^2 *approximation rate is limited to* $N^{-1/2}$ *in the linear case, while in the nonlinear case we have the "spectral estimate"*

$$\mathrm{dist}_{L^2}(f, S_N) \le C_s N^{-s}, \tag{4.2.36}$$

for any $s \le n+1$ *where* n *is the order of polynomial reproduction associated with the wavelet basis.*

Remark 4.2.6 *The space* $B^t_{q,q}$ *is contained in* $B^s_{p,p}$ *since these spaces are also defined as* ℓ^q *and* ℓ^p *of the sequences* $\|c_\lambda\psi_\lambda\|_{B^s_{p,p}}$*. Note that this embedding is not compact, and that neither* $B^{t-\varepsilon}_{q,q}$ *nor* $B^t_{q-\varepsilon,q}$ *is contained in* $B^s_{p,p}$ *for any* $\varepsilon > 0$*. This critical situation is graphically summarized in Figure 4.2.1: any function space corresponding to "*s *derivative in* L^p*" is represented by the point* $(1/p, s)$*, and the space* $B^t_{q,q}$ *sits on the critical line of slope* d *initiated from* $B^s_{p,p}$*.*

Remark 4.2.7 *All the results in this section can be adapted in a straightforward manner to the case of Besov spaces with boundary conditions, using the characterization (3.10.26) of Theorem 3.10.5 in place of (3.7.32). This remark also applies to the results of the next section.*

4.3 Nonlinear wavelet approximation in L^p

The results of §4.2 are rather easy to prove, due to the simple links that exist between Besov spaces and ℓ^p spaces through wavelet decompositions. For example, Theorem 4.2.2 can be viewed as the rewriting in terms of Besov spaces of the following simple norm equivalence for sequences: if $(c_n)_{n>0}$ is a positive decreasing sequence and $d_n = \|(c_k)_{k\ge n}\|_{\ell^p}$, then for $q < p$, one has

$$\|(c_n)_{n>0}\|_{\ell^q} \sim [\sum_{n>0} n^{-1}(n^{1/q-1/p}d_n)^q]^{1/q} = \|(n^{-1/p}d_n)_{n>0}\|_{\ell^q}. \tag{4.3.1}$$

The study of best N-term approximation in L^p norm is more difficult since, as it was pointed out in §3.8 of Chapter 3, we cannot identify L^p with

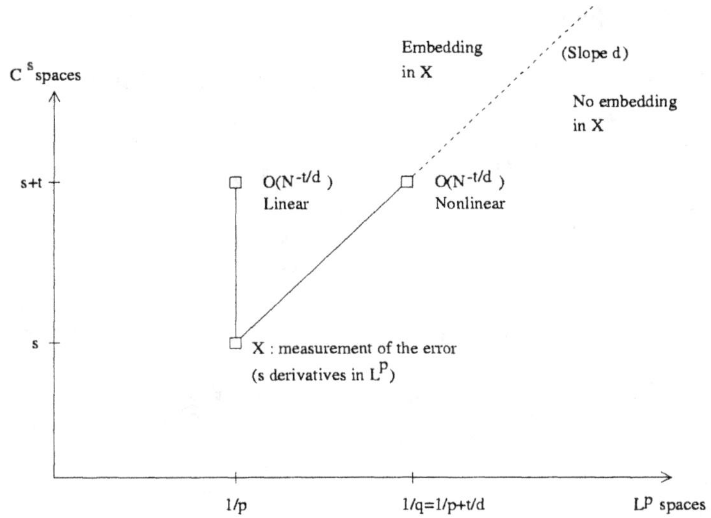

Figure 4.2.1: Pictorial interpretation of linear and nonlinear approximation

$B^0_{p,p}$ for $p \neq 2$. In this section, we shall see that switching from $B^0_{p,p}$ (which is treated by Theorem 4.2.2) to L^p does not affect the results of N-term approximation for $1 < p < \infty$.

Let us start with a basic result that allows us to estimate the L^p norms of a linear combination of wavelets according to the size of the coefficients.

Lemma 4.3.1 *Let $1 \leq p < \infty$, and assume that the wavelet basis $\{\psi_\lambda\}_{\lambda \in \nabla}$ is generated from scaling functions in L^p. If Λ is a 272subset of ∇ of cardinality $|\Lambda| < \infty$ and if $(c_\lambda)_{\lambda \in \Lambda}$ is a sequence such that $\|c_\lambda \psi_\lambda\|_{L^p} \leq 1$, $i = 1, \cdots, N$, then*

$$\|\sum_{\lambda \in \Lambda} c_\lambda \psi_\lambda\|_{L^p} \leq C|\Lambda|^{1/p}, \tag{4.3.2}$$

where C is independent of $|\Lambda|$.

Proof By assumption, we have for $\lambda \in \Lambda$

$$\|c_\lambda \psi_\lambda\|_{L^\infty} \lesssim 2^{|\lambda|d/p} \|c_\lambda \psi_\lambda\|_{L^p} \leq 2^{|\lambda|d/p}. \tag{4.3.3}$$

Let us define $\Lambda_j = \Lambda \cap \nabla_j$ and $N_j = |\Lambda_j|$ the number of wavelets at level j in Λ. By the controlled overlapping property (3.1.15), we have

$$\|\sum_{\lambda \in \Lambda_j} c_\lambda \psi_\lambda\|_{L^\infty} \lesssim 2^{jd/p}. \tag{4.3.4}$$

For $x \in \Omega$, we define

$$j(x) = \max\{j \geq -1 \; ; \; x \in \mathrm{Supp}(\psi_\lambda) \text{ for some } \lambda \in \Lambda_j\}, \tag{4.3.5}$$

and

$$\Omega_j := \{x \in \Omega \; ; \; j(x) = j\}. \tag{4.3.6}$$

Thus $\Omega = \cup_{j \geq -1} \Omega_j$. We next remark that

$$|\Omega_j| \leq |\cup_{\lambda \in \Lambda_j} \mathrm{Supp}(\psi_\lambda)| \lesssim N_j 2^{-dj}, \tag{4.3.7}$$

and that for $x \in \Omega_j$, we have by (4.3.4)

$$|\sum_{\lambda \in \Lambda} c_\lambda \psi_\lambda| \leq \sum_{l \leq j} |\sum_{\lambda \in \Lambda_l} c_\lambda \psi_\lambda| \lesssim \sum_{l \leq j} 2^{ld/p} \lesssim 2^{jd/p}. \tag{4.3.8}$$

It follows that

$$\begin{aligned} \|\textstyle\sum_{\lambda \in \Lambda} c_\lambda \psi_\lambda\|_{L^p}^p &= \textstyle\sum_{j \geq 1} \int_{\Omega_j} |\sum_{\lambda \in \Lambda} c_\lambda \psi_\lambda|^p \\ &\lesssim \textstyle\sum_{j \geq 1} |\Omega_j| 2^{jd} \\ &\lesssim \textstyle\sum_{j \geq 1} N_j = |\Lambda| \end{aligned}$$

which concludes the proof. ◇

Note that, by rescaling, this result gives for a general finite linear combination

$$\|\sum_{\lambda \in \Lambda} c_\lambda \psi_\lambda\|_{L^p} \lesssim |\Lambda|^{1/p} \sup_{\lambda \in \Lambda} \|c_\lambda \psi_\lambda\|_{L^p}. \tag{4.3.9}$$

By duality, we shall obtain an analogous result with reversed inequalities. We assume here in addition that the dual basis $\{\psi_\lambda\}_{\lambda \in \nabla}$ is generated from scaling functions in $L^{p'}$, where $1/p + 1/p' = 1$.

Lemma 4.3.2 *Let $1 < p \leq \infty$. If Λ is a finite subset of ∇ of cardinality $|\Lambda| < \infty$ and if $(c_\lambda)_{\lambda \in \Lambda}$ is a sequence such that $\|c_\lambda \psi_\lambda\|_{L^p} \geq 1$, $i = 1, \cdots, N$, then*

$$\|\sum_{\lambda \in \Lambda} c_\lambda \psi_\lambda\|_{L^p} \geq C|\Lambda|^{1/p}, \tag{4.3.10}$$

where C is independent of $|\Lambda|$.

Proof If $f = \sum_{\lambda \in \Lambda} c_\lambda \psi_\lambda$, we define

$$g = \sum_{\lambda \in \Lambda} |c_\lambda|^{-2} \overline{c_\lambda} \tilde{\psi}_\lambda, \tag{4.3.11}$$

so that

$$|\Lambda| = \langle f, g\rangle \leq \|f\|_{L^p}\|g\|_{L^{p'}}. \tag{4.3.12}$$

Remarking that

$$\begin{aligned} \|\overline{c_\lambda}|c_\lambda|^{-2}\psi_\lambda\|_{L^{p'}} &\lesssim |c_\lambda|^{-1}2^{d|\lambda|(1/2-1/p')} \\ &\lesssim \|c_\lambda\psi_\lambda\|_{L^p}^{-1} \lesssim 1, \end{aligned}$$

we derive from Lemma 4.3.1 that

$$\|g\|_{L^{p'}} \lesssim |\Lambda|^{1/p'}. \tag{4.3.13}$$

Combining (4.3.12) and (4.3.13) yields (4.3.10). ◇

We also get, by rescaling, for a general finite linear combination

$$|\Lambda|^{1/p} \inf_{\lambda\in\Lambda} \|c_\lambda\psi_\lambda\|_{L^p} \lesssim \|\sum_{\lambda\in\Lambda} c_\lambda\psi_\lambda\|_{L^p}. \tag{4.3.14}$$

A first consequence of (4.3.9) and (4.3.14) is the following result originally proved in TEMLYAKOV [1998]: a near best N-term approximation in L^p can be achieved by a simple thresholding procedure.

Theorem 4.3.1 *We assume here that the basis $\{\psi_\lambda\}_{\lambda\in\nabla}$ is generated from scaling functions which are C^ε for some $\varepsilon > 0$. Let $f = \sum_{\lambda\in\nabla} c_\lambda\psi_\lambda \in L^p$ with $1 < p < \infty$. Then, we have*

$$\text{dist}_{L^p}(f, S_N) \lesssim \|f - \sum_{\lambda\in\Lambda_N} c_\lambda\psi_\lambda\|_{L^p}, \tag{4.3.15}$$

where $\Lambda_N = \Lambda_N(f, L^p)$ is the set of indices corresponding to the N largest contributions $\|c_\lambda\psi_\lambda\|_{L^p}$ or equivalently the N largest $2^{(d/2-d/p)|\lambda|}|c_\lambda|$.

Proof We first remark that if V_Λ denotes the space generated by ψ_λ, $\lambda \in \Lambda$, for some $\Lambda \subset \nabla$, then we have

$$\text{dist}_{L^p}(f, V_\Lambda) \lesssim \|f - \sum_{\lambda\in\Lambda} c_\lambda\psi_\lambda\|_{L^p}, \tag{4.3.16}$$

with a constant that does depend on Λ. This is a straightforward consequence of Lemma 2.3.1 and the uniform boundedness of the projectors

$$T_\Lambda f := \sum_{\lambda\in\Lambda} \langle f, \tilde{\psi}_\lambda\rangle\psi_\lambda, \tag{4.3.17}$$

that we already observed in §3.8 of Chapter 3.

It is thus sufficient to prove that if $|\Lambda| = N$, one has

$$\|f - T_{\Lambda_N} f\|_{L^p} \lesssim \|f - T_\Lambda f\|_{L^p}. \tag{4.3.18}$$

For this, we define $A_N = \Lambda \setminus \Lambda_N$ and $B_N = \Lambda_N \setminus \Lambda$. Clearly we have $|A_N| = |B_N|$. By definition of Λ_N, we have

$$\sup_{\lambda \in A_N} \|c_\lambda \psi_\lambda\|_{L^p} \leq \inf_{\lambda \in B_N} \|c_\lambda \psi_\lambda\|_{L^p}. \tag{4.3.19}$$

Combining this with (4.3.9) and (4.3.14), we obtain

$$\|\sum_{\lambda \in A_N} c_\lambda \psi_\lambda\|_{L^p} \lesssim \|\sum_{\lambda \in B_N} c_\lambda \psi_\lambda\|_{L^p}. \tag{4.3.20}$$

From this, we derive

$$\begin{aligned} \|T_\Lambda f - T_{\Lambda_N} f\|_{L^p} &\leq \|T_{A_N} f\|_{L^p} + \|T_{B_N} f\|_{L^p} \\ &\lesssim \|T_{B_N} f\|_{L^p} \\ &= \|T_{B_N} T_{\nabla \setminus \Lambda} f\|_{L^p} \\ &\lesssim \|T_{\nabla \setminus \Lambda} f\|_{L^p} \\ &= \|f - T_\Lambda f\|_{L^p}, \end{aligned}$$

which implies (4.3.18). ◇

We shall now use (4.3.9) and (4.3.14) in order to prove the Jackson and Bernstein estimates for N-term approximation in L^p.

Theorem 4.3.2 *Let $1 < p < \infty$ and $\{\psi_\lambda\}_{\lambda \in \nabla}$ be a wavelet basis constructed from scaling functions in L^p. Assuming that the space $B^s_{q,q}$, $1/q = 1/p + s/d$ admits a wavelet characterization of the type (3.7.32), we have the Jackson estimate*

$$\text{dist}_{L^p}(f, S_N) \lesssim N^{-s/d} \|f\|_{B^s_{q,q}}, \tag{4.3.21}$$

and, for $f \in S_N$, the Bernstein estimate

$$\|f\|_{B^s_{q,q}} \lesssim N^{s/d} \|f\|_{L^p}. \tag{4.3.22}$$

Proof Let $f \in B^s_{q,q}$. As in the proof of Theorem 4.2.2, we remark that the norm equivalence (3.7.32) also writes

$$\|f\|_{B^s_{q,q}} \sim \|(\|c_\lambda \psi_\lambda\|_{L^p})_{\lambda \in \nabla}\|_{\ell^q}, \tag{4.3.23}$$

In particular, we can control the ℓ^q_w norm of the sequence $(\|c_\lambda\psi_\lambda\|_{L^p})$ by $\|f\|^q_{B^s_{q,q}}$: for all $\eta > 0$

$$\#\{\lambda\ ;\ \|c_\lambda\psi_\lambda\|_{L^p} \geq \eta\} \lesssim \eta^{-q}\|f\|^q_{B^s_{q,q}}. \tag{4.3.24}$$

It follows that there exists a constant $C > 0$ (depending on the constant in the equivalence (4.3.23)) such that if we define

$$A_j = \{\lambda\ ;\ C2^{-j/q}\|f\|_{B^s_{q,q}} \leq \|c_\lambda\psi_\lambda\|_{L^p} \leq C2^{-(j-1)/q}\|f\|_{B^s_{q,q}}\}, \tag{4.3.25}$$

we then have

$$|A_j| \leq 2^j. \tag{4.3.26}$$

From (4.3.9), we can evaluate the L^p norm of $T_{A_j}f = \sum_{\lambda\in A_j} c_\lambda\psi_\lambda$ by

$$\|T_{A_j}\|_{L^p} \lesssim 2^{-j/q}\|f\|_{B^s_{q,q}}|A_j|^{1/p} \leq 2^{j(1/p-1/q)}\|f\|_{B^s_{q,q}}. \tag{4.3.27}$$

Now define $B_j = \cup_{l=0}^{j-1} A_l$. By (4.3.26), we have $|B_j| \leq 2^j$. For $\Sigma_j := S_N$, we thus have

$$\begin{aligned} \text{dist}_{L^p}(f,\Sigma_j) &\leq \|f - T_{B_j}f\|_{L^p} \\ &\leq \textstyle\sum_{l\geq j}\|T_{A_l}f\|_{L^p} \\ &\lesssim \textstyle\sum_{l\geq j} 2^{l(1/p-1/q)}\|f\|_{B^s_{q,q}} \\ &\lesssim 2^{j(1/p-1/q)}\|f\|_{B^s_{q,q}} = 2^{-js/d}\|f\|_{B^s_{q,q}}. \end{aligned}$$

By the monotonicity of $\text{dist}_{L^p}(f, S_N)$, this implies the direct estimate (4.3.21) for all N.

In order to prove the Bernstein estimate, we distinguish two cases: $p \geq 2$ or $p \leq 2$. If $p \geq 2$, we have

$$\|f\|_{B^0_{p,p}} \sim \|(\|c_\lambda\psi_\lambda\|)_{L^p})_{\lambda\in\nabla}\|_{\ell^p} \lesssim \|f\|_{L^p}. \tag{4.3.28}$$

One way of checking (4.3.28) is to use an interpolation argument: the property holds when $p = 2$ for which one actually has the equivalence $\|f\|_{B^0_{2,2}} \sim \|f\|_{L^2}$ and $p = \infty$ since

$$\|c_\lambda\psi_\lambda\|_{L^\infty} \lesssim 2^{|\lambda|d/2}|c_\lambda| = 2^{|\lambda|d/2}|\langle f, \tilde{\psi}_\lambda\rangle| \lesssim \|f\|_{L^\infty}. \tag{4.3.29}$$

One can also directly check (4.3.28) using the characterization of L^p by the square function Sf as shown in (3.8.38). For this range of p, the Bernstein estimate is thus a straightforward consequence of the same estimate derived for $B^0_{p,p}$ in the previous section.

In the case where $p \leq 2$, let $f = \sum_{\lambda\in\Lambda} c_\lambda\psi_\lambda \in S_N$, i.e. such that $|\Lambda| \leq N$. We take here the same definitions of Λ_j, N_j, $j(x)$ and Ω_j as in the proof of Lemma 4.3.1. We then estimate $\|f\|_{B^s_{q,q}}$ as follows.

$$\begin{aligned}\|f\|^q_{B^s_{q,q}} &\lesssim \textstyle\sum_{\lambda\in\Lambda} \|c_\lambda\psi_\lambda\|^q_{L^p}\\ &= \textstyle\sum_{\lambda\in\Lambda} |c_\lambda|^q\|\psi_\lambda\|^p_{L^p}\|\psi_\lambda\|^{q-p}_{L^p}\\ &\lesssim \textstyle\int_\Omega\sum_{\lambda\in\Lambda} |c_\lambda|^q|\psi_\lambda|^p 2^{d(1/2-1/p)(q-p)|\lambda|}\\ &\lesssim \textstyle\int_\Omega\sum_{\lambda\in\Lambda} |c_\lambda|^q|\psi_\lambda|^q[2^{d(1/p-1/2)|\lambda|}|\psi_\lambda|]^{p-q}\\ &\lesssim \textstyle\int_\Omega [Sf(x)]^q R_\Lambda(x)dx,\end{aligned}$$

where we have applied Hölder's inequality on sequences to obtain the last line. Here, $Sf(x)$ is the square function defined by (3.8.37) and $R_\Lambda(x)$ can be estimated by

$$\begin{aligned}R_\Lambda(x) &= \left(\textstyle\sum_{\lambda\in\Lambda}[2^{d(1/p-1/2)|\lambda|}|\psi_\lambda(x)|]^{2(p-q)/(2-q)}\right)^{(2-q)/2}\\ &\lesssim \left(\textstyle\sum_{\lambda\in\Lambda,\psi_\lambda(x)\neq 0} 2^{2d|\lambda|(p-q)/(2p-qp)}\right)^{(2-q)/2}\\ &\lesssim 2^{j(x)d(1-q/p)}.\end{aligned}$$

Using Hölder's inequality, we thus obtain

$$\begin{aligned}\|f\|^q_{B^s_{q,q}} &\lesssim \|Sf\|^{q/p}_{L^p}\left(\textstyle\int_\Omega 2^{j(x)d}dx\right)^{1-q/p}\\ &\lesssim \|f\|_{L^p}[\textstyle\sum_{j\geq -1}|\Omega_j|2^{jd}]^{1-q/p}\\ &\lesssim \|f\|_{L^p}[\textstyle\sum_{j\geq -1}N_j]^{1-q/p}\\ &= N^{1-q/p}\|f\|_{L^p},\end{aligned}$$

i.e. the Bernstein estimate. ◇

Remark 4.3.1 *The direct estimate shows in particular that $B^s_{q,q}$ is embedded in L^p, as announced in §3.8 of the previous chapter.*

Remark 4.3.2 *For $p \leq 2$, one can also check that $\|f\|_{L^p} \lesssim \|f\|_{B^0_{p,p}}$: the direct estimate is then a straightforward consequence of the same estimate established in Theorem 4.2.2 for $B^0_{p,p}$.*

We are now ready to prove the main result of this section.

Theorem 4.3.3 *Let $1 < p < \infty$ and $\{\psi_\lambda\}_{\lambda\in\nabla}$ be a wavelet basis constructed from scaling functions in L^p. Assuming that for $0 < s < t$, the space $B^s_{q,q}$, $1/q = 1/p + s/d$ admits a wavelet characterization of the type (3.7.32), then one also has the norm equivalence*

$$\|f\|_{B^s_{q,q}} \sim \|f\|_{L^p} + \|(2^{js}\mathrm{dist}_{L^p}(f,\Sigma_j))_{j\geq 0}\|_{\ell^q}. \tag{4.3.30}$$

Proof We want to prove here that $\mathcal{A}^s_{p,q} = B^s_{q,q}$. We first note that the direct estimate in Theorem 4.3.2 is equivalent to

$$B^s_{q,q} \subset \mathcal{A}^s_\infty(L^p). \tag{4.3.31}$$

On the other hand, if $f \in A^s_{q,q}$ let $f_j \in \Sigma_j$ be such that

$$\|f - f_j\|_{L^p} \leq 2\text{dist}_{L^p}(f, \Sigma_j), \tag{4.3.32}$$

and let $r = \min\{1, q\}$. We then have

$$\begin{aligned} \|f\|_{B^s_{q,q}} &\leq \left(\|f_0\|^r_{B^s_{q,q}} + \sum_{j\geq 0} \|f_{j+1} - f_j\|^r_{B^s_{q,q}}\right)^{1/r} \\ &\lesssim \left(\|f\|^r_{L^p} + \sum_{j\geq 0} 2^{js}\|f_{j+1} - f_j\|^r_{L^p}\right)^{1/r}. \end{aligned}$$

where we have used the inverse estimate of Theorem 4.3.2.

We thus have

$$\mathcal{A}^s_{p,r} \subset B^s_{q,q} \subset \mathcal{A}^s_{p,\infty}. \tag{4.3.33}$$

Recalling the interpolation properties (3.5.41) for the $\mathcal{A}^s_{p,r}$ spaces (which are easily extended to the case of nonlinear approximation using Theorem 4.2.1) and (4.2.33) for the Besov spaces $B^s_{q,q}$, $1/q = 1/p + s/d$, we conclude by reiteration (Theorem 3.5.1) that $B^s_{q,q} = \mathcal{A}^s_{p,q}$ with equivalent norms. ◇

Remark 4.3.3 *The above result shows that N-term approximation in L^p, $1 < p < +\infty$ is also well described by Figure 38.1. Moreover, all the results of this section can be extended to the case where the error is measured in $W^{m,p}$, $1 < p < \infty$. Here one obtains a norm equivalence of the type*

$$\|f\|_{B^s_{q,q}} \sim \|f\|_{W^{m,p}} + \|(2^{j(s-m)}\text{dist}_{W^{m,p}}(f, \Sigma_j))_{j\geq 0}\|_{\ell^q}, \tag{4.3.34}$$

for $s > m$ and $1/q - 1/p = (s - m)/d$. The extension to this case is rather straightforward. One simply notices that, due to the characterization of L^p by the square function Sf and of $W^{m,p}$ by the modified square function $S_m f$, the transformation D_m defined by

$$\langle D_m f, \tilde{\psi}_\lambda\rangle = 2^{m|\lambda|}\langle f, \tilde{\psi}_\lambda\rangle, \tag{4.3.35}$$

is an isomorphism between $W^{m,p}$ and L^p. From this simple remark, one can easily adapt all the results of this section to the $W^{m,p}$ setting and derive (4.3.34).

Remark 4.3.4 *The cases $p = 1$ and $p = \infty$ are not covered by the results of this section. For $p = 1$, a basic difficulty is that the spaces $B^s_{q,q}$, $1/q =$*

$1 + s/d$, are not simply characterized by the inner products with wavelet coefficients. If one accepts measuring the error in the Hardy space H^p, then the case $p \leq 1$ can be treated with the approach of KYRIASIS *[1996] which allows us to characterize Besov spaces by wavelet decompositions for a wider range of indices. For $p = \infty$, results on N-term approximation in the class $C^0(\mathbb{R}^d)$ can be found in* DEVORE, PETRUSHEV *and* YU *[1992].*

Remark 4.3.5 *Also not covered by our results is the case where the function f that we want to approximate sits in the space $W^{t,q}$ which differs from $B^t_{q,q}$ when t is an integer and $q \neq 2$. If $q \geq 2$, one has the embedding $W^{t,q} \subset B^t_{q,q}$ and therefore we can still derive from our results that $f \in W^{t,q}$ and $X = W^{s,p}$ or $B^s_{p,p}$ with $t - s = d/q - d/p$, we then have*

$$\|(2^{j(t-s)}\mathrm{dist}_X(f, \Sigma_j))_{j\geq 0}\|_{\ell^q} \lesssim \|f\|_{W^{t,q}}. \tag{4.3.36}$$

In the case $q < 2$, this embedding does not hold. The following result was however proved in COHEN, DEVORE, PETRUSHEV *and* XU *[1998], which corresponds to the case $q = 1$ and $t = 1$: if $f \in W^{1,1}$, then its normalized wavelet coefficient sequence $(\|c_\lambda\psi_\lambda\|_{W^{1,1}})_{\lambda\in\nabla}$ belongs to the weak space ℓ^1_w (this result also holds for the slightly larger space BV of functions with bounded variation). Using similar ideas, one can prove that for $1 \leq q < 2$ and $t > 0$ an integer, $f \in W^{t,q}$ implies that $(\|c_\lambda\psi_\lambda\|_{W^{t,q}})_{\lambda\in\nabla}$ belongs to the weak space ℓ^q_w. Following the same method as in the proof of Theorems 4.2.2 and 4.3.2, this weak summability property is sufficient to establish the direct estimate*

$$\mathrm{dist}_X(f, S_N) \lesssim N^{-(t-s)/d} \tag{4.3.37}$$

for $X = W^{s,p}$ or $B^s_{p,p}$ with $t - s = d/q - d/p$.

4.4 Adaptive finite element approximation

Adaptive finite element approximation corresponds to the situation where the S_N are defined by (4.1.3), i.e. the union of all finite element spaces of a fixed type, associated with partitions of at most N elements. For such spaces, the answer to the two basic questions raised in §3.1 - what are the properties of f which govern the decay of the best approximation error $\sigma_N(f)$ in some norm X and how to compute in a simple way a near-optimal approximation of f in S_N - is less complete than for best N-term wavelet approximation. The goal of this section is to give an account of the existing results in this area and compare them to the wavelet results.

We first consider the case of finite elements in one space dimension, with $\mathcal{T}$ a partition of an interval $[a, b]$. In this case, we can view adaptive finite elements as *free knot splines*, i.e. spline functions with non-uniformly spaced knots on $[a,b]$ that are allowed to be adapted to the function that we want to approximate. More precisely, with any subdivision

$$a = x_0 < x_1 < \cdots < x_N < x_{N+1} = b, \tag{4.4.1}$$

of $[a,b]$ into N intervals, we associate for $0 \le m \le n$ the spaces

$$S(m,n,(x_i)) := \{f \in C^{m-1}([a,b]) \;\; f_{|]x_i,x_{i+1}[} \in \Pi_n\}, \tag{4.4.2}$$

of piecewise polynomials functions of degree n with smoothness $m-1$ at the knots x_i. If $a = -\infty$ or $b = +\infty$, we simply modify the definition by imposing that f is identically zero on $]\infty, x_1]$ or $[x_N, +\infty[$. For $m = n$, these are the classical spline functions. For $m = 0$, no smoothness is imposed and we obtain simple piecewise polynomials functions. Functions in the above spaces with $m < n$ can also be obtained as limits of spline functions with $m = n$ by merging knots by groups of at most $n - m + 1$ points. For this reason they are also called splines of degree n and multiplicity $n - m + 1$ (they were already evoked in Example 5 of §2.11 in Chapter 2). We then define the space of free-knot splines of degree n and multiplicity $n - m + 1$ on N intervals as the union

$$S_N(m,n) = \cup_{(x_i)\in R_N} S(m,n,(x_i)), \tag{4.4.3}$$

where R_N is the set of all possible subdivisions of $[a,b]$ into N-intervals. For fixed parameters m and n, we are thus interested in approximation by functions of the nonlinear space $S_N = S_N(m,n)$. Note that (4.2.20) does hold for such nonlinear spaces since we clearly have $S_N + S_N \subset S_{2N}$.

The first results on free-knot spline approximation appeared in the 1960's (KAHANE [1961] and BIRMAN and SOLOMJAK [1967]) and a rather complete understanding of this theory (in the case of univariate functions) was reached at the end of the 1980's with works by OSWALD [1990] and PETRUSHEV [1988]. These results were strongly influencial in the subsequent study of N-term wavelet approximation as presented in §4.2 and §4.3 of this chapter, due to the strong analogies between both tools: in §1.5 of Chapter 1, we already observed that the thresholding process of N-term approximation in the Schauder basis amounts to the selection of an adaptive set of N knots. In the univariate case, it is actually known that the approximation spaces are the same for both techniques. More precisely, the following result was proved in PETRUSHEV [1988] for the free knot splines of degree n and multiplicity n (i.e. piecewise polynomials of degree n) on a univariate interval I.

Theorem 4.4.1 *For $1 \le p < \infty$, $0 < s < n+1$ and $1/q = 1/p + s$, the approximation space $\mathcal{A}_q^s(L^p)$ for the free knot splines coincides with the Besov space $B_{q,q}^s(I)$.*

As in the case of N-term wavelet approximation, a natural way of proving such a result consists in establishing first a direct and inverse estimate for the spaces S_N and concludes by interpolation arguments, using the fact that we have $S_N + S_N \subset S_{2N}$.

While we leave the proof of the inverse estimate as an exercise to the reader, it is particularly instructive to follow the arguments of an elementary proof for the direct estimate

$$\inf_{g \in S_N} \|f - g\|_{L^p(I)} \lesssim N^{-s} |f|_{B_{q,q}^s(I)}, \tag{4.4.4}$$

in the case where $S_N = S_N(0, n)$, i.e. with no smoothness requirement on the spline approximants. From the embedding of $B_{q,q}^s$ in L^p, we derive that for a given interval J,

$$\inf_{g \in \Pi_n} \|f - g\|_{L^p(J)} \lesssim \inf_{g \in \Pi_n} \|f - g\|_{B_{q,q}^s(J)}. \tag{4.4.5}$$

Then, using general results on polynomial approximation (for example the local Whitney estimate in Theorem 3.2.2), we conclude that

$$\inf_{g \in \Pi_n} \|f - g\|_{L^p(J)} \le C |f|_{B_{q,q}^s(J)}. \tag{4.4.6}$$

In this last estimate, a scaling argument immediately shows that the constant C does not depend on the size of J. From the superadditivity of $|f|_{B_{q,q}^s(J)}^q$ with respect to unions of disjoint intervals (see Remark 3.3.5), we can find for all N a disjoint partition $I = \cup_{k=1}^N I_k$ such that

$$|f|_{B_{q,q}^s(I_k)} \lesssim N^{-1/q} |f|_{B_{q,q}^s(I)}, \tag{4.4.7}$$

with a constant that does not depend on N. Combining (4.4.6) and (4.4.7), we finally obtain

$$\begin{aligned} \inf_{g \in S_N} \|f - g\|_{L^p(I)} &\le [\textstyle\sum_{k=1}^N \inf_{h \in \Pi_n} \|f - h\|_{L^p(I_k)}^p]^{1/p} \\ &\lesssim N^{1/p - 1/q} |f|_{B_{q,q}^s(I)} = N^{-s} |f|_{B_{q,q}^s(I)}. \end{aligned}$$

This proof shows that a natural subdivision of the interval for the approximation of f by free knot splines corresponds to a well-balanced splitting of the semi-norm $B_{q,q}^s(I)$ between the knots. Note that the critical Sobolev and

Besov embeddings, which were already identified in the context of N-term wavelet approximation, now play an explicit role in this proof.

How does one practically implement free knot spline approximation ? The above reasoning suggests equilibrating the Besov semi-norm $|f|_{B^s_{q,q}}$ over the intervals of the adaptive partition. However, such a strategy is not practically convenient since it relies on our knowledge of the Besov smoothness of the function to be approximated, in contrast to wavelet thresholding. A more realistic approach is to fix some tolerance $\varepsilon > 0$ and design the N intervals I_k in such a way that ***the local polynomial approximation error is well equilibrated***, in the sense that $\inf_{h\in\Pi_n} \|f-h\|_{L^p(I_k)} \sim \varepsilon$ for all k. More precisely let us assume that we have for all k

$$\varepsilon/2 \le \inf_{h\in\Pi_n} \|f-h\|_{L^p(I_k)} \le \varepsilon, \tag{4.4.8}$$

(the number of intervals N depends then on ε). The resulting free knot spline approximation clearly satisfies $\|f-g\|_{L^p(I)} \le N^{1/p}\varepsilon$, and furthermore if $f \in B^s_{q,q}$, $1/q = 1/p + s$, we can use (4.4.6) to derive

$$\begin{aligned}\|f-g\|_{L^p(I)} &\le N^{1/p}\varepsilon \\ &\le N^{-s}N^{1/q}\varepsilon \\ &\lesssim [\textstyle\sum_{k=1}^N \inf_{h\in\Pi_n} \|f-h\|^q_{L^p(I_k)}]^{1/q} N^{-s} \\ &\lesssim N^{-s}|f|_{B^s_{q,q}(I)}.\end{aligned}$$

If we now turn to adaptive finite elements on a bounded domain $\Omega \subset \mathbb{R}^d$ in dimension $d \ge 2$, several results are available if one chooses to consider adaptive partitions with ***shape constraints*** in terms of a bound on the aspect ratio of the elements

$$\max_{K\in\mathcal{T}}\Big([\mathrm{Diam}(K)]^d/\mathrm{vol}(K)\Big) \le C, \tag{4.4.9}$$

where C is uniform over all the allowed partitions. Such a restriction means that the local refinement is isotropic, in a similar way to wavelets. With such constraints, it thus seems reasonable that the resulting approximation spaces coincide with those of best N-term wavelet approximation: defining $\Sigma_j = S_{2^{dj}}$, we expect $\mathcal{A}^s_{p,q} = B^s_{q,q}$ for $1/q = 1/p + s/d$ and $1 < p < \infty$, assuming that we use finite elements which are piecewise polynomials of degree n with $n+1 \ge s$.

However, one immediately faces several difficulties when trying to apply the general mechanism of proof - direct estimates, inverse estimates and interpolation theory - in this situation. A first difficulty is that the sum $h = f + g$ of two elements in S_N is not generally contained in S_{aN} for a fixed a for the following reason: we need to define for h a partition with elements contained in both partitions associated with f or g. Indeed, while it can be seen that the shape constraint (4.4.9) implies the existence of such a partition with aN elements, these new elements might not anymore fulfill (4.4.9) (some very thin overlap between the elements of f and g results in very thin elements for h) and therefore need to be further subdivided. In turn, while inverse estimates can be proved rather easily for elements of $f_j \in \Sigma_j$, they cannot be applied to functions of the type $f_j - f_{j-1}$.

If we next try to generalize the proof of the direct estimate that we gave in the case $d = 1$, we first establish an estimate on the local polynomial approximation error similar to (4.4.6) namely

$$\inf_{g\in\Pi_n} \|f - g\|_{L^p(K)} \leq C|f|_{B^s_{q,q}(K)}. \tag{4.4.10}$$

with $1/q = 1/p + s/d$, and by using an affine change of variable onto a reference element, we observe that the constant C is independent of the element K as long as it satisfies the shape constraint (4.4.9). It follows that if the N elements are designed in such a way that they equilibrate the $B^s_{q,q}$ semi-norm

$$|f|_{B^s_{q,q}(K)} \lesssim N^{-1/q}|f|_{B^s_{q,q}(\Omega)}, \tag{4.4.11}$$

or the local polynomial approximation error

$$\varepsilon/2 \leq \inf_{h\in\Pi_n} \|f - h\|_{L^p(K)} \leq \varepsilon, \tag{4.4.12}$$

we obtain by the same arguments as in the case $d = 1$ that the resulting piecewise polynomial approximation g (again without smoothness conditions at the interfaces between the elements) satisfies

$$\|f - g\|_{L^p(\Omega)} \lesssim N^{-s/d}|f|_{B^s_{q,q}(I)}. \tag{4.4.13}$$

However, the construction of a partition which equilibrates the $B^s_{q,q}$ semi-norm or the local polynomial approximation error is not as obvious as in the case $d = 1$. In practice, one typically relies on one of the following more practical but sub-optimal strategies:

1. **Adaptive coarsening:** starting from a spline or finite element approximation on a very fine uniform discretization, mesh points are iteratively removed at the location where the local error is below the tolerance ε. For univariate splines, this technique is known as "knot removal" and a survey can be found in LYCHE [1993]. For multivariate finite elements, this technique is rather known as "thinning" and it involves a step of local retriangulation in the vicinity of the point which has been removed, see e.g. DYN, FLOATER and ISKE [2002].

2. **Adaptive refinement:** starting from a coarse piecewise polynomial approximation, one refines it adaptively by splitting the elements as long as the local error of polynomial approximation remains higher than the prescribed tolerance ε. The analysis of such algorithms has been done in DEVORE and YU [1990] in the one dimensional case where it is shown that they reach the rates N^{-s}, as best approximation in S_N, with a slight extra smoothness assumption on the function, e.g. $f \in B^{s+\alpha}_{q,q}$ or $f \in B^s_{q+\alpha,\infty}$ for some arbitrarily small $\alpha > 0$. In this sense this procedure almost achieves the approximation rates predicted by Theorem 4.4.1.

Remark 4.4.1 *If we impose continuity or higher smoothness across the elements of the partition, we need to modify the proof of the direct estimate, since we cannot build the approximant by taking the best polynomial on each element. Using instead a quasi-interpolant operator, we are typically led to equilibrate the local error* $\inf_{h\in\Pi_n} \|f - h\|_{L^p(N(K))}$ *where* $N(K)$ *consists of the element* K *and its immediate neighbours. While this does not affect the estimates for* $d = 1$, *we should definitely expect that the performance of adaptive finite elements with smoothness constraints are not as good as those of wavelets in higher dimension: continuity imposes that the partition is conformal, which together with shape constraints induces a "grading" property - very small elements cannot be adjacent to a very large one - which is a limitation to adaptivity. However, we still expect to obtain the rate* $N^{-s/d}$ *with a slight extra smoothness assumption on the function, e.g.* $f \in B^{s+\alpha}_{q,q}$ *or* $f \in B^s_{q+\alpha,\infty}$ *for some arbitrarily small* $\alpha > 0$.

Remark 4.4.2 *In the one dimensional case, an alternative approach to the construction of a free knot spline approximation is to simply apply a thresholding procedure on the decomposition of the function* f *in a spline wavelet basis. One can then exploit the results on* N*-term approximation that were discussed in §4.2 and §4.3.*

4.5 Other types of nonlinear approximations

Nonlinear approximation is a rich theory that covers a wider range of techniques beside wavelets and adaptive splines and finite elements. In this section, we shall give an account of the other main items in this theory. Generally speaking, we are interested in approximation in spaces S_N (or $\Sigma_j = S_{2^{dj}}$) that are not necessarily linear, but such that $f \in S_N$ can still be parametrized by N (or CN with a fixed C) degrees of freedom, and such that $S_N \subset S_{N+1}$.

Example 1. Rational approximation in one dimension
Here S_N denotes the set of rational functions of degree N, i.e. functions of the type $r(x) = p(x)/q(x)$ where $p, q \in \Pi_N$. In the case of univariate functions, elements in this set are characterized by $2N$ parameters. As for free knot splines, we have $S_N + S_N \subset S_{2N}$.

Approximation by rational functions has been considered since the beginning of the last century. In particular, the works of Padé provided a systematic approach to the approximation of an analytic function f in the complex domain by rational functions p/q, $p \in \Pi_N$, $q \in \Pi_M$ which are determined by maximizing the degree of cancellation of $qf - p$ at the origin. We refer to BAKER [1975] and BREZINSKI and VAN ISEGHEM [1994] for detailed surveys on Padé approximants. These methods typically apply to the approximation of analytic functions in the complex domain.

More recently came the idea that rational functions have the ability to produce adaptive approximations for functions that contain singularities in the real domain. One of the first and most striking results in this direction was given in NEWMAN [1964] and deals with the approximation of the function $f(x) = |x|$ in the uniform metric on $I = [-1, 1]$.

Theorem 4.5.1 *There exists a constant C and a sequence of polynomials $p_N, q_N \in \Pi_N$ such that*

$$\|f - \frac{p_N}{q_N}\|_{L^\infty(I)} \le Ce^{-\pi\sqrt{N}}, \tag{4.5.1}$$

and thus $\inf_{g \in S_N} \|f - g\|_{L^\infty(I)} \lesssim e^{-\pi\sqrt{N}}$.

Such a result contrasts deeply with polynomial approximation of f which is limited by the presence of the singularity at the origin: one can easily check that $\inf_{g \in \Pi_N} \|f - g\|_{L^\infty(I)} \sim \mathcal{O}(1/N)$.

This result also reveals the links between rational and free knot spline approximation, since an easy corollary is that a continuous piecewise polynomial function on a bounded interval can be approximated at the rate $e^{-\pi\sqrt{N}}$ in the uniform metric by rational functions in S_N. We can thus expect that a function which is approximated with rate N^{-s} by free knot splines can also be approximated with the same rate by rational functions. Such links were studied in detail and exploited in PETRUSHEV [1988] where the following result is proved.

Theorem 4.5.2 *For $1 \le p < \infty$, $s > 0$ and $1/q = 1/p + s$, the approximation space $\mathcal{A}_q^s(L^p)$ for rational functions coincides with the Besov space $B_{q,q}^s(I)$.*

An analogous result with the metric BMO (bounded mean oscillation) as a close substitute for L^∞ was proved in PELLER [1985]. In other words, the approximation spaces for rational functions coincide with those of free knot splines and wavelets. Here again, the way of proof for such a result consists in establishing direct and inverse estimates for the S_N spaces. However, a simple algorithm that would directly produce an optimal rational approximation (without going through an intermediate free knot spline approximant) is still to be found.

The adaptive potential of rational functions can also be intuitively understood by the following observations: if $\psi(x)$ is a rational function in $L^2(\mathbb{R})$ such that $\int \psi = 0$, e.g. $\psi(x) = x/(x^2+1)$, then the family

$$\psi_{j,k}(x) = 2^{j/2}\psi(2^j x - k) = 2^{-j/2}\frac{(x - 2^{-j}k)}{(x-2^{-j}k)^2 + 2^{-2j}}, \tag{4.5.2}$$

has the appearance of a wavelet basis, while all its elements are in S_2. This reveals the analogy between nonlinear rational and wavelet approximations since an N-term combination of the above functions is in S_{2N}. Note that as j grows, the poles $2^{-j}k \pm i2^{-j}$ of $\psi_{j,k}$ get closer to the real line, which corresponds to the idea that an adaptive approximation of f by $f_N \in S_N$ requires a particular placement of the poles of f_N in the vicinity of the singularities of f.

Remark 4.5.1 *It should be well understood that the above analogy between wavelets and rational functions is only heuristic since the family in (4.5.2) is not known to constitute a basis. A general open problem has been to find a Riesz basis for $L^2(I)$ of a form generalizing (4.5.2), i.e. consisting of rational functions of degree less than N for some prescribed $N \geq 0$, which could provide the optimal approximation orders by simple thresholding procedures. A positive answer to this problem was recently given in* PETRUSHEV *[2000], which reinforces the links between rational functions and wavelets. It might also provide practical algorithms for adaptive rational approximation.*

Theorems 4.4.1 and 4.5.2 show that, in the univariate setting, free knot splines and rational functions have approximation properties which are very similar to N-term wavelet approximation. The same basically holds for adaptive finite elements with shape constraints in several dimensions. We now want to point out several instances of nonlinear approximation for which adaptivity is pushed to a higher level. This results in better approximation rates than wavelets for certain types of functions, but also in a more difficult practical implementation as well as a still incomplete theoretical analysis of these methods, which are sometimes referred to as "highly nonlinear".

Example 2. Anisotropic approximation
Consider the L^2-approximation of the characteristic function $f = \chi_\Omega$ of a smooth domain $\Omega \subset [0,1]^d$. Due to the singularity on the boundary $\partial\Omega$, one can easily check that the linear approximation in the multiresolution spaces V_j cannot behave better than

$$\sigma_N(f) = \|f - P_j f\|_{L^2} \sim \mathcal{O}(2^{-j/2}) \sim \mathcal{O}(N^{-1/2d}), \tag{4.5.3}$$

where $N = \dim(V_j) \sim 2^{dj}$. Turning to nonlinear approximation, we notice that since $\int \tilde{\psi}_\lambda = 0$, all the coefficients $d_\lambda = \langle f, \tilde{\psi}_\lambda \rangle$ are zero except those such that the support of $\tilde{\psi}_\lambda$ overlaps the boundary. At scale level j there are thus at most $K2^{d-1}j$ non-zero coefficients, where K depends on the support of the ψ_λ and on the $d-1$ dimensional measure of $\partial\Omega$. For such coefficients, we have the estimate

$$|d_\lambda| \le \|\tilde{\psi}_\lambda\|_{L^1} \le C2^{-dj/2}. \tag{4.5.4}$$

In the univariate case, i.e. when Ω is a simple interval, the number of non-zero coefficients up to scale j is bounded by jK. Therefore, using N non-zero coefficients at the coarsest levels gives an error estimate with exponential decay

$$\sigma_N(f) \le [\sum_{j \ge N/K} K|C2^{-dj/2}|^2]^{1/2} \le \tilde{C}2^{-dN/2K}, \tag{4.5.5}$$

which is a spectacular improvement on the linear rate. In the multivariate case, the number of non-zero coefficients up to scale j is bounded by $\sum_{l=0}^{j} K2^{(d-1)l}$ and thus by $\tilde{K}2^{(d-1)j}$. Therefore, using N non-zero coefficients at the coarsest levels gives an error estimate

$$\sigma_N(f) \le [\sum_{\tilde{K}2^{(d-1)j} \ge N} K2^{(d-1)j}|C2^{-dj/2}|^2]^{1/2} \le \tilde{C}N^{-1/(2d-2)}, \tag{4.5.6}$$

which is much less of an improvement. For example, in the bidimensional case, we only go from $N^{-1/4}$ to $N^{-1/2}$ by switching to nonlinear wavelet approximation.

This simple example illustrates the *curse of dimensionality* for best N-term wavelet approximation of functions which have singularities concentrated on lower dimensional manifolds. The main reason for the degradation of the approximation rate is the large number $K2^{(d-1)j}$ of wavelets which are needed to refine the boundary from level j to level $j+1$. On the other hand, if we view the boundary itself as the graph of a smooth function, it is clear that approximating this graph with accuracy 2^{-j} should require many less parameters than $K2^{(d-1)j}$. This reveals the fundamental limitation of wavelet bases: they fail to exploit the smoothness of the boundary and therefore cannot capture the simplicity of f in a small number of parameters. Another way of describing this limitation is by remarking that nonlinear wavelet approximation allows local refinement of the approximation, but imposes some ***isotropy*** in this refinement process. In order to capture the boundary with a small number of parameters, one would typically need to refine more in the normal direction than in the tangential directions, i.e. apply ***anisotropic local refinement.*** Such anisotropic refinements are made possible by adaptive finite elements provided that we do not impose the shape constraint (4.4.9). For instance, with the above example in the case $d = 2$, one can easily check that the use of piecewise constant functions on an adaptive partition of N triangles allows one to recover the rate $\sigma_N(f) \sim \mathcal{O}(N^{-1})$, precisely because one is allowed to use arbitrarily thin triangles to match the boundary, as illustrated in Figure 4.5.1.

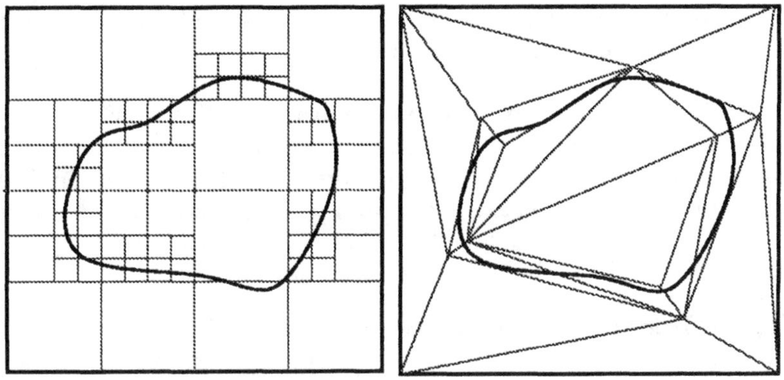

Figure 4.5.1: Isotropic and anisotropic refinements

Rational functions in several dimensions also have the ability to capture anisotropic singularities with a small number of parameters: this is also made possible since the poles are now algebraic curves that can be calibrated in order to match the singularity manifolds with high precision. An even more spectacular result is therefore conjectured : if $\partial\Omega$ is C^∞, then $\sigma_N(f) \leq C_r N^{-r}$ for any $r > 0$. Apart from these intuitive remarks, the approximation theory for anisotropic finite elements and multivariate rational functions is not well understood. These approximation families are "more nonlinear" than the examples that we have considered so far, in the sense that they do no longer satisfy the property (4.2.20). In turn, the corresponding approximation spaces $\mathcal{A}^s_q(X)$ are not ensured to be linear: two functions f and g might be approximated with the rate N^{-s} while this rate is not achieved for $f+g$. There is therefore no hope of describing these approximation spaces through a classical notion of smoothness.

From a more practical point of view, a major limitation of anisotropic finite elements and multivariate rational functions is so far the lack of existing efficient algorithms to build an optimal approximation. In particular, the adaptive coarsening or refinement algorithms that we mentioned in §4.4 can be applied with anisotropic finite elements but are not proved to build an optimal partition.

The development of new approximation and representation tools, which could both capture anisotropic features such as edges with a very small number of parameters and be implemented by fast and robust procedures, is currently the object of active research.

Example 3: best N-term approximation in a dictionary
A second instance of highly nonlinear approximation is encountered in the following situation: one wants to approximate $f \in L^2(\Omega)$ by an N-term linear combination $\sum_{k\in\Lambda} c_k e_k$, where $(e_k)_{k\geq 0}$ is a "dictionary" of functions which spans $L^2(\Omega)$ in a redundant fashion. Several choices of such redundant families have been the object of particular interest.

1. **Time-frequency atoms:** these are families of functions of the form $g_{m,n}(x) = g(x-ma)e^{inbx}$, $m,n \in \mathbb{Z}$, where g is a fixed shape function and $a,b>0$ fixed parameters. A well studied case is the that of Gabor functions corresponding to the choice of a Gaussian function for g. In such a case (see DAUBECHIES [1992, chapter 4]), for $a,b>0$ small enough, these functions constitute a frame of $L^2(\mathbb{R})$ and are in turn a complete family. Interesting variants are the *local cosine bases* of H. Malvar (see e.g. MEYER [1992]), which in contrast to Gabor functions, allow us to generate orthonormal bases.

2. **Wavelet packets:** these functions, introduced in COIFMAN, MEYER, QUAKE and WICKERHAUSER [1993], are a natural generalization of wavelet bases, produced by allowing more flexibility in the multiscale decomposition algorithms described in §2.6 of Chapter 2. More precisely, we allow ourselves to iterate the basic two-band decomposition (the block diagram of Figure 2.6.3) on the result of the high-pass fil-

tering of the previous iteration (while wavelet decompositions only iterate this action on the result of the low-pass filtering). Depending on the successive choices of applying this basic decomposition or of leaving the resulting subsignals unchanged, one expands the signal into a specific basis corresponding to a particular subband decomposition, one of them being the wavelet basis (corresponding to the choice of only splitting the low-pass subsignal). Wavelet packets can thus be viewed as a collection of bases, having many common elements.

3. **Ridge functions:** these are families of multivariate functions of the type $g_u(x) := g(\langle u, x\rangle)$, where g is a univariate function chosen among a finite or infinite set and u a direction vector chosen among a finite or infinite set. Approximation by such functions has been studied in relation to the problem of modelling neural networks (see BARRON [1994]).

The redundancy of these dictionaries is intended to allow a better level of adaptivity, with the hope of describing complicated functions by very few parameters. However, difficulties arise when it comes to the practical implementation of an optimal N-term approximation: while keeping the N largest coefficients is the natural procedure to apply in the case of a basis, the selection of the best N-term combination within a redundant family is a highly complex problem (it can be shown to have NP-hard complexity in very simple instances, see AVELLANEDA, DAVIS and MALLAT [1997]). For this reason, one in practice makes use of sub-optimal algorithms:

1. **Best bases algorithms**, proposed by COIFMAN and WICKERHAUSER [1992], operate in the context of a dictionary made out of the union of several bases. In a first step, the algorithm compares the action of the different bases on a given function f, using an "entropy criterion", i.e. a measurement of the sparsity of f for each basis. According to the discussion in §4.2, if one is interested in concentrating the L^2 norm on a small number of coefficients, a possibility is to take the ℓ^p norm of the coefficients, for some $p < 2$. The basis which minimizes the criterion is selected and used to compress the function by a simple N-term approximation. In the case of wavelet packets, it is interesting to note that a best basis can be obtained in $\mathcal{O}(N \log N)$ operations, making use of the particular tree-structured algorithm which generates the expansions of a discretized function in these bases.

2. **Matching pursuit algorithms**, as studied in AVELLANEDA, DAVIS and MALLAT [1997], operate on any type of dictionary. The simplest (and most often used) of these approximation schemes is the so-called

greedy algorithm. We assume here that the e_k are normalized in L^2. As a first step, the algorithm compares the values of the coefficients $\langle f, e_k \rangle$ and selects the largest one $\langle f, e_{k_0} \rangle$. A first approximation of f is chosen to be $f_0 = \langle f, e_{k_0} \rangle e_{k_0}$. In a second step, the residual $r_0 = f - f_0$ is analyzed in a similar way, yielding an updated approximation of f by $f_1 = f_0 + \langle r_0, e_{k_1} \rangle e_{k_1}$. Iterating this procedure N times yields an N term approximation of f.

Although these algorithms perform very well in several practical situations, most known theoretical results about their nonlinear approximation properties are negative in the following sense: if the function f is such that $\|f - f_N\|_{L^2} \lesssim N^{-s}$ for a sequence of N-term approximations, the algorithm generally fails to produce such a rate of convergence. An interesting open problem is to establish reasonable assumptions on the dictionary or on the class of functions to be analyzed that would restore the optimality of these algorithms.

Example 4. Data compression
Among the different techniques of nonlinear approximation that we have discussed, wavelets offer the simplest algorithmic approach: (i) the function is decomposed by a fast algorithm from a fixed finest discretization level (typically imposed by the sampling rate in the case of a digitized signal), (ii) its largest coefficients are retained by imposing either their number or a threshold, (iii) the function can be reconstructed in a compressed form by a fast algorithm up to the finest discretization level.

This particular simplicity explains the interest in wavelets methods when dealing with data compression problems for certain types of signals. In §1.5 of Chapter 1, we already made clear that wavelet thresholding behaves particularly well on images, since it corresponds with reducing the number of parameters by focusing on the description of significant structures such as edges. However, another aspect of nonlinear approximation appears in data compression: these parameters (here the unthresholded wavelet coefficients) need to be *quantized* in order to be *encoded* by a finite number of bits. In turn, the success of a wavelet-based compression is strongly dependent on the chosen quantization and encoding strategy.

In the most standard approach, quantization is performed by a *scalar* quantizer which independently assigns an approximate value $\tilde{c}_\lambda$ to each coefficient c_λ. A natural generalization is *vector* quantization, which assigns approximate values to blocks of coefficients (see e.g. GERSHO and GRAY [1992]). The approximate values - or symbols - are then encoded by an *entropic* encoder which typically allocates more bits to the symbols which appear with low probability. This approach was successfully applied to wavelet decompositions of digital pictures in ANTONINI, BARLAUD, DAUBECHIES and MATHIEU [1992].

More recently, an alternate approach emerged, closer in spirit to nonlinear approximation: each coefficient c_λ is allocated a number of bits $n(\lambda, |c_\lambda|)$ that will directly depend on the size of c_λ. Typically, one benefits from the fact that there are only a few large coefficients by ignoring the small coefficients below some threshold η and allocating j bits to the coefficients in the range $2^j\eta \le |c_\lambda| < 2^{j+1}\eta$, which in the fixed precision $|c_\lambda - \tilde{c}_\lambda| \le \eta$. One of the main interests of this approach is that it leads to *progressive encoders*: the finer reconstruction obtained by dividing η by 2 can be encoded by updating the previous bit stream of 1 bit for each coefficient such $\eta/2 \le |c_\lambda|$. Note that this approach implies the additional encoding of the addresses of those coefficients within a given order of magnitude. In practice, this encoding is facilitated by using the existing correlations between these addresses. A first strategy, developed in SHAPIRO [1993] and SAID and PEARLMAN [1996], exploits the correlations across scales: if a coefficient d_λ is large, it is then likely that the support of $\tilde{\psi}_\lambda$ contains an edge singularity and that d_μ is also large for the indices μ such that $|\mu| < |\lambda|$ and $\text{Supp}(\tilde{\psi}_\lambda) \subset \text{Supp}(\tilde{\psi}_\mu)$. Therefore d_μ should be automatically quantized with at least as many bits as d_λ. A second strategy, developed in TAUBMAN [1998], exploits the correlations within each scale: large coefficients d_λ tend to be organized along curves which match the edges of the image. This approach was chosen in the image compression standard JPEG 2000. On a more mathematical point of view, it has been shown in COHEN, DAHMEN, DAUBECHIES and DEVORE [2001] that for various function classes taken as models for the signal, such algorithms lead to optimal asymptotic behaviour of the L^2-error with respect to the number of bits.

As an example we have displayed in Figure 4.5.2, the result of a wavelet-based compression strategy for the image that we already manipulated in §1.5 of Chapter 1, with a compression factor of 1/40. We have used here the biorthogonal spline wavelet pair $(\psi_{3,5}, \tilde{\psi}_{3,5})$ introduced in §2.11 of Chapter 2: reconstruction is ensured by piecewise bilinear functions which allows us to avoid the visual artefacts of the Haar system.

Remark 4.5.2 *Other important applications of wavelet-based parameter reduction are statistical estimation and signal denoising: a natural restoration strategy consists of thresholding the coefficients of the noisy signal at a level that will remove most of the noise, but preserve the few significant coefficients in the signal. This is in contrast to classical linear low pass filtering, which tends to blur the edges while removing the noise. Such nonlinear procedures have been the object of important theoretical work and have motivated recent progresses in nonlinear approximation, see e.g.* DONOHO,

Figure 4.5.2: Adaptive compression by a factor 1/40

JOHNSTONE, KERKYACHARIAN *and* PICARD *[1992]. It is fair to say that during the 1990's, wavelets have stimulated a general evolution of signal and image processing toward nonlinear and adaptive methods. The reader interested in a "wavelet tour of signal processing" will certainly enjoy the reading of* MALLAT *[1998].*

4.6 Adaptive approximation of operators

The goal of this section is both to analyze the sparse structure that results from the multiscale discretization of most operators involved in partial differential and integral equations, and to study the application of such operators to functions that also have a sparse multiscale representation.

We already discussed an instance of "matrix compression" in §1.5 of Chapter 1. In order to reach a better theoretical understanding of such a process in more general situations, we need more information on the size of the matrix elements in the wavelet discretizations of general classes of operators. In some sense, we need to analyze the size properties of these elements in the same way as we did in Chapter 3 for the wavelet coefficients of general classes of functions.

Given an operator A acting on functions defined on a domain $\Omega \subset \mathbb{R}^d$

and two wavelet bases $(\psi_\lambda)_{\lambda\in\nabla}$ and $(\tilde{\psi}_\lambda)_{\lambda\in\nabla}$ adapted to such a domain, we are interested in evaluating the entries

$$m_{\lambda,\mu} = \langle A\psi_\lambda, \tilde{\psi}_\mu\rangle. \tag{4.6.1}$$

When the two bases are biorthogonal, $M = (m_{\lambda,\mu})_{\lambda,\mu\in\nabla}$ is exactly the matrix of A in the basis $(\psi_\lambda)_{\lambda\in\nabla}$. Note however that such a coupling between the two bases is not necessary in the general framework of Galerkin and Petrov-Galerkin discretizations of operator equations. Therefore, we do not impose this constraint here. We shall first derive some basic estimates for specific classes of operators.

Example 1. Partial differential operators
For the sake of simplicity, let us first consider a one dimensional differential operator of the type $Af(x) = \sum_{m\leq r} a_m(x)f^{(m)}(x)$, where the coefficients $a_m(x)$ are C^∞ functions. From the local nature of this operator, we can derive the trivial estimate $m_{\lambda,\mu} = 0$ if ψ_λ and $\tilde{\psi}_\mu$ have disjoint support.

If these supports overlap, we can estimate the entries as follows: for $|\lambda| \leq |\mu|$, assuming that the wavelets ψ_λ have $C^{r+\alpha}$ smoothness and that the wavelets $\tilde{\psi}_\mu$ are orthogonal to polynomials of degree $n \geq \alpha-1$, we derive the estimate

$$\begin{aligned} |m_{\lambda,\mu}| &= \textstyle\int A\psi_\lambda(x)\, \tilde{\psi}_\mu(x)dx \\ &\leq \|\tilde{\psi}_\mu\|_{L^1} \inf_{g\in\Pi_n} \|A\psi_\lambda - g\|_{L^\infty(\mathrm{Supp}(\tilde{\psi}_\mu))} \\ &\lesssim 2^{-|\mu|(1/2+\alpha)}\|A\psi_\lambda\|_{C^\alpha(\mathrm{Supp}(\tilde{\psi}_\mu))} \\ &\lesssim 2^{-|\mu|(1/2+\alpha)}2^{|\lambda|(r+\alpha+1/2)} \\ &= 2^{(1/2+\alpha)(|\lambda|-|\mu|)}2^{r|\lambda|}. \end{aligned}$$

In the case where $|\lambda| \geq |\mu|$, we can integrate by part and take advantage of the smoothness of the coefficients $a_m(x)$ to obtain a symmetric estimate for $m_{\lambda,\mu}$, under the asumption that the wavelets $\tilde{\psi}_\mu$ have $C^{r+\alpha}$ smoothness and that the wavelets ψ_λ are orthogonal to polynomials of degree $n \geq \alpha-1$. In summary, we have obtained

$$|m_{\lambda,\mu}| \lesssim 2^{-(1/2+\alpha)||\lambda|-|\mu||}2^{r\inf\{|\lambda|,|\mu|\}} \tag{4.6.2}$$

which also can be written

$$|m_{\lambda,\mu}| \lesssim 2^{-(1/2+r/2+\alpha)||\lambda|-|\mu||}, 2^{r(|\lambda|+|\mu|)/2}. \tag{4.6.3}$$

Remark 4.6.1 *The case $|\lambda| = -1$ or $|\mu| = -1$ which corresponds to the scaling functions at the coarsest level is not covered by the above considerations since, in contrast to the wavelets, these functions are not orthogonal to* *polynomials. However, we remark that (4.6.3) only required the vanishing moment properties for the wavelet of highest level. Thus the only case not covered remains $|\lambda| = |\mu| = -1$ for which (4.6.3) becomes trivial.*

The case of a more general partial differential operator of order r, i.e. of the type $Af(x) = \sum_{|m|\leq r} a_m(x)\partial^m f(x)$ with $x \in \mathbb{R}^d$ can be treated in a similar way: if both wavelet bases have $C^{r+\alpha}$ smoothness and are orthogonal to polynomials of degree $n \geq \alpha - 1$, one obtains the estimate

$$|m_{\lambda,\mu}| \lesssim 2^{-(d/2+r/2+\alpha)||\lambda|-|\mu||}2^{r(|\lambda|+|\mu|)/2}, \tag{4.6.4}$$

for entries corresponding to overlapping supports. Note that we can also use the Cauchy-Schwarz inequality in the estimation of $|m_{\lambda,\mu}|$, according to

$$\begin{aligned} |m_{\lambda,\mu}| &\leq \inf_{g\in\Pi_n} \|A\psi_\lambda - g\|_{L^2(\mathrm{Supp}(\tilde{\psi}_\mu))} \\ &\lesssim 2^{-\alpha|\mu|}\|A\psi_\lambda\|_{H^\alpha} \\ &\lesssim 2^{-\alpha|\mu|}\|\psi_\lambda\|_{H^{\alpha+r}} \\ &\lesssim 2^{\alpha(|\lambda|-|\mu|)}2^{r|\lambda|}, \end{aligned}$$

allowing us to make the weaker assumption that both wavelet bases have $H^{r+\alpha}$ smoothness, and are orthogonal to polynomials of total degree $n \geq \alpha - 1$. We then obtain the weaker estimate

$$|m_{\lambda,\mu}| \lesssim 2^{-(r/2+\alpha)||\lambda|-|\mu||}2^{r(|\lambda|+|\mu|)/2}. \tag{4.6.5}$$

One can actually further weaken the assumptions on the smoothness of the wavelet bases: if both wavelet bases have $H^{r/2}$ smoothness, the entries $|m_{\lambda,\mu}|$ still make sense. However, we see from (4.6.5) that there is a gain in increasing the smoothness of the wavelets, together with their number of vanishing moments, since the factor $2^{-(\alpha+r/2)||\lambda|-|\mu||}$ improves the decay of the matrix entries away from the diagonal blocks corresponding to $|\lambda| = |\mu|$.

Example 2. Integral operators
We first consider the very simple example of the integral operator that defines the solution of the two point boundary value problem $-\frac{d^2u}{dx^2} = f$ with $u(0) = u(1) = 0$. This operator has the explicit form

$$Af(x) = \int_0^x (1-x)yf(y)dy + \int_x^1 (1-y)xf(y)dy = \int_0^1 K(x,y)f(y)dy.$$

In such a case, we remark that the wavelet coefficients

$$m_{\lambda,\mu} = \int_{[0,1]^2} K(x,y)\tilde{\psi}_\mu(x)\psi_\lambda(y)dx\,dy, \tag{4.6.6}$$

are zero if ψ_λ and $\tilde{\psi}_\mu$ have disjoint supports, provided that ψ_λ or $\tilde{\psi}_\mu$ is orthogonal to polynomials of degree $n = 1$. In order to estimate the remaining entries, we can proceed in the same way as we did in the previous example: for $|\lambda| \geq |\mu|$, assuming that the wavelets $\tilde{\psi}_\mu$ have C^α smoothness and that the wavelets ψ_λ are orthogonal to polynomials of degree $n+2$ with $n \geq \alpha - 1$, we derive the estimate

$$\begin{aligned} |m_{\lambda,\mu}| &= \int A\psi_\lambda(x)\, \tilde{\psi}_\mu(x)dx \\ &\leq \|A\psi_\lambda\|_{L^1} \inf_{g\in\Pi_n} \|\tilde{\psi}_\mu - g\|_{L^\infty(\mathrm{Supp}(A\psi_\lambda))} \\ &\lesssim 2^{-|\lambda|(5/2+\alpha)} \|\tilde{\psi}_\mu\|_{C^\alpha(\mathrm{Supp}(A\psi_\lambda))} \\ &\lesssim 2^{-|\lambda|(5/2+\alpha)} 2^{|\mu|(\alpha+1/2)} \\ &= 2^{(1/2+\alpha)(|\mu|-|\lambda|)} 2^{-2|\lambda|}. \end{aligned}$$

By symmetry, we finally obtain the estimate

$$|m_{\lambda,\mu}| \lesssim 2^{-(\alpha+1/2)||\lambda|-|\mu||} 2^{-2\sup\{|\lambda|,|\mu|\}} = 2^{-(\alpha+5/2)||\lambda|-|\mu||} 2^{-(|\lambda|+|\mu|)}.$$

This simple example reveals a striking similarity between the estimates for differential and integral operators. Here, the vanishing moments of ψ_λ are also used to ensure that $A\psi_\lambda$ has a local support of size $\sim 2^{-|\lambda|}$ (the same as ψ_λ itself). The same comments as in Remark 4.6.1 apply here to the particular case of scaling functions that do not oscillate.

The case that we have just treated is a very particular instance of a multiscale discretization where most of the matrix entries are zero. One cannot expect this property to hold for more general operators. On the other hand, a fairly large class of integral operators gives rise to "almost sparse" matrices, i.e. containing only a few numerically significant entries.

This is the case of the Calderon-Zygmund type operators, i.e. L^2 bounded operators which satisfy estimates of the type (3.8.30) and (3.8.31) (see §3.8 of Chapter 3). A prototype of Calderon-Zygmund is the univariate Hilbert transform given by the kernel $K(x,y) = (x-y)^{-1}$ if $x \neq y$. In this case, we have the more general estimate

$$|\partial_x^m K(x,y)| + |\partial_y^m K(x,y)| \leq |x-y|^{-1-m}. \tag{4.6.7}$$

From (4.6.7), we can again estimate the entries $m_{\lambda,\mu}$ in the case where the supports of ψ_λ and $\tilde{\psi}_\mu$ are strongly disjoint in the sense that

$$\mathrm{dist}(\mathrm{Supp}(\psi_\lambda), \mathrm{Supp}(\tilde{\psi}_\mu)) \geq C 2^{-\inf\{|\lambda|,|\mu|\}}, \tag{4.6.8}$$

where C is a fixed constant. We first define a function that will play the role of a spatial distance between indices in ∇ by

$$d(\lambda,\mu) := 1 + 2^{\min\{|\lambda|,|\mu|\}} \mathrm{dist}(\mathrm{Supp}(\psi_\lambda), \mathrm{Supp}(\psi_\mu)). \tag{4.6.9}$$

For $|\lambda| \leq |\mu|$, assuming that $\tilde{\psi}_\mu$ is orthogonal to all polynomials of degree n, we can exploit (4.6.7) to derive the estimate

$$\begin{aligned}|m_{\lambda,\mu}| &= |\int K(x,y)\psi_\lambda(y),\tilde{\psi}_\mu(x)dxdy \\ &\leq \|\psi_\lambda\|_{L^1}\|\tilde{\psi}_\mu\|_{L^1}\sup_{y\in \mathrm{Supp}(\psi_\lambda)}\inf_{g\in\Pi_n}\|K(\cdot,y)-g\|_{L^\infty(\mathrm{Supp}(\tilde{\psi}_\mu))} \\ &\lesssim 2^{-(|\lambda|+|\mu|)/2}2^{-n)|\mu|}[2^{-|\lambda|}+\mathrm{dist}(\mathrm{Supp}(\psi_\lambda),\mathrm{Supp}(\tilde{\psi}_\mu))]^{-(n+1)} \\ &\lesssim 2^{-(|\lambda|+|\mu|)/2}2^{-n|\mu|}2^{(n+1)|\lambda|}d(\lambda,\mu)^{-(n+1)} \\ &= 2^{(n+1/2)(|\lambda|-|\mu|)}d(\lambda,\mu)^{-(n+1)}.\end{aligned}$$

By symmetry, if ψ_λ is also orthogonal to all polynomials of degree n, we finally obtain

$$|m_{\lambda,\mu}| \lesssim 2^{-(n+1/2)||\lambda|-|\mu||}d(\lambda,\mu)^{-(n+1)}. \qquad (4.6.10)$$

In the case where the supports overlap, estimates of the type (4.6.10) can be preserved if in addition to the $n+1$ vanishing moments one assumes smoothness properties on the basis functions. We shall illustrate this point in the simple case of the Hilbert transform, with two standard wavelet bases generated from translations and dilations of single functions ψ and $\tilde{\psi}$. In this case, we can use the Parseval formula to derive

$$\begin{aligned}|m_{\lambda,\mu}| &= |\int \overline{\hat{\psi}_\lambda(\omega)\hat{\tilde{\psi}}_\mu(\omega)}\mathrm{Sign}(\omega)d\omega| \\ &\lesssim 2^{-(|\lambda|+|\mu|)/2}\int|\hat{\psi}(2^{-|\lambda|}\omega)\hat{\tilde{\psi}}(2^{-|\mu|}\omega)|d\omega \\ &\lesssim 2^{(|\lambda|-|\mu|)/2}\int|\hat{\psi}(\omega)\hat{\tilde{\psi}}(2^{|\lambda|-|\mu|}\omega)|d\omega.\end{aligned}$$

Here, we shall assume that ψ and $\tilde{\psi}$ are smooth enough so that we have $|\hat{\psi}(\omega)|+|\hat{\tilde{\psi}}(\omega)| \lesssim (1+|\omega|)^{-n-1}$, where n is the degree of cancellation, so that we also have $|\hat{\psi}(\omega)|+|\hat{\tilde{\psi}}(\omega)| \lesssim |\omega|^{n+1}$ near the origin. In the case where $|\lambda| \leq |\mu|$, we can then evaluate the above integral by splitting it between $|\omega| \leq 2^{|\mu|-|\lambda|}$ and $|\omega| > 2^{|\mu|-|\lambda|}$. For the first part, we obtain

$$\begin{aligned}\int_{|\omega|\leq 2^{|\mu|-|\lambda|}}|\hat{\psi}(\omega)\hat{\tilde{\psi}}(2^{|\lambda|-|\mu|}\omega)|d\omega &\lesssim \int_{|\omega|\leq 2^{|\mu|-|\lambda|}}\left(\frac{|2^{|\lambda|-|\mu|}\omega|}{1+|\omega|}\right)^{n+1}d\omega \\ &\lesssim 2^{n(|\lambda|-|\mu|)}.\end{aligned}$$

For the second part, we also obtain

$$\begin{aligned}\int_{|\omega|>2^{|\mu|-|\lambda|}}|\hat{\psi}(\omega)\hat{\tilde{\psi}}(2^{|\lambda|-|\mu|}\omega)|d\omega &\lesssim \int_{|\omega|>2^{|\mu|-|\lambda|}}(1+|\omega|)^{-n-1}d\omega \\ &\lesssim 2^{n(|\lambda|-|\mu|)}.\end{aligned}$$

It follows that (4.6.10) holds. By a symmetric argument if $|\mu| \leq |\lambda|$, we finally obtain that (4.6.10) holds for all entries.

Note that the example of the single layer operator acting on 1-periodic functions, which was evoked in §1.5 of Chapter 1, is strongly related to the Hilbert transform since a singularity of the type $1/(x-y)$ appears if we differentiate $K(x,y) = \log|e^{i2\pi x} - e^{i2\pi y}|$. This allows us to evaluate the corresponding matrix entries $m_{\lambda,\mu} = \int K(x,y)\psi_\lambda(y), \tilde{\psi}_\mu(x)dxdy$ using integration by part in the variable x (resp. y) if $|\lambda| \le |\mu|$ (resp. $|\lambda| \ge |\mu|$), which yields

$$\begin{aligned} |m_{\lambda,\mu}| &\lesssim 2^{-(n+1/2)||\lambda|-|\mu||} d(\lambda,\mu)^{-(n+1)} 2^{-\sup\{|\lambda|,|\mu|\}} \\ &= 2^{-(n+1)||\lambda|-|\mu||} d(\lambda,\mu)^{-(n+1)} 2^{-(|\lambda|+|\mu|)/2}. \end{aligned}$$

In the general context of pseudodifferential operators on periodic domains, the reader will find similar estimates in DAHMEN, PRÖSSDORF and SCHNEIDER [1993].

It is interesting to note from these examples that, while finite element discretizations of partial differential and integral operators result in very different types of matrices - banded matrices for partial differential operators and densely populated matrices for integral operators - multiscale discretizations give rise to the same structure in both cases (this is the "finger-like" structure that was illustrated in Figure 1.5.10 in §1.5 of Chapter 1). In order to treat these different examples within a unified framework that describes this structure, we shall now introduce general classes of matrices associated with operators through wavelet bases.

Definition 4.6.1 *Let $s \in \mathbb{R}$ and $\alpha, \beta > 0$. A matrix M belongs to the class $\mathcal{M}^s_{\alpha,\beta}$ if and only if its entries satisfy the estimate*

$$|m_{\lambda,\mu}| \le C_M 2^{s(|\lambda|+|\mu|)} 2^{-(d/2+\alpha)||\lambda|-|\mu||} d(\lambda,\mu)^{-(d+\beta)}. \tag{4.6.11}$$

We denote by $\mathcal{M}_{\alpha,\beta}$ this class when $s = 0$.

Before going further, let us analyze the meaning of the various factors in the right side of (4.6.11).

The factor $2^{s(|\lambda|+|\mu|)}$ describes the growth or decay (depending on the sign of s) of the entries of M along the diagonal, i.e. the multiplicative effect of the operator on the range of scales. The parameter s thus indicates the order of the operator: for instance $s = 2$ if $A = \Delta$, $s = 0$ for the Hilbert transform, $s = -1/2$ for the single layer operator, etc. Note that the diagonal matrix $D_s = (2^{s|\lambda|}\delta_{\lambda,\mu})_{\lambda,\mu}$ allow us to renormalize M in the sense that $\widetilde{M} = D_s^{-1} M D_s^{-1}$ satisfies the estimate (4.6.11) with $s = 0$, i.e. belongs to the class $\mathcal{M}_{\alpha,\beta}$. Note that such a renormalization is exactly the preconditioning process described on finite matrices in §3.11 of Chapter 3.

The factor $2^{-(d/2+\alpha)||\lambda|-|\mu||}$ describes the decay of the entries away from the diagonal blocks corresponding to $|\lambda| = |\mu|$. In all examples that we have discussed, the value of α is both influenced by the nature of the operator (e.g. the smoothness of its coefficients in the case of a nonconstant partial differential operator or the regularity of the kernel for an integral operator), as well as by the smoothness and number of vanishing moments of the wavelet basis.

Finally, the factor $d(\lambda,\mu)^{-(d+\beta)}$ describes the decay of the entries away from the diagonal within each block corresponding to fixed values of $|\lambda|$ and $|\mu|$. In the case of a local operator such as a partial differential operator, we note that β can be taken arbitrarily large, up to a modification in the constant C_M in (4.6.11). In the case of an integral operator, β is influenced like α by the operator and the wavelet basis.

A basic tool for the study of the classes $\mathcal{M}^s_{\alpha,\beta}$ is the Schur lemma that we recall below, leaving its proof as an exercise to the reader.

Lemma 4.6.1 *Let $M = (m_{\lambda,\mu})_{\lambda,\mu\in\nabla}$ be a matrix indexed by ∇. Assume that there exists a sequence of positive numbers $(\omega_\lambda)_{\lambda\in\nabla}$ and a constant C such that*

$$\sum_{\mu\in\nabla} \omega_\mu |m_{\lambda,\mu}| + \sum_{\mu\in\nabla} \omega_\mu |m_{\mu,\lambda}| \le C\omega_\lambda, \tag{4.6.12}$$

for all $\lambda \in \nabla$. Then M defines a bounded operator in $\ell^2(\nabla)$ with $||M|| \le C$.

A first application of the Schur lemma is the following simple result.

Theorem 4.6.1 *If $\alpha,\beta > 0$, then any $M \in \mathcal{M}_{\alpha,\beta}$ defines a bounded operator in $\ell^2(\nabla)$. In turn, any matrix $M \in \mathcal{M}_{\alpha,\beta}$ together with a Riesz basis $(\psi_\lambda)_{\lambda\in\nabla}$ of $L^2(\Omega)$ defines an L^2 bounded operator A represented by M in this basis.*

Proof We shall use the Schur lemma with $\omega_\lambda = 2^{-d|\lambda|/2}$. From (4.6.11), we first obtain

$$\begin{aligned}
\sum_{\mu\in\nabla} \tfrac{\omega_\mu}{\omega_\lambda} |m_{\lambda,\mu}| &\lesssim \sum_{\mu\in\nabla} 2^{d\frac{|\lambda|-|\mu|}{2}} 2^{-(\frac{d}{2}+\alpha)||\lambda|-|\mu||} d(\lambda,\mu)^{-(d+\beta)} \\
&\lesssim \sum_{j\ge -1} 2^{d\frac{|\lambda|-j}{2}} 2^{-(\frac{d}{2}+\alpha)||\lambda|-j|} \sum_{|\mu|\in\nabla_j} d(\lambda,\mu)^{-(d+\beta)}.
\end{aligned}$$

Since $\beta > 0$ the last factor $\sum_{|\mu|\in\nabla_j} d(\lambda,\mu)^{-(d+\beta)}$ is bounded by a uniform constant if $j \le |\lambda|$ and by $2^{d(j-|\lambda|)}$ if $j \ge |\lambda|$. Splitting the sum in j according to these two cases, we finally obtain

$$\begin{aligned}
\omega_\lambda^{-1} \sum_{\mu\in\nabla} \omega_\mu |m_{\lambda,\mu}| &\lesssim \sum_{j\ge -1} 2^{d||\lambda|-j|/2} 2^{-(d/2+\alpha)||\lambda|-j|} \\
&\le 2\sum_{l\ge 0} 2^{-\alpha l} = C < \infty,
\end{aligned}$$

which shows that (4.6.12) holds with such weights. ◇

An immediate consequence of this result concerns the classes $\mathcal{M}^s_{\alpha,\beta}$ for $s \neq 0$. We already noted that for M in this class, the preconditioned matrix

$$\widetilde{M} = D_s^{-1} M D_s^{-1}, \tag{4.6.13}$$

with $D_s = (2^{s|\lambda|}\delta_{\lambda,\mu})_{\lambda,\mu}$ belongs to the class $\mathcal{M}_{\alpha,\beta}$. Introducing the weighted spaces

$$\ell_t^2(\nabla) := \{(c_\lambda)_{\lambda\in\nabla}\ ;\ \|(c_\lambda)\|_{\ell_t^2}^2 := \sum_{\lambda\in\nabla} 2^{2t|\lambda|}|c_\lambda|^2 < \infty\}, \tag{4.6.14}$$

(which are associated with the Sobolev spaces H^s by the results of Chapter 3, e.g. Theorem 3.10.5), we remark that D_s defines an isomorphism from $\ell_t^2(\nabla)$ to $\ell_{t+s}^2(\nabla)$. Combining these remarks with Theorem 4.6.1, we can describe the action of $M = D_s\widetilde{M}D_s$ as follows.

Corollary 4.6.1 *If $\alpha, \beta > 0$, then any $M \in \mathcal{M}^s_{\alpha,\beta}$ defines a bounded operator from $\ell_s^2(\nabla)$ to $\ell_{-s}^2(\nabla)$. In turn, any matrix $M \in \mathcal{M}^s_{\alpha,\beta}$ together with a wavelet basis $(\psi_\lambda)_{\lambda\in\nabla}$ that both characterizes $H^s(\Omega)$ and $H^{-s}(\Omega)$ (possibly with boundary conditions) defines a bounded operator A from H^s to H^{-s}, represented by M in this basis.*

Remark 4.6.2 *One can address the problem of the multiplication of two matrices in the classes $\mathcal{M}^s_{\alpha,\beta}$. With a slight modification - appending the additional factor $[1 + (|\lambda| - |\mu|)^2]^{-1}$ in the estimate (4.6.11) - it can be proved by direct matrix product computations (see e.g.* MEYER *[1990]) that the resulting classes $\widetilde{\mathcal{M}}_{\alpha,\beta}$ (again for $s = 0$) are algebras of operators. In contrast, these classes are not stable by inversion: it is false in general that an invertible matrix in $\mathcal{M}_{\alpha,\beta}$ or $\widetilde{\mathcal{M}}_{\alpha,\beta}$ has its inverse in the same class (see* TCHAMITCHIAN *[1996] for some explicit counterexamples).*

Our next step is to show that the estimate (4.6.11) allows us to compress the matrices in the class $\mathcal{M}^s_{\alpha,\beta}$ by discarding certain entries. Again, we first consider the case $s = 0$.

Theorem 4.6.2 *Let $M \in \mathcal{M}_{\alpha,\beta}$ and $t < \min\{\alpha/d, \beta/d\}$. For all $N \geq 0$ one can discard the entries of M in such a way that the resulting matrix M_N has N non-zero entries per row and column and satisfies*

$$\|M - M_N\| \lesssim N^{-t}, \tag{4.6.15}$$

in the operator norm of $\ell^2(\nabla)$.

Proof We first truncate the matrix M in scale: for a given $J > 0$, we discard $m_{\lambda,\mu}$ if $||\lambda| - |\mu|| \geq J$. Denoting by A_J the resulting matrix, we can use the same technique as in the proof of Theorem 4.6.1 (Schur lemma with weights $2^{d|\lambda|/2}$) to measure the error $\|M - A_J\|$ in operator norm. By a very similar computation, we obtain

$$\|M - A_J\| \lesssim \sum_{l \geq J} 2^{-\alpha l} \lesssim 2^{-\alpha J}. \tag{4.6.16}$$

We next trunctate A_J in space, by preserving in each remaining block of A_J the entries $m_{\lambda,\mu}$ such that $d(\lambda,\mu) \geq k(||\lambda| - |\mu||)$ where the function k is to be determined. We denote by B_J the resulting matrix. Again using the Schur lemma in the same way as in the proof of Theorem 4.6.1, we evaluate the error $\|A_J - B_J\|$ by the supremum in λ of

$$\omega_\lambda^{-1} \sum_{j=|\lambda|-J}^{|\lambda|+J} \sum_{\mu \in \nabla_j} \omega_\mu |b^J_{\lambda,\mu} - m_{\lambda,\mu}|, \tag{4.6.17}$$

and we obtain an estimated contribution of $2^{-\alpha J}$ for each term in j by taking $k(l) = 2^{J\alpha/\beta} 2^{l(1-\alpha/\beta)}$. The total error is thus estimated by

$$\|M - B_J\| \lesssim J 2^{-\alpha J}, \tag{4.6.18}$$

while the number of non-zero entries per row and column in B_J is estimated by $N(J) \lesssim \sum_{l=0}^{J} k(l)^d$. In the case where $\alpha > \beta$ (resp. $\beta < \alpha$), this sum is dominated by the first term $k(0)^d = 2^{Jd\alpha/\beta}$ (resp. last term $K(J) = 2^{dJ}$). In the case $\alpha = \beta$ we obtain $N(J) \lesssim J 2^{dJ}$. In all cases, it follows from the evaluation of $N(J)$ and of the error $\|M - B_J\|$ by (4.6.18) that

$$\|M - B_J\| \lesssim N(J)^{-t}, \tag{4.6.19}$$

if t is such that $t < \min\{\alpha/d, \beta/d\}$. Since J ranges over all positive integers, this is enough to conclude the proof. ◇

It is important to note that in the above proof, the construction of the truncated matrix is independent of the value of t. We can derive simple consequences of this result concerning the sparsity of the operators in the classes $\mathcal{M}^s_{\alpha,\beta}$, by the same considerations as for the study of their boundedness properties: for $M \in \mathcal{M}^s_{\alpha,\beta}$, we apply the compression process of Theorem 4.6.2 to the preconditioned matrix $\widetilde{M} \in \mathcal{M}_{\alpha,\beta}$ defined by (4.6.13). Denoting by $\widetilde{M}_N$ the compressed matrix, we then define $M_N = D_s \widetilde{M}_N D_s$. This new matrix also has N non-zero entries per row and column and approximates M in the sense expressed by the following corollary.

Corollary 4.6.2 *Let $M \in \mathcal{M}^s_{\alpha,\beta}$ and $t < \min\{\alpha/d, \beta/d\}$. For all $N \geq 0$ one can discard the entries of M in such a way that the resulting matrix M_N has N non-zero entries per row and column and satisfies*

$$\|M - M_N\| \lesssim N^{-t}, \tag{4.6.20}$$

in the norm of operators from $\ell^2_s(\nabla)$ to $\ell^2_{-s}(\nabla)$.

Our last results deal with the application of sparse matrices of the type that we have introduced in this section on sparse vectors that result of the multiscale discretization of functions. In the context of nonlinear approximation theory, the level of sparsity can be described by the rate of decay of the error of N-term approximation: an infinite vector U has approximation rate $t > 0$ in some metric X, if there exists a sequence of vectors $(U_N)_{N\geq 0}$ such that U_N has N non-zero coordinates and such that

$$\|U - U_N\|_X \lesssim N^{-t}. \tag{4.6.21}$$

In the case where $X = \ell^2$, the vectors U_N are simply obtained by keeping the N largest coordinates in U. In this case, we have observed in Remark 4.2.4 that the property (4.6.21) is equivalent to

$$\#\{\lambda\,;\ |u_\lambda| > \eta\} \lesssim \eta^{-p}, \tag{4.6.22}$$

with $1/p = 1/2 + t$. Therefore, the level of sparsity of U is also described by the smallest p such that $U \in \ell^p_w$. The following result shows that this level of sparsity is preserved by the action of a sparse matrix.

Theorem 4.6.3 *A matrix $M \in \mathcal{M}_{\alpha,\beta}$ defines a bounded operator in $\ell^2 \cap \ell^p_w$ if $1/p = 1/2 + t$ and $t < \min\{\alpha/d, \beta/d\}$. In other words a vector U with approximation rate t in ℓ^2 is mapped by M onto a vector $V = MU$ with at least the same approximation rate.*

Proof Our proof will consist of directly constructing an N-term approximation to $V = MU$ from the N-term approximation of U. For $j \geq 0$, we denote by U_j the vector that consists of the 2^j largest coordinates of U. From the assumptions we know that

$$\|U - U_j\| \lesssim 2^{-tj}, \tag{4.6.23}$$

where for the sake of notational simplicity we denote by $\|\cdot\|$ the ℓ^2 norm. Using Theorem 4.6.2, we can define truncated operators M_j such that M_j

has at most $2^j\alpha_j$ non-zero entries per row and column for some summable sequance $(\alpha_j)_{j\geq 0}$ such as $\alpha_j = (j+1)^{-2}$ and such that for all $r < \min\{\alpha/d, \beta/d\}$,

$$\|M - M_j\| \lesssim 2^{-rj}, \tag{4.6.24}$$

with a constant that may depend on the closeness of r to $\min\{\alpha/d, \beta/d\}$. We define an approximation to $V = MU$ by

$$\begin{aligned} V_j &:= M_jU_0 + M_{j-1}(U_1 - U_0) + \cdots M_0(U_j - U_{j-1}) \\ &= \textstyle\sum_{l=0}^{j} M_{j-l}(U_l - U_{l-1}), \end{aligned} \tag{4.6.25}$$

where we have used the notation $U_{-1} = 0$. We can thus evaluate the number of non-zero entries of V_j by

$$N(j) \leq 2^j\alpha_j + \sum_{l=1}^{j} \alpha_{j-l}2^{j-l}2^{l-1} \lesssim 2^j. \tag{4.6.26}$$

Taking $r \in]t, \min\{\alpha/d, \beta/d\}[$, we can evaluate the error of approximation as follows.

$$\begin{aligned} \|V - V_j\| &= \|M(U - U_j) + \textstyle\sum_{l=0}^{j}(M - M_l)(U_{j-l} - U_{j-l-1})\| \\ &\leq \|M\|\,\|U - U_j\| + \textstyle\sum_{l=0}^{j}\|M - M_l\|\,\|U_{j-l} - U_{j-l-1}\| \\ &\lesssim 2^{-tj} + 2^{-tj}\textstyle\sum_{l=0}^{j-1} 2^{(t-r)l} + 2^{-rj} \\ &\lesssim 2^{-tj} \lesssim N(j)^{-t}, \end{aligned}$$

with the constant proportional to $\|U\|_{\ell^p_w}$. Since j ranges over all possible integers, we have thus proved that $V \in \ell^p_w$. ◇

We can again derive an immediate corollary, by the same considerations as for Corollaries 4.6.1 and 4.6.2.

Corollary 4.6.3 *Let $M \in \mathcal{M}^s_{\alpha,\beta}$ and U be a vector of approximation rate t in ℓ^2_s with $t < \inf\{\alpha/d, \beta/d\}$. Then $V = MU$ has approximation rate t in the dual space ℓ^2_{-s}.*

It is interesting to note that the simple choice $V_j = M_jU_j$ in the above proof does not give the optimal decay estimate. The particular construction of the approximation by (4.6.25) corresponds to the intuitive idea that the operator should be truncated according to the size of the component of U to which it is applied. In the case where U is already a finitely supported coefficient vector, we can use this construction to build a *sparse matrix-vector multiplication algorithm* which computes the product $\mathcal{M}U$ up to some prescribed tolerance ε:

Definition 4.6.2 *Let $M \in \mathcal{M}_{\alpha,\beta}$ and U be a finite vector. If $\varepsilon > 0$ is a given tolerance, using the same notation as in the proof of Theorem 4.6.3, we define the approximation*

$$\mathbf{APPROX}(MU, \varepsilon) = \sum_{l=0}^{j} M_{j-l}(U_l - U_{l-1}), \tag{4.6.27}$$

where j is the smallest integer such that quantity

$$r_j := \|M\| \, \|U - U_j\| + \sum_{l=0}^{j} \|M - M_l\| \, \|U_{j-l} - U_{j-l-1}\|,$$

is less than ε.

In this algorithm, we can use the a-priori estimates derived from Theorem 4.6.2 in place of the quantities $\|M\|$ and $\|M - M_j\|$. This algorithm will be invoked in §4.9 for solving operator equations by adaptive wavelet methods. By definition, we know that $\|\mathbf{APPROX}(MU, \varepsilon) - MU\| \leq \varepsilon$. From Theorem 4.6.3, we can derive that this algorithm has optimal complexity in the following sense.

Corollary 4.6.4 *If $U \in \ell^p_w$ for $1/p = 1/2 + t$ and $t < \min\{\alpha/d, \beta/d\}$, and if $V = \mathbf{APPROX}(MU, \varepsilon)$, then the number of non-zero entries of V satisfies*

$$\#(V) \lesssim \varepsilon^{-1/t} \|U\|_{\ell^p_w}^{1/t}, \tag{4.6.28}$$

and therefore

$$\|V - MU\| \lesssim \#(V)^{-t} \|U\|_{\ell^p_w}. \tag{4.6.29}$$

The total number of arithmetic operations for computing V is also bounded by

$$\mathcal{N}(V) \lesssim \varepsilon^{-1/t} \|U\|_{\ell^p_w}^{1/t}, \tag{4.6.30}$$

provided that the quantities $\|U - U_j\|$ and $\|U_{j-l} - U_{j-l-1}\|$ are known. We also have

$$\|V\|_{\ell^p_w} \lesssim \|U\|_{\ell^p_w}. \tag{4.6.31}$$

The constants in the above estimates only depend on the closeness of t to $\min\{\alpha/d, \beta/d\}$.

Proof The estimate (4.6.28) is proved by the same arguments as in the proof of Theorem 4.6.4: it follows from

$$\#(V) = N(j) \lesssim 2^j \tag{4.6.32}$$

combined with

$$e < r_{j-1} \lesssim \|U\|_{\ell^p_w} 2^{-tj}. \tag{4.6.33}$$

The estimate (4.6.30) is derived in a similar way, remarking that the total number of multiplications involved in the computation of V is bounded by

$$\mathcal{N}(V) \le 2^j \alpha_j + \sum_{l=1}^{j} \alpha_{j-l} 2^{j-l} 2^{l-1} \lesssim 2^j. \tag{4.6.34}$$

The estimate (4.6.31) is less immediate. For this, we define by W the best N-term approximation of MU with $N = \#(W)$. Using triangle and H'older's inequality, we have

$$\begin{aligned} \|V\|_{\ell^p_w} &= \|W\|_{\ell^p_w} + \|V - W\|_{\ell^p_w} \\ &\le \|MU\|_{\ell^p_w} + \|V - W\|_{\ell^p} \\ &\le \|MU\|_{\ell^p_w} + (2N)^t \|V - W\| \\ &\le \|MU\|_{\ell^p_w} + (2N)^t (\|V - MU\| + \|W - MU\|). \end{aligned}$$

From the approximation properties of V and W and since M defines a bounded operator in ℓ^p_w, it thus follows that

$$\|V\|_{\ell^p_w} \lesssim \|MU\|_{\ell^p_w} + \|U\|_{\ell^p_w} \lesssim \|U\|_{\ell^p_w}, \tag{4.6.35}$$

which concludes the proof. ◇

Remark 4.6.3 *In this section, we have addressed sparsity only in the sense where the error is measured in ℓ^2 or ℓ^2_s. In a similar way it is possible to derive similar results with ℓ^p and ℓ^p_s spaces (which corresponds to analyzing the error in Besov spaces that differ from H^s). The Hilbertian theory is of particular interest, due to its direct application to the discretization of elliptic problems, as will be seen in §4.8).*

Another technique of matrix compression is based on the so-called non-standard representation, which was introduced in BEYLKIN, COIFMAN and ROKHLIN [1993], in relation to multipole algorithms (GREENGARD and ROKHLIN [1987]) and panel clustering techniques (HACKBUSCH and NOWAK [1989]). The non-standard representation is derived from an initial (usually full) matrix M_J representing the discretization of an operator in a nodal basis at resolution 2^{-J}. One views this matrix as the "restriction" of an operator A onto a space V_J, in the sense that its entries $m^J_{\lambda,\mu}$ are identified with the inner products $\langle A\varphi_\mu, \tilde{\varphi}_\lambda \rangle$, $\lambda, \mu \in \Gamma_J$. Thus M_J represents the operator $P_J A P_J$, which can be decomposed according to

$$\begin{aligned} P_J A P_J &= (P_{J-1} + Q_{J-1}) A (P_{J-1} + Q_{J-1}) \\ & P_{J-1} A P_{J-1} + P_{J-1} A Q_{J-1} + Q_{J-1} A P_{J-1} + Q_{J-1} A Q_{J-1} \end{aligned}$$

Iterating at coarser scales, we therefore obtain

$$P_J A P_J = P_0 A P_0 + \sum_{j=0}^{J-1} [P_{j-1} A Q_{j-1} + Q_{j-1} A P_{j-1} + Q_{j-1} A Q_{j-1}]. \quad (4.6.36)$$

The first term of this decomposition is represented by the coarse scale matrix $(\langle A\varphi_\mu, \tilde{\varphi}_\lambda\rangle)_{\lambda,\mu\in\Gamma_0}$, and the other terms by the matrices $(\langle A\psi_\mu, \tilde{\varphi}_\lambda\rangle)_{\lambda\in\Gamma_j,\mu\in\nabla_j}$, $(\langle A\varphi_\mu, \tilde{\psi}_\lambda\rangle)_{\lambda\in\nabla_j,\mu\in\Gamma_j}$ and $(\langle A\psi_\mu, \tilde{\psi}_\lambda\rangle)_{\lambda\in\nabla_j,\mu\in\nabla_j}$. Note that these matrices are obtained from M_J by applying a standard (isotropic) bivariate wavelet decomposition: the resulting representation has the same organization as the decompostion of an image (see Figures 1.5.6 and 1.5.7 in §1.5 of Chapter 1). We can thus expect that, similarly to the standard wavelet discretization that was discussed in the present section, many entries are numerically small and can be thresholded in order to compress the action of A, provided that it has some smoothness properties. It should be well understood that such a representation is not a change of basis since its application on a vector requires the redundant data of both its approximation and wavelet coefficients at levels $0, \cdots, J-1$. In turn one cannot hope to combine this type of operator compression with a nonlinear approximation of the vector (as we did in the case of standard multiscale discretization) since the approximation coefficients are in general fully populated. On the other hand, a particular advantage of the non-standard form is that in many practical cases, e.g. Calderon-Zygmund operators, the structure of the compressed matrices consists of a set of uniformly banded square matrices, simpler and more sparse than the finger-like structure encountered with standard wavelet discretizations. We refer to MEYER [1997] for a detailed analysis of operator compression algorithms based on the non-standard representation.

Remark 4.6.4 *Wavelet-based multiscale representations of operators can also be applied to matrix reduction problems in the context of homogeneization: an operator acting on coarse scale is derived from a fine scale operator by multilevel reduction technique, in order to solve a problem on a coarse grid with a "memory" of fine grid information that may affect the behaviour of the coarse solution, e.g. highly oscillating coefficients. These methods have been introduced in* BEYLKIN *and* BREWSTER *[1995] (see also* GILBERT *[1998] for a comparison with classical homogeneization schemes).*

4.7 Nonlinear approximation and PDE's

The goal of this section is to present instances of PDE's for which solutions can be proved to have a high smoothness in the scale which governs the rate of nonlinear approximation, while its classical smoothness is low due to the presence of isolated singularities. In such instances, there is an important potential gain in switching from uniform to adaptive discretizations.

Example 1. Hyperbolic conservation laws
It is well known that the solutions $u(x,t) = (u_1(x,t), \cdots, u_m(x,t))$, of initial value problems of the type

$$\begin{array}{l} \frac{\partial u_i}{\partial t} + \mathrm{Div}_x[A(u)] = 0, \quad x \in \mathbb{R}^n,\ t \geq 0 \\ u_i(x,0) = u_{i,0}(x), \quad i = 1, \cdots, m, \end{array} \tag{4.7.1}$$

might develop singularities in finite time even when the functions $u_{i,0}$ and A are smooth, if the flux A is not a simple linear transformation (in which case the solution is simply advected). The simplest prototype of such a situation is furnished by the Burger equation for which $m = n = 1$ and $A(u) = u^2/2$.

When $u(x,t)$ is no longer C^1, the solution of (4.7.1) can only be meant in a distributional sense and is not necessarily unique. In the case of scalar conservation laws, i.e. when $m = 1$, uniqueness can be restored by appending suitable *entropy conditions* to the equation. We refer to the monograph by LAX [1972] for a synthetic survey of these fundamental issues.

The generation of shocks is clearly a problem for the approximation of solutions by numerical schemes (see LEVEQUE [1992] for an introduction to this aspect of conservation laws): a uniform discretization with mesh size h can at most yield $\|u(\cdot,t) - u_h(\cdot,t)\|_{L^1} \lesssim h$ (with $u_h(\cdot,t)$ the approximate solution at time t), since solutions generally fail to have $W^{1,1}$ smoothness, but still lie in the space BV (functions with bounded variation, i.e. such that $\|u(\cdot,t) - u(\cdot + h,t)\|_{L^1} \leq C|h|$) under reasonable assumptions in the case of scalar conservation laws. If we now intend to approximate u on an adaptive grid, smoothness should be reconsidered in $B^s_{p,p}$, $1/p = 1 + s/n$. In the case of scalar univariate conservation laws, i.e. $m = n = 1$, this issue was addressed in DEVORE and LUCIER [1990] with the following striking result.

Theorem 4.7.1 *Assume that the flux function A is C^∞ and strictly convex, and that the initial data u_0 is in $B^s_{p,p}$ for some $s > 0$ and $1/p = 1+s$. Then for all time $t > 0$, the function $u(\cdot,t)$ remains in the space $B^s_{p,p}$.*

The method of proof of the theorem of DeVore and Lucier can be understood at the intuitive level by considering the simple case of the Burger's equation (i.e. $A(u) = u^2/2$) with the smoothness parameter $s \leq 2$. For such range of smoothness, $B^s_{p,p}$ can be exactly characterized by the decay property of approximation by free knot piecewise affine functions (without smoothness constraints) according to Theorem 4.4.1. One then remarks that the associated evolution operator $\mathcal{E}_t$ which maps the initial condition u_0 to the entropy solution of Burger's equation $u(\cdot,t)$ at time $t > 0$ is an L^1 contraction, and that $\mathcal{E}_t$ maps a piecewise affine function with N pieces into another piecewise affine function with at most $2N$ pieces. The second property is due to the fact that the speed $A'(u) = u$ is itself affine in u and the constant 2 accounts for the possible appearance of rarefaction waves if the initial piecewise affine function has positive jumps. Thus if u_0 is approximated at a certain rate by piecewise affine functions $v_N \in S_N$, one obtains the same rate for $u(\cdot,t)$ with the affine functions $\mathcal{E}_t v_N \in S_{2N}$, i.e. $u(\cdot,t)$ has the same smoothness as u_0 in the scale Besov spaces associated with nonlinear approximation by piecewise polynomials in L^1.

It is interesting to note here that the proof goes via the study of a specific approximation procedure. However, having once established the Besov smoothness property, we can also conclude from other nonlinear approximation results that other tools such as wavelets or rational functions might yield high order schemes for conservation laws. A multivariate analog of this result is still an open problem (one should expect a less optimistic statement, due to the anisotropic structures of shocks in several dimensions).

Example 2. Elliptic operator equations
We already saw in §3.11 of Chapter 3 that a large class of elliptic equations can be preconditioned through multiscale Galerkin discretization. We again consider this general setting, recalling (3.11.1) to (3.11.7). In particular, we assume that the energy norm $\|\cdot\|_a$ is equivalent to a Sobolev norm $\|\cdot\|_H$, with the multiscale characterizations (3.11.6) and (3.11.7).

The use of adaptive multiscale methods is related to the following questions: (i) what are the possible sources of singularities in elliptic equations ? (ii) Are these singularities highly regular in the scale of nonlinear approximation for some specific metric ?

Since the Galerkin method produces a near-best approximation of u by a function of the trial space in the norm $\|\cdot\|_H$, it is natural to compare the linear and nonlinear approximation spaces associated with the specific Sobolev space H. Consider for instance a second order elliptic operator A such as $A = -\Delta$, in which case H is the space H^1 possibly appended with some boundary condition. Then, the results that we have stated in Chapter 3 as well as in §4.2 of the present chapter allow us to give a precise description of these approximation spaces.

1. **Uniform case:** if u has smoothness H^{1+t} then uniform discretizations produce solutions u_h described by $N = N(h)$ parameters such

that

$$\|u - u_h\|_{H^1} \lesssim N^{-t/d}, \tag{4.7.2}$$

where u_h is the Galerkin solution on the finite element space V_h or on the approximation space V_j with $h \sim 2^{-j}$.

2. **Adaptive case:** if u has smoothness $B^{1+t}_{p,p}$, $1/p = 1/2 + t/d$, then for all N, there exists an adaptive space $V_\Lambda = \mathrm{Span}\{\psi_\lambda \ ; \ \lambda \in \Lambda\}$ with $|\Lambda| = N$ such that

$$\|u - u_\Lambda\|_{H^1} \lesssim N^{-t/d}, \tag{4.7.3}$$

where u_Λ is the Galerkin solution on V_Λ.

In the second case, an additional problem is to construct the adaptive index set $\Lambda = \Lambda(N)$ such that (4.7.3) holds. We postpone this crucial point to §4.9. For the time being, our main point is to compare the smoothness of the solution u in both senses above. We list below the various possible sources of singularities for the solutions of linear elliptic equations and the known facts about their linear and nonlinear approximation properties.

1. **Singularities generated by the interior data:** if the data f has poor smoothness, say f does not belong to H^t for $t \geq t_0$, the solution of $Au = f$ will not be in H^t for $t \geq t_0 + 2s$ where $2s$ is the order of the operator A. It may occur that the data f has higher smoothness in the sense of nonlinear approximation in some metric, in which case one might also expect a higher smoothness for the solution in this sense. In other words, we raise the question: do sparse data generate sparse solutions ? The answer to this question relies on the sparsity of the inverse operator A^{-1} in its multiscale discretization. In the most favorable cases, estimates on the kernel of A^{-1} show that the matrix of this operator in a wavelet basis belongs to the class $\mathcal{M}^{-s}_{\alpha,\beta}$ (defined in §4.6), where α and β can be made arbitrarily large by increasing the smoothness and number of vanishing moments of the basis functions. This is the case (with $s = 1$) when $A = \Delta$ with homogeneous Dirichlet conditions on $\partial\Omega$ and Ω is a C^∞ domain. In such cases, we derive from Corollary 4.6.3 that if the data f is sparse in H' in the sense that $\mathrm{dist}_{H'}(f, S_N) \lesssim N^{-t}$, then the solution u will be sparse in H in the sense that $\mathrm{dist}_H(f, S_N) \lesssim N^{-t}$ (assuming here that the wavelet basis has enough smoothness and oscillations to ensure the estimate (4.6.11) on the coefficients of A^{-1} with $t < \inf\{\alpha/d, \beta/d\}$). In less favorable cases, other sources of singularities (that we describe below) might perturb the sparsity of A^{-1}.

2. **Singularities generated by the boundary data:** if we append a non homogeneous boundary condition $Bu = g$ on $\partial\Omega$, a singular behaviour of g can perturb the smoothness of the solution u. Intuitively, one expects to remedy this problem by local refinement of the interior mesh near the location of the singularity on the boundary. This approach can be precisely analyzed in a simple situation: consider the example of the Laplace equation $-\Delta u = f$ on the square $\Omega = [0,1]^2$ with Dirichlet data $u = g$ on $\partial\Omega$. In such a case, it is shown in COHEN and MASSON [1998] that a tensor product multiscale decomposition of $[0,1]$ can be coupled together with a multiscale decomposition adapted to the boundary, in the sense where each wavelet ψ^b_λ is the trace of an interior wavelet $\psi^i_\mu(\lambda)$. Such a coupling allows us to derive a natural lifting of $g = \sum_{\lambda\in\nabla^b} g_\lambda\psi^b_\lambda$ into $Lg = \sum_{\lambda\in\nabla^b} g_\lambda\psi^i_\lambda$. From the function space characterization results for both interior and boundary bases, one can derive that L is stable from H^s to $H^{s+1/2}$ (for a range of s that depends on the wavelet bases), and that L preserves sparsity: if $\|g - g_N\|_{H^{1/2}} \lesssim N^{-t}$ for some N-term approximation on the boundary, then Lg_N is an N-term approximation of Lg which satisfies $\|Lg - Lg_N\|_{H^1} \lesssim N^{-t}$. Since $u = Lg + v$ where v is a solution of $-\Delta v = f + \Delta(Lg)$ with homogeneous Dirichlet boundary condition, we are back to the previous case, i.e. facing a (sparse) singularity in the interior data.

3. **Singularities generated by the coefficients in the operator:** the presence of singularities in the coefficients, e.g. a jump in $a(x)$ for an operator $u \mapsto \nabla(a(x)\nabla u)$ also perturbs the smoothness of u. The analysis of the sparsity of u, in particular when the coefficients have themselves a sparse wavelet decomposition is so far not well understood, although some first results are available in DAHLKE [1998]. An additional difficulty is that such singularities in the coefficients also affect the sparsity of the operator A and the possibility of compressing it (by reducing the coefficient α in the estimate (4.6.11)).

4. **Singularities generated by the geometry of the domain:** it is well known that the solution u of an elliptic problem with smooth data f and g and coefficients $a_i(x)$ in A can still be singular if the geometry of the domain is not smooth. Are these singularities still highly regular in the sense of nonlinear approximation ? In certain cases for which the nature of the singularity can be precisely elucidated, one can give a clear answer to this question. As an example, consider the Laplace equation $-\Delta u = f$ on a polyhedral domain by measuring the smoothness of these singularities in the scale of Besov spaces related to

nonlinear approximation in H^1. Then it is known (see e.g. GRISVARD [1983]) that for a data $f \in H^m$, the solution can be decomposed into $u = u_r + u_s$, where the regular part u_r belongs to H^{m+2} while u_s is a finite linear combination of functions s_{n,v_i}, $n = 1, \cdots, n_{\max} = n_{\max}(m)$, which are singular at the corresponding vertex v_i of $\partial\Omega$ in the sense that they do not belong to H^{m+2}. With a conformal change of coordinate that maps v_i to the origin and such that Ω locally agrees with the cone $C = \{(r, \omega) \; ; \; r > 0 \text{ and } 0 < \omega < \theta\}$, the singular functions are given in polar coordinates by $s_n(r, \omega) = r^{n\pi/\theta} \sin(n\pi\omega/\theta)$ near the origin. One easily checks that such functions always belong to $B^{3/2+t}_{p,p}$ for all $t > 0$ and p such that $1/p = 1/2 + t/2$. In turn they also belong to all spaces $B^{1+t}_{p,p}$, $1/p = 1/2 + s/2$, and thus do not perturb the rate of nonlinear approximation of the solution in H^1 (see DAHLKE [1999] for a more detailed analysis). Less favorable cases include polyhedrons which develop anisotropic singularities along the edges (and thus require better adapted anisotropic refinements), as well as general Lipschitz domains which may develop singularities along the whole boundary. Several results dealing with this last case can be found in DAHLKE and DEVORE [1996].

We have thus identified several instances where the use of adaptive discretization methods is mathematically justified. We insist however on the substantial gap between the evidence of a high order nonlinear approximation rate for the solution and the effective implementation of an adaptive scheme which indeed displays the predicted rate. Nonlinear approximation provides therefore an *ideal benchmark* for an adaptive scheme: if $\|\cdot\|_X$ is the norm in which we measure the error between the solution of a PDE and its numerical approximation, and if the solution u is such that $\|u - u_N\|_X \lesssim N^{-s}$ for a family of nonlinear approximations $u_N \in S_N$ (such as the best N-term approximations of u), then an *optimal adaptive scheme* should produce N-term approximate solutions $\tilde{u}_N \in S_N$ such that one also has $\|u - \tilde{u}_N\|_X \lesssim N^{-s}$. From a computational point of view, optimality also means that the operation cost in evaluating $\tilde{u}_N$ should remain of order $\mathcal{O}(N)$. Note that this last property was fulfilled for the multiscale elliptic solver described in Example 4 of §3.11, however in the context of uniform discretizations. Our present goal here is somehow more ambitious since the algorithm should also provide the optimal adaptive discretization. It should also be noted that the norm X in which one can hope for an optimal error estimate is often dictated by the problem at hand: for example, in the case of an elliptic problem, this will typically be a Sobolev norm equivalent to the energy norm (e.g. the H^1 norm for a second order problem).

In most instances, the above ideal goals are far from being reached, either by the wavelet-based schemes that we shall evoke here, or by more classical finite differences or finite element adaptive schemes. In the latter, a crucial tool for adaptivity is ***a-posteriori analysis*** which consists of the derivation of ***local error indicators*** from the currently computed solution. These indicators are then used to locally refine (or derefine) the discretization at the next computation step. Wavelet-based schemes have been developed since the early 1990's following the empirical idea that local error indicators are directly given by the size of the currently computed wavelet coefficients: a large coefficient indicates important fluctations of the solution on the support of the corresponding wavelet, and suggests refining the approximation by adding wavelets at finer scales in this region. Most existing wavelet adaptive schemes therefore have in common the following general structure: at some step n of the computation, a set Λ^n is used to represent the numerical solution

$$u_n = u_{\Lambda^n} = \sum_{\lambda\in\Lambda^n} d_\lambda^n \psi_\lambda. \tag{4.7.4}$$

This information is then used to compute the new set Λ^{n+1} and the coefficients d_λ^{n+1} describing the numerical solution at the next step.

In the case of an *evolution problem* of the type

$$\partial_t u = \mathcal{E}(u), \tag{4.7.5}$$

such as hyperbolic conservation laws, the numerical solution at step n is typically an approximation to u at time $n\Delta t$ where Δt is the time step of the resolution scheme.

In the case of a ***stationary problem*** of the type

$$\mathcal{F}(u) = 0, \tag{4.7.6}$$

such as elliptic operator equations, the numerical solution at step n is typically an approximation to u which should converge to the exact solution as n tends to $+\infty$.

At this level of generality, note the first class of problems could be embedded in the second one by setting $\mathcal{F} = \partial_t - \mathcal{E}$. However, the above distinction is clearly important from a numerical point of view. In the first case the approximate solution is computed incrementally from a time step to the next one rather than globally in time and space: the storage of its value is always limited to one or a few time steps, and an approximation error at some time n influences the error at later times in an irreversible way. In the second case, the approximation of the solution is global in all variables and we shall rely on the residual of the equation $\mathcal{F}(u_n)$ to improve the accuracy at the next step. This has led to two distinct classes of adaptive wavelet schemes, with fairly different numerical analysis, which are discussed in the next two sections.

4.8 Adaptive multiscale processing

In this section, we consider evolution problems of the type (4.7.5). In the adaptive numerical treatment of such problems, a specific difficulty is that the singularities might not only appear but also move as time progresses, so that the adaptive mesh should be updated at each time step. While "moving mesh methods" are feasible, the numerical analysis of their performance is delicate, as well as their practical implementation in more than one dimension. In this context, the use of *multiresolution methods* has appeared since the 1980's as an interesting alternative for developing simple and efficient adaptive numerical schemes. The rough idea is to use a hierarchy of fixed nested grids at different resolutions, which offers the possibility of locally selecting an appropriate level of discretization. In the context of finite volume schemes for Computational Fluid Dynamics, this approach has been developed since the early 1980's with pioneering contributions such as BERGER, M. and J. OLIGER [1984] and BERGER, M. and P. COLLELA [1989].

In the context of wavelet discretizations, the selection of the appropriate adaptive grid corresponds to properly updating the set of indices Λ^n at each time step $n\Delta t$. This has led to the introduction of *dynamically adaptive schemes* in MADAY, PERRIER and RAVEL [1991], in which the derivation of $(\Lambda^{n+1}, u_{\Lambda^{n+1}})$ from $(\Lambda^n, u_{\Lambda^n})$ typically goes in three basic steps:

1. **Refinement:** a larger set $\tilde{\Lambda}^{n+1}$ with $\Lambda^n \subset \tilde{\Lambda}^{n+1}$ is derived from an *a-posteriori* analysis of the computed coefficients d_λ^n, $\lambda \in \Lambda^n$.

2. **Evolution:** a first numerical solution $u_{\tilde{\Lambda}^{n+1}} = \sum_{\lambda\in\tilde{\Lambda}^{n+1}} d_\lambda^{n+1}\psi_\lambda$ is computed from u_n and the data of the problem.

3. **Coarsening:** the smallest coefficients of $\tilde{u}_{n+1}$ are thresholded, resulting in the numerical solution $u_{\Lambda^{n+1}} = \sum_{\lambda\in\Lambda^{n+1}} d_\lambda^{n+1}\psi_\lambda$ supported on the smaller set $\Lambda^{n+1} \subset \tilde{\Lambda}^{n+1}$.

Before discussing the derivation of $(u_{\Lambda^{n+1}}, \Lambda^{n+1})$ from $(u_{\Lambda^n}, \Lambda^n)$, we need to briefly discuss the initialization of the scheme: ideally, we can obtain an optimal adaptive expansion u_{Λ^0} of the intial value data u_0 into a linear combination of wavelets by a thresholding procedure on its global expansion, i.e.

$$u_{\Lambda^0} = \sum_{\lambda\in\Lambda^0} d_\lambda^0\psi_\lambda, \quad \Lambda^0 := \{\lambda \text{ s.t. } \|d_\lambda^0\psi_\lambda\|_X \geq \eta\}, \tag{4.8.1}$$

where X is some prescribed norm in which we are interested in controlling the error, η a prescribed threshold and $d_\lambda^0 := \langle u_0, \tilde{\psi}_\lambda \rangle$ are the wavelet coefficients of u_0. In practice, we cannot compute all the values of these coefficients, and one thus needs a more reasonable access to a compressed representation. This is typically done through some *a-priori* analysis of the initial value u_0. In particular, if u_0 is provided by an analytic expression, or if we have some information on the local size of its derivatives, estimates on the decay of wavelet coefficients, such as in Remark 3.6.5, can be used to avoid the computation of most details which are below threshold. With such a strategy, we expect to obtain Λ^0 and $(u_\lambda^0)_{\lambda \in \Lambda^0}$ with a memory and computational cost which is proportional to $\#(\Lambda^0)$.

Then, assuming that at time $n\Delta t$ the approximate solution u_{Λ^n} has the form (4.7.4) for some set Λ^n of coefficients, the problem is thus both to select a correct set of indices Λ^{n+1} and to compute the new coefficients d_λ^{n+1} for $\lambda \in \Lambda^{n+1}$. As we already explained, this is done by (i) refining Λ^n into an intermediate set $\tilde{\Lambda}^{n+1}$ which is well fitted to describing the solution at time $(n+1)\Delta t$, (ii) computing $u_{\tilde{\Lambda}^{n+1}}$ supported by $\tilde{\Lambda}^{n+1}$ and (iii) deriving $(u_{\Lambda^{n+1}}, \Lambda^{n+1})$ from $u_{\tilde{\Lambda}^{n+1}}$ by a thresholding process similar to (4.8.1). The selection of the intermediate set $\tilde{\Lambda}^{n+1}$ should thus take into account the effect of the evolution operator $\mathcal{E}$ on the sparse expansion (4.7.4), integrated between $n\Delta t$ and $(n+1)\Delta t$. Once a procedure for the refinement of Λ^n into $\tilde{\Lambda}^{n+1}$ has been prescribed, several strategies are available for computing $u_{\tilde{\Lambda}^{n+1}}$ from u_{Λ^n}, such as Petrov-Galerkin methods in MADAY, PERRIER and RAVEL [1991] or collocation methods in BERTOLUZZA [1995b,1997]. All these strategies are based on the computation of the inner products $\langle \mathcal{E}(u_{\Lambda^n}), \tilde{\psi}_\lambda \rangle$ for $\lambda \in \tilde{\Lambda}^{n+1}$ up to some precision. In the case where the evolution operator $\mathcal{E}$ is linear, this amounts to a matrix-vector product, and one can make use of the sparse multiplication algorithm which was proposed at the end of §4.6. However, in many cases of interest, the evolution operator $\mathcal{E}$ is nonlinear, making this computation more difficult and costly. Generally speaking, the discretization of nonlinear operators is a less simple task in the wavelet coefficient domain than in the physical domain.

In the following, we shall present a systematic approach which allows us to solve this problem, by a suitable combination of both wavelet and physical representations of the numerical solution. This approach is based on the *discrete framework* described in §2.13. Recall that such a framework allows us to consider wavelet decompositions of point value or cell average discretizations. In this context, one can hope to combine the above ideas of dynamic adaptivity with existing finite difference or finite volume schemes which are proved or known to perform well: the main idea is to start from such a scheme which operates on discretizations U_J^n of $u(\cdot, n\Delta t)$ at scale 2^{-J}, and accelerate the scheme by applying dynamic adaptivity on a discrete multiscale decomposition of U_J^n at intermediate scales $j = 0, \cdots, J$. Before explaining this strategy in more detail, let us point out that imposing a limitation on the scale in an explicit dynamically adaptive scheme is usually necessary for ***stability***: in the context of linear or nonlinear convection terms, the time-step Δt is tied to the highest level J of refinement contained in the sets Λ^n according to the so-called CFL condition which imposes that $\Delta t \lesssim c\Delta x \sim c2^{-J}$ where c is the speed of convection. For a given time step, this constraint typically imposes a limitation on the scales contained in Λ^n, i.e. fixes some highest discretization level J.

As in the case of continuous wavelet bases, approximations of a vector $U_J \in \ell^\infty(\Gamma_J)$ are performed by truncating its wavelet coefficients obtained by the discrete wavelet transform

$$\mathcal{M}U_J = (d_\lambda)_{\lambda \in \nabla^J}, \tag{4.8.2}$$

onto a smaller subset $\Lambda \subset \nabla^J$: denoting by $\mathcal{R}_\Lambda$ the operator which maps d_λ to 0 if $\lambda \notin \Lambda$ and leaves it unchanged otherwise, the corresponding approximation is thus defined as

$$R_\Lambda U_J := \mathcal{M}^{-1}\mathcal{R}_\Lambda \mathcal{M} U_J. \tag{4.8.3}$$

In particular Λ can be defined by a thresholding procedure in order to obtain an adaptive approximation of the vector U_J. For this, we remark that we have

$$U_J = \sum_{|\lambda|<J} d_\lambda \Psi_{J,\lambda}, \tag{4.8.4}$$

where $\Psi_{J,\lambda}$ corresponds with the reconstruction from a single coefficient $\delta_\lambda = 1$. Given a discrete norm $\|\cdot\|_X$ of interest, it is thus natural to consider the set

$$\Lambda = \Lambda(\eta) = \{\lambda \text{ s.t. } \|d_\lambda \Psi_{J,\lambda}\|_X > \eta\}, \tag{4.8.5}$$

and the corresponding thresholding operator $T_\eta = R_\Lambda$. Note that keeping those λ such that $\|d_\lambda \Psi_{J,\lambda}\|_X > \eta$ typically corresponds with using a level dependent threshold $|d_\lambda| > \eta_{|\lambda|}$: in the case where X is the discrete ℓ^p norm

$$\|U_J\|_X := \Big((\#\Gamma_J)^{-1} \sum_{\gamma \in \Gamma_J} |u_\gamma|^p\Big)^{1/p}, \tag{4.8.6}$$

one can check that $\|\Psi_{J,\lambda}\|_X \sim 2^{-jd/p}$ so that $\eta_j = 2^{jd/p}\eta$.

In addition, it will be important to assume that the set Λ has a *tree structure* in the following sense: we assume that if a coefficient λ is contained in Λ, all the coefficients at coarser scales which are used to predict the physical value (point values or cell averages) which will be corrected by d_λ are also contained in Λ. Such a tree structure is not automatically ensured by straightforward thresholding, but we can always replace in the definition of R_Λ and T_η the set Λ by the smallest tree which contains it. The interest in using tree structures is that one can then define an isomorphism between the coefficients of the compressed multiscale representation $(d_\lambda)_{\lambda\in\Lambda}$ and the physical values (point values or cell averages) $U_J(\gamma)$ on an *adaptive mesh* $\Gamma(\Lambda)$ which corresponds to the values which can be exactly reconstructed. We illustrate in Figure 4.8.1, in the one dimensional point value case, the adaptive physical and multiscale grids $\Gamma(\Lambda)$ and Λ. Note that in the cell average case, we mean by $U_J(\gamma)$ the average of U_J on the cell γ which resolution level is not necessarily the finest one. The adaptive physical and multiscale grids $\Gamma(\Lambda)$ and Λ are connected by a fast multiscale transform $\mathcal{M}_\Lambda$ such that

$$(d_\lambda)_{\lambda\in\Lambda} = \mathcal{M}_\Lambda(U_J(\gamma))_{\gamma\in\Gamma(\Lambda)}. \tag{4.8.7}$$

The complexity of the adaptive decomposition and reconstruction algorithms $\mathcal{M}_\Lambda$ and $\mathcal{M}_\Lambda^{-1}$ is in $\mathcal{O}(\#(\Lambda))$.

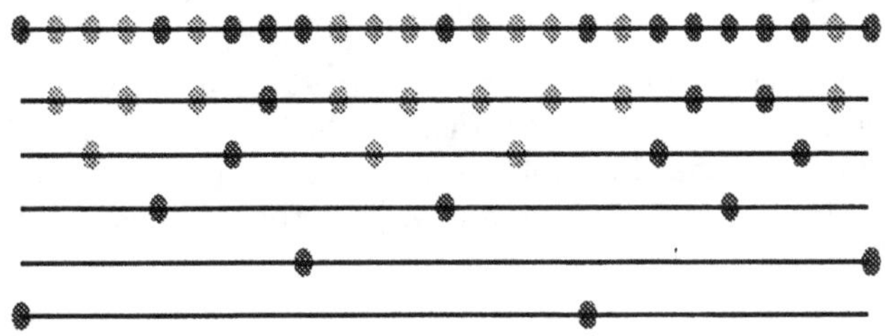

Figure 4.8.1: Adaptive grid $\Gamma(\Lambda)$ (dark points - top) and tree-structured set Λ (dark points - bottom)

We assume that we are given a reference scheme on the finest grid Γ_J: approximations of $u(x, n\Delta t)$ by $U_J^n = (U_J^n(\gamma))_{\gamma\in\Gamma_J}$ are computed by a one step formula of the type

$$U_J^{n+1}(\gamma) = U_J^n(\gamma) + F(U_J^n(\mu)\ ;\ \mu \in S(\gamma)), \tag{4.8.8}$$

where $S(\gamma)$ is the local stencil associated with γ. In the case of finite volume conservative schemes, F has the form of a *flux balance* over the edges $\Gamma_{\gamma,\mu}$ surrounding the cell γ

$$U_J^{n+1}(\gamma) = U_J^n(\gamma) + \sum_{\mu \text{ s.t. } |\Gamma_{\gamma,\mu}|\neq 0} F_{\gamma,\mu}^n,$$

where $F^n_{\gamma,\mu} = -F^n_{\mu,\gamma}$ is a function of the $U^n_J(\nu)$ for ν in a local stencil surrounding γ and μ. We shall use the notation

$$U^{n+1}_J = E_J U^n_J, \tag{4.8.9}$$

where E_J denotes the discrete (linear or nonlinear) evolution operator.

The goal of the adaptive multiscale processing is to compute approximations of $u(x, n\Delta t)$ by V^n_J, where $V^n_J = (V^n_J(\gamma))_{\gamma\in\Gamma_J}$ can be represented by its coefficients $(d^n_\lambda)_{\lambda\in\Lambda^n_\eta}$ in an adaptive set Λ^n_η (all other coefficients being zero) or equivalently by its physical values (point values or cell averages) on the adaptive mesh $(V^n_J(\gamma))_{\gamma\in\Gamma(\Lambda^n_\eta)}$ (we shall systematically impose the tree structure on the adaptive set Λ^n_η).

The parameter η is a level dependent threshold which will monitor the loss of accuracy of the adaptive scheme with respect to the reference scheme. Ideally the set Λ^n_η would be the tree structured set of coefficients obtained by thresholding the reference numerical solution U^n_J, i.e. the smallest tree containing $\{\lambda,\ |d_\lambda(U^n_J)| \geq \eta_{|\lambda|}\}$, and η is a level dependent threshold which ensures a prescribed accuracy ε in the sense that

$$\|U^n_J - \mathcal{T}_\eta U^n_J\|_X \leq \varepsilon. \tag{4.8.10}$$

However, this set will not be accessible in the adaptive algorithm since we do not want to compute the exact U^n_J. Just for the initialization of the adaptive scheme, we do apply the thresholding procedure on U^0_J as explained in the beginning of this section, i.e. define Λ^0_η as the smallest tree containing $\{\lambda,\ |d_\lambda(U^0_J)| \geq \eta_{|\lambda|}\}$ and set

$$V^0_J := T_\eta U^0_J. \tag{4.8.11}$$

For $n > 0$, the adaptive solution V^n_J will differ from $T_\eta U^n_J$. Our goal is to control the error $\|U^n_J - V^n_J\|_X$.

As already explained, the derivation of $(V^{n+1}_J, \Lambda^{n+1})$ from (V^n_J, Λ^n), is executed by three basic steps:

1. **Refinement:** a superset $\Lambda^n_\eta \subset \tilde{\Lambda}^{n+1}_\eta$ adapted to describe the solution at time $n+1$ is predicted from the available data (V^n_J, Λ^n). Ideally this set should be the smallest tree such that $|d_\lambda(E_J V^n_J)| < \eta_{|\lambda|}$ if $\lambda \notin \tilde{\Lambda}^{n+1}_\eta$. We extend the numerical solution on this larger set by simply taking $d^n_\lambda = 0$ for $\lambda \in \tilde{\Lambda}^{n+1}_\eta \setminus \Lambda^n_\eta$, i.e. by leaving V^n_J unchanged.

2. **Evolution:** a first value $\tilde{V}^{n+1}_J(\gamma)$ is computed from V^n_J for $\gamma \in \Gamma(\tilde{\Lambda}^{n+1}_\eta)$. Ideally, this evolution step should compute the restriction of $E_J V^n_J$ on the coefficient set $\tilde{\Lambda}^{n+1}_\eta$, i.e. $\tilde{V}^{n+1}_J = R_{\tilde{\Lambda}^{n+1}_\eta} E_J V^n_J$.

3. **Coarsening:** we apply the level dependent thresholding operator T_η to the coefficients $(\tilde{d}_\lambda^{n+1})_{\lambda\in\tilde{\Lambda}_\eta^{n+1}}$ of $\tilde{V}_J^{n+1}$. This results in the new set $\Lambda_\eta^{n+1} \subset \tilde{\Lambda}_\eta^{n+1}$ and the numerical solution V_J^{n+1} at time step $n+1$.

The loss of accuracy with respect to the reference scheme is monitored by the threshold η: if $\eta = 0$ we have $V_J^n = U_J^n$, i.e. the adaptive scheme coincides with the reference scheme. This threshold should typically be tuned in order to remain within the accuracy which is achieved by the reference scheme while reducing the CPU time and memory space.

Remark 4.8.1 *One should clearly make the distinction between the discretization of the reference scheme and of the multiresolution approximation, which should be thought of as decoupled:*

1. *One can apply a multiresolution decomposition with high order precision together with a low order reference scheme.*

2. *One can apply multiresolution decompositions based on point values together with a finite volume scheme based on a cell average discretization, i.e. the cell average values are decomposed as if they were point values. In the context where the conservation of mass is an important issue, it should be well understood that such a strategy results in a loss of conservation, since we subsample the numerical values on the adaptive grid instead of averaging them. However, we can hope to control the error with respect to a conservative reference scheme.*

As we shall analyze further in more detail, the loss of accuracy with respect to the reference scheme is influenced by each of the three basic steps which can be viewed as perturbations to the reference scheme. For the coarsening step, this perturbation is controlled by the level of the threshold η. For the refinement step, the predicted set $\tilde{\Lambda}_\eta^{n+1}$ should take into account the action of the discrete evolution operator E_J on the size of the coefficients in the multiscale decomposition. Finally, the evolution step needs to be performed accurately on the adaptive grid. In particular, we cannot (and do not want to) exactly compute $E_J V_J^n$ since this would require reconstructing V_J^n on the finest grid. Concerning this last step, let us distinguish between two different strategies.

Strategy 1: Direct application of the numerical scheme on the adaptive grid $\Gamma(\tilde{\Lambda}_\eta^{n+1})$. This is the usual adaptive mesh refinement approach, which consists of computing $V_J^{n+1}(\gamma)$, for $\gamma \in \Gamma(\tilde{\Lambda}_\eta^{n+1})$, by direct application of the numerical scheme E_j at the local scale 2^{-j} in the vicinity of γ. Here we exploit the fact that the adaptive discretization is locally uniform, except in the regions of transition between two scales j and $j+1$ (the tree structure assumption on $\tilde{\Lambda}_\eta^{n+1}$ does not allow a transition of more than one scale). In such regions, the discretization can still be made locally uniform at the scale $j+1$ by applying the prediction operator to the coarse grid data. This "artificial refinement" is illustrated in Figure 4.8.2 in the cell-average framework. The computation has then the same complexity as the adaptive discretization, i.e. $\mathcal{O}(\#(\tilde{\Lambda}_\eta^{n+1}))$. This approach gives satisfactory results if the reference scheme has high order accuracy comparable to the prediction operator. However, in the case where the reference scheme is low order, it produces a significant loss of accuracy in the regions where the adaptive mesh $\Gamma(\tilde{\Lambda}_\eta^{n+1})$ is coarse. In such a case, a better strategy is described next.

Strategy 2: Exact computation of $R_{\tilde{\Lambda}_\eta^{n+1}} E_J V_J^n$ by local adaptive reconstruction. The idea is to exactly reconstruct at fine scale $V_J^n(\mu)$ for those $\mu \in \Gamma_J$ which are needed to compute exactly $E_J V_J^n$ on the adaptive grid $\Gamma(\tilde{\Lambda}_\eta^{n+1})$. In the point value case, this means that we reconstruct $V_J^n(\mu)$ for $\mu \in \Gamma_J$ contained in one of the stencils $S(\gamma)$ for $\gamma \in \tilde{\Lambda}_\eta^{n+1}$. In this case, the additional computational time for this reconstruction remains essentially proportional to the complexity of the adaptive discretization $\mathcal{O}(\#(\tilde{\Lambda}_\eta^{n+1}))$. This is illustrated in Figure 4.8.3 for the one dimensional case when the local reconstruction is performed by iteration of the 4-point cubic prediction rule from the local scale j of the adaptive grid near γ to the finest scale J. For each $\gamma \in \tilde{\Lambda}_\eta^{n+1}$, this amounts to applying to its neighbours on the adaptive grids the $J-j$ power of a 5×5 matrix which can be stored once and for all. A similar strategy applies in higher dimensions. In the cell average case and if the reference scheme is a conservative finite volume scheme, we need to reconstruct $V_J^n(\mu)$ for $\mu \in \Gamma_J$ such that μ is used in the flux evaluations for the fine grid interfaces which are part of the interfaces of the adaptive grid. In dimension 1, this process is essentially similar to the point value case. However, as illustrated in Figure 4.8.4, it becomes more complex in higher dimensions since we need to reconstruct many fine grid cells for each coarse grid cell of the adaptive grid: in dimension d, the complexity of this process grows therefore like $\mathcal{O}(\sum_{j=1}^{J} 2^{(d-1)(J-j)}\#\{\lambda \in \tilde{\Lambda}^{n+1}, |\lambda| = j\}$, which is still less than $\mathcal{O}(\#(\Gamma_J))$, but makes this strategy too costly in all the practical test cases that were investigated.

In order to evaluate the error $\|U_J^n - V_J^n\|_X$ in some prescribed norm, we shall consider here the second evolution strategy, i.e. exact evaluation by local reconstruction. In this case, the vector $\tilde{V}_J^{n+1}$ is exactly given by

$$\tilde{V}_J^{n+1} = R_{\tilde{\Lambda}_\eta^{n+1}} E_J V_J^n, \tag{4.8.12}$$

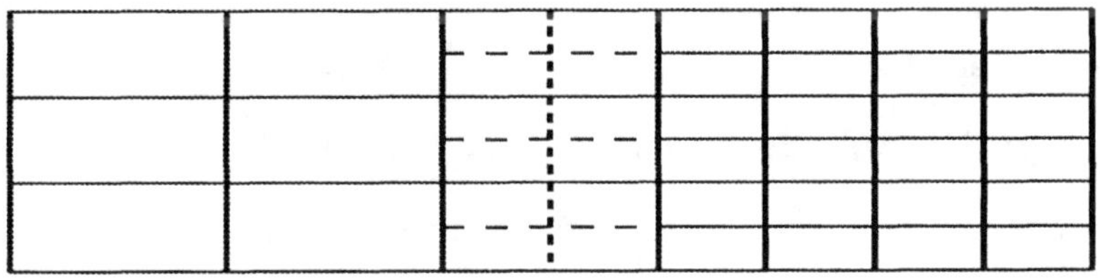

Figure 4.8.2: Artificial refinement (dashed cells) in a transition region

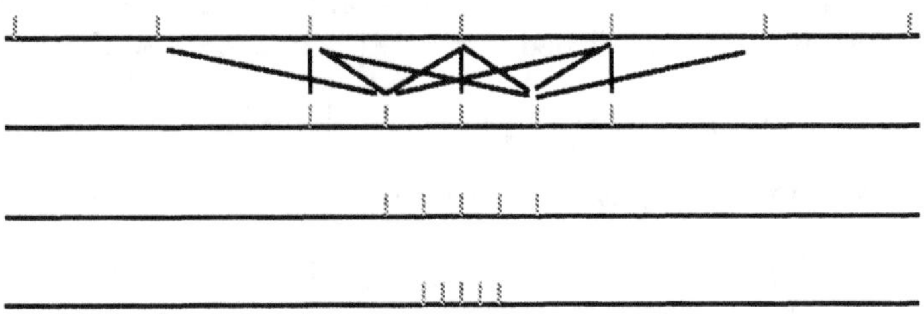

Figure 4.8.3: Local reconstruction in one dimension for point values

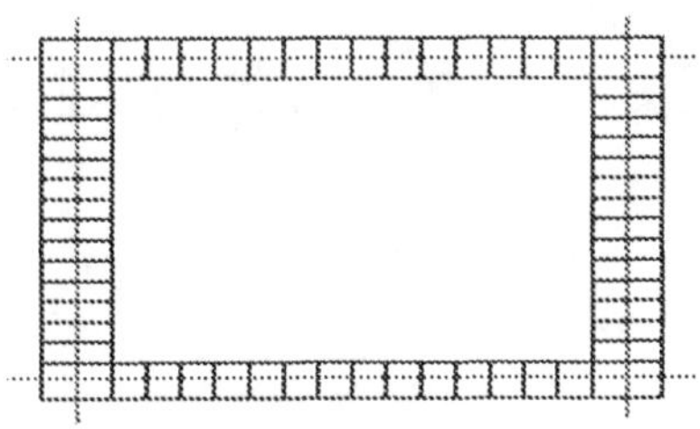

Figure 4.8.4: Local reconstruction in two dimensions for cell averages

so that we have

$$V_J^{n+1} = R_{\Lambda_\eta^{n+1}} R_{\tilde{\Lambda}_\eta^{n+1}} E_J V_J^n. \quad (4.8.13)$$

We can therefore write

$$\|U_J^{n+1} - V_J^{n+1}\|_X \le \|E_J U_J^n - E_J V_J^n\|_X + t_n + r_n, \quad (4.8.14)$$

where t_n is the error produced by the thresholding step

$$t_n := \|T_\eta \tilde{V}_J^{n+1} - \tilde{V}_J^{n+1}\|, \quad (4.8.15)$$

and r_n is the error produced by the refinement step

$$r_n := \|R_{\tilde{\Lambda}^{n+1}} E_J V_J^n - E_J V_J^n\|_X. \quad (4.8.16)$$

The refinement and thresholding strategies should be designed in order to control both terms with a prescribed precision ε. Assuming stability of the reference scheme in the prescribed norm $\|\cdot\|_X$ in the sense that

$$\|E_J U - E_J V\| \le (1 + c\Delta t)\|U - V\|_X, \quad (4.8.17)$$

this yields the cumulative estimate at time $T = (n+1)\Delta t$,

$$\|U_J^{n+1} - V_J^{n+1}\|_X \le (1 + c\Delta t)\|U_J^n - V_J^n\|_X + 2\varepsilon \le \cdots \le C(T)n\varepsilon. \quad (4.8.18)$$

Remark 4.8.2 *In many practical instances, such as hyperbolic conservation laws, this estimate turns out to be too pessimistic in the sense that thresholding and refinement error do not really accumulate, so that ε can be chosen larger than the value prescribed by this analysis.*

Let us finally explain the refinement and thresholding strategies which allow us to control r_n and t_n. For the thresholding error, we obviously have with $\eta_j = 2^{dj}\eta_0$,

$$t_n \le \sum_{\|d_\lambda^{n+1}\Psi_{J,\lambda}\|_X < \eta} \|d_\lambda^{n+1}\Psi_{J,\lambda}\|_X. \quad (4.8.19)$$

A first crude estimate of the right hand side is by $\eta\#(\nabla^J) \sim \eta 2^{dJ}$, which suggests the choice $\eta = \varepsilon 2^{-dJ}$. A sharper estimate is by $\eta\#(\tilde{\Lambda}^{n+1})$. This estimate suggests a *time dependent threshold*

$$\eta = \varepsilon[\#(\tilde{\Lambda}^{n+1})]^{-1}. \quad (4.8.20)$$

For the refinement error, we would like to define $\tilde{\Lambda}^{n+1}$ in such a way that $|d_\lambda(E_J V_J^n)| < \eta_{|\lambda|}$ if $\lambda \notin \tilde{\Lambda}_\eta^{n+1}$. We typically expect that $\tilde{\Lambda}^{n+1}$ should be larger yet *close* to Λ^n. Indeed, we already observed in §4.6 that a large class of *linear operators* have indeed a local action in the wavelet domain, in the sense that this action can be approximated by the sparse matrix-vector multiplication (4.6.25) in Theorem 4.6.3. If $\mathcal{E}$ is one such linear operator, one can thus make use of a sparse matrix-vector multiplication algorithm such as APPROX defined in the end of §4.6 in order to design a local refinement procedure which ensures the required estimate on r_n. For nonlinear evolution operators, we need to rely on more specifically adapted refinement rules. In the context of hyperbolic problems and with cell average multiresolution decompositions, the following rules were proposed in HARTEN [1995]:

- If $|d_\lambda^n| > \eta_{|\lambda|}$, include in $\tilde{\Lambda}_\eta^{n+1}$ the immediate neighbors of λ at the same level.
- If $|d_\lambda^n| > 2^{r-1}\eta_{|\lambda|}$, with r the order of accuracy of the prediction operator, also include the children of λ at the finer level (i.e. refine by one level the corresponding cell of the adaptive grid).

These rules are derived heuristically, assuming that the reference scheme is explicit and satisfies a CFL condition $\Delta t \leq C2^{-J}$, which limits the development and propagation of high gradients from a time step to the next one. Although they are satisfactory in practice, these rules do not seem sufficient to prove that $|d_\lambda(E_J V_J^n)| < \eta_{|\lambda|}$ if $\lambda \notin \tilde{\Lambda}_\eta^{n+1}$. In COHEN, KABER, MUELLER and POSTEL [2003], a more severe refinement rule is proposed: refine by n levels if $2^{n(s-1)}\eta_{|\lambda|} \leq |d_\lambda| < 2^{(n+1)(s-1)}\eta_{|\lambda|}$, with s the Hölder smoothness of the underlying continuous wavelet system (ψ_λ) obtained as the limit of $(\Psi_{J,\lambda})$ if we let $J \to +\infty$. With such a refinement rule, it is possible to prove that $|d_\lambda(E_J V_J^n)| < \eta_{|\lambda|}$ if $\lambda \notin \tilde{\Lambda}_\eta^{n+1}$. In practice, it is nevertheless observed that Harten's rule is sufficient and that the thresholding error t_n tends to dominate the refinement error r_n.

Remark 4.8.3 *The approach that we have presented in this section allows us to derive adaptive wavelet schemes for a fairly large class of equations, with the possibility of controlling the error with respect to a reference scheme up to some prescribed accuracy. However, a complete analysis of the exact trade-off between this accuracy and the complexity - represented by the number of parameters $\#(\Lambda_\eta^n)$ used in the approximation - is still not available, even for simple equations such as one dimensional scalar conservation laws, for which theoretical results such as Theorem 4.7.1 indicate the adaptive methods should behave significantly better than non-adaptive methods from this point of view.*

Remark 4.8.4 *In the case of hyperbolic equations derived from conservation laws (as considered in §4.7), the multiresolution approach developed in* HARTEN *[1995] can be qualified as "semi-adaptive" in the following sense: while the solution* U_J^n *is still described and evolved on a uniform grid, the increment* B_J^n *(corresponding here with the numerical flux balance vector) is replaced by its adaptive approximation* $T_{\Lambda_\eta^n} B_J^n$ *where* Λ_η^n *now represents the set of wavelet coefficients of* U_J^n *above some threshold. Such an approach allows some substantial saving on the computational cost in the case where the evaluation of* B_J^n *is the heavy part of the reference scheme.*

4.9 Adaptive space refinement

Let us now turn to stationary problems of the type (4.7.6). Here, we shall embed in the following general framework: we assume that $\mathcal{F}$ is a continuous mapping from a Hilbert space H onto its dual H', which admits a *non-singular solution* $u \in H$ in the sense that the Frechet derivative $D\mathcal{F}(u)$ is an isomorphism from H to H'. Therefore u is also the solution of the variational formulation of the problem: find $u \in H$ such that for all $v \in H$,

$$\langle \mathcal{F}(u), v \rangle = 0, \tag{4.9.1}$$

where $\langle \cdot, \cdot \rangle$ represents the duality product between H and H'. In the linear case, this means that $\mathcal{F}$ is of the form

$$\mathcal{F}(u) = \mathcal{A}u - f \tag{4.9.2}$$

with $f \in H'$ and $\mathcal{A}$ an isomorphism from H to H'. Let us give some classical instances:

1. Elliptic problems of positive order such as the Laplace equation $-\Delta u = f$ on a domain $\Omega \subset \mathbb{R}^d$ with homogeneous boundary condition $u = 0$ on $\partial\Omega$. In this case $\mathcal{A} = -\Delta$ and $H = H_0^1$.

2. Saddle point problems such as the Stokes equation $-\Delta u + \nabla p = f$ on a domain $\Omega \subset \mathbb{R}^d$ subject to the constraint $\mathrm{Div} u = 0$ and the boundary condition $u = 0$ on $\partial\Omega$. In this case, with the solution and data defined as (u, p) and $(f, 0)$, we have $\mathcal{A} = (-\Delta + \nabla, \mathrm{Div})$ and $H = (H_0^1)^d \times L_0^2$ where L_0^2 is the space of L^2 functions with vanishing integral.

3. Boundary integral equations on the contour of a domain $\Omega \subset \mathbb{R}^d$ such as the single layer potential $\int_{\partial\Omega} K(\cdot, y)u(y)dy = f$ with $K(x, \cdot)$ the fundamental solution of $-\Delta v = \delta_x$ on $\mathbb{R}^d$. In this case, $\mathcal{A}$ is the integral operator defined by the kernel K and $H = H^{1/2}(\partial\Omega)$.

Instances in the nonlinear case include equations such as

$$-\Delta u + u|u|^{p-1} = f, \tag{4.9.3}$$

with homogeneous Dirichlet boundary condition $u = 0$ and under the restriction $p \leq \frac{d+2}{d-2}$ if $d \geq 3$, which allows us to take $H = H_0^1$.

The *classical* approach to numerically solving this type of equation by the finite element method is typically concerned with the following issues:

1. Well-posedness of the equation, i.e. existence, uniqueness and stability of the solution.

2. Discretization into a finite element problem by the Galerkin method - find $u_{\mathcal{T}} \in V_{\mathcal{T}}$ such that $\langle \mathcal{F}(u_{\mathcal{T}}), v_{\mathcal{T}} \rangle = 0$ for all $v_{\mathcal{T}} \in V_{\mathcal{T}}$ - analysis of well-posedness and of the approximation error $\|u - u_{\mathcal{T}}\|_X$.

3. Numerical resolution of the finite dimensional system resulting from the Galerkin discretization.

4. Mesh refinement based on a-posteriori error estimators in the case of adaptive finite element methods.

Several difficulties are associated with each of these steps. First of all, note that the well-posedness of the finite element problem is in general *not* a consequence of the well-posedness of the continuous problem. Typical examples even in the linear case are saddle point problems for which it is well known that, for Galerkin discretizations to be stable, the finite element spaces for the different solution components have to satisfy certain compatibility conditions (LBB or Inf-Sup condition), which are also crucial in the derivation of optimal error estimates. Thus the discrete problem does not necessarily inherit the "nice properties" of the original infinite dimensional problem. Concerning the numerical resolution of the discrete system, a typical source of trouble is its possible *ill-conditioning*, which interferes with the typical need to resort to iterative solvers in high dimension. An additional difficulty which occurs in the case of integral equations is the manipulation of matrices which are densely populated.

Finally, let us elaborate more on the adaptivity step. Adaptive refinement schemes in a finite element context have been developed since the end of the 1970's. Key ingredients in such adaptive algorithms are *a-posteriori error estimators* which are typically derived from the current residual $\mathcal{F}(u_{\mathcal{T}})$: in the case where the Frechet derivative $D\mathcal{F}(u)$ is an isomorphism between Banach function spaces X to Y, one can hope to estimate the error $\|u - u_{\mathcal{T}}\|_X$ by the evaluation of $\|\mathcal{F}(u_{\mathcal{T}})\|_Y$. The rule of thumb is then to decompose $||\mathcal{F}(u_{\mathcal{T}})||_Y$ into computable local error indicators η_K which aim to describe as accurately as possible the local error on each element $K \in \mathcal{T}$. In the case of elliptic problems, these indicators typically consist of local residuals and other quantities such as jumps of derivatives across the interface between adjacent triangles or simplices.A typical refinement algorithm will subdivide those elements K for which the error indicator η_K is larger than a prescribed tolerance ε resulting in a new mesh $\tilde{\mathcal{T}}$. Note that this strategy is theoretically in accordance with our remarks in §4.4 on adaptive finite element approximation, since it tends to *equilibrate* the local error. Two other frequently used strategies consist of refining a fixed proportion of the elements corresponding to the largest η_K, or the smallest number of elements K for which the η_K contribute to the global error up to a fixed proportion. It is therefore hoped that the iteration of this process from an initial mesh $\mathcal{T}_0$ will produce optimal meshes $(\mathcal{T}_n)_{n\geq 0}$ in the sense that the associated solutions $u_n := u_{\mathcal{T}_n} \in V_{\mathcal{T}_n}$ converge to u at the optimal rate:

$$\sigma_N(u) \leq CN^{-r} \Rightarrow ||u - u_n||_X \leq C[\#(\mathcal{T}_n)]^{-r}, \tag{4.9.4}$$

up to a change in the constant C. Unfortunately, severe obstructions appear when trying to prove (4.9.4) even in the simplest model situations. One of them is that η_K is in general not an estimate from above of the local error, which reduces the chances of deriving the optimal rate. For most adaptive refinement algorithms, the theoretical situation is actually even worse in the sense that it cannot even be proved that the refinement step actually results in a reduction of the error by a fixed amount and that u_n converges to u as n grows. Only recently in DOERFLER [1996] and in MORIN, NOCETTO and SIEBERT [2000] have proofs of convergence appeared for certain type of adaptive finite element methods, yet without convergence rate and therefore no guaranteed advantage over their non-adaptive counterparts.

Wavelet methods vary from finite element methods in that they can be viewed as solving systems that are finite sections of one fixed infinite dimensional system corresponding with the discretization of the equation in the full basis. This observation has led to a *new paradigm* which has been explored in COHEN, A., W. DAHMEN and R. DEVORE [2002], for linear variational problems. It aims at closely intertwining the analysis - discretization - solution process. The basic steps there read as follows:

1. Well-posedness of the variational problem.

2. Discretization into an *equivalent* infinite dimensional problem which is well posed in ℓ^2.

3. Devising of an iterative scheme for the ℓ^2-problem that exhibits a fixed error reduction per iteration step.

4. Numerical realization of the iterative scheme by means of an *adaptive application* of the involved infinite dimensional operators within some dynamically updated accuracy tolerances.

Thus the starting point is the same as in the classical approach. The main difference is that one aims at staying as long as possible with the infinite dimensional problem. Only at the very end, when it comes to applying the operators in the ideal iteration scheme, one enters the finite dimensional realm. However, the finite number of degrees of freedom is determined at each stage by the adaptive application of the operator, so that at no stage is any specific trial space fixed.

The discretization process in the infinite dimensional setting can be understood as follows: if we express the solution u in a wavelet basis

$$u = \sum_{\lambda\in\nabla} u_\lambda \psi_\lambda \tag{4.9.5}$$

such that the basis functions ψ_λ are elements of H, we can rewrite the variational formulation (4.9.1) as an infinite system

$$\langle \sum_{\lambda\in\nabla} u_\lambda \psi_\lambda, \psi_\mu \rangle = 0, \quad \mu \in \nabla, \tag{4.9.6}$$

which we summarize as

$$F(U) = 0 \tag{4.9.7}$$

where $U = (u_\lambda)_{\lambda\in\nabla}$ is the coordinate vector and the mapping F is defined by

$$F(U) := (\langle \sum_{\lambda\in\nabla} u_\lambda \psi_\lambda, \psi_\mu \rangle)_{\mu\in\nabla}. \tag{4.9.8}$$

In addition, we shall make the assumption that the Hilbertian Sobolev space H is characterized by a norm equivalence of the type established in Chapter 3. According to Remark 3.6.4, this means that if we renormalize the ψ_λ according to

$$\|\psi_\lambda\|_H = 1, \tag{4.9.9}$$

we obtain a Riesz basis of H which defines an isomorphism between H and ℓ^2

$$\|u\|_H \sim [\sum_{\lambda\in\nabla} |u_\lambda|^2]^{1/2} = \|U\| \tag{4.9.10}$$

where $||\cdot||$ denotes the ℓ^2 norm. According to the results of §3.10, we also know that the dual basis $(\psi_\lambda)_{\lambda\in\nabla}$ characterizes the dual space H', which simply means here that

$$||f||_{H'} \sim [\sum_{\lambda\in\nabla} |\langle f, \psi_\lambda\rangle|^2]^{1/2} = ||F||. \tag{4.9.11}$$

It follows that the above defined F is a continuous mapping from ℓ^2 to ℓ^2. We also see that if u is a nonsingular solution, $DF(U)$ should be an isomorphism in ℓ^2.

In this context, the simplest iterative algorithms that may converge toward the solution of the problem (4.9.7), are of the form

$$U^{n+1} = U^n + D_n F(U^n) \tag{4.9.12}$$

where F is the *residual* of the equation and D_n some linear operator. Typical instances include:

1. Richardson iteration: $D_n := \tau$ a fixed scalar.
2. Newton method: $D_n := [DF(U^n)]^{-1}$.
3. Least square gradient descent: $D_n := [DF(U^n)]^T$.

Of course, $U^n = (u^n_\lambda)$ corresponds to an approximation

$$u^n = \sum_{\lambda\in\nabla} u^n_\lambda \psi_\lambda, \tag{4.9.13}$$

to the solution u. Note that the residual allows us to measure the size of the approximation error in the H norm: since $DF(U)$ is an isomorphism, we have

$$||u^n - u||_H \sim ||U^n - U|| \sim ||F(U^n)|| \tag{4.9.14}$$

for U^n sufficiently close to U (and without this restriction for linear problems).

The adaptive algorithm will consist of applying this iteration in an approximate fashion in order to construct vectors U^n which are supported on finite sets of indices Λ^n. We shall give an example of this in the simplest setting of a ***linear elliptic problem***

$$\mathcal{A}u = f \tag{4.9.15}$$

for some $f \in H'$ in which case (4.9.7) rewrites

$$AU = F, \tag{4.9.16}$$

where the right hand side vector

$$F := (\langle f, \psi_\lambda \rangle)_{\lambda\in\nabla} \tag{4.9.17}$$

is in ℓ^2 and the infinite stiffness matrix

$$A(\lambda, \mu) = \langle \mathcal{A}\psi_\lambda, \psi_\mu \rangle \tag{4.9.18}$$

is elliptic in ℓ^2. In this case, a converging infinite dimensional algorithm can be simply obtained by the Richardson iteration

$$U^n := U^{n-1} + \tau(F - AU^{n-1}) \tag{4.9.19}$$

with $0 < \tau < 2[\lambda_{\max}(A)]^{-1}$ and $U^0 = 0$, which guarantees the reduction rate $\|U - U^n\| \le \rho\|U - U^{n-1}\|$ with $\rho = \max\{1 - \tau\lambda_{\min}(A), \tau\lambda_{\max}(A) - 1\}$. Note that renormalizing the wavelet system amounts to applying a multiscale preconditioning, similar to those techniques discussed in §3.11 yet operated at the infinite dimensional level.

At this stage, we may enter finite dimensional adaptive computation by modifying the Richardson iteration up to a prescribed tolerance according to

$$U^n := U^{n-1} + \tau(\mathbf{COARSE}(F,\varepsilon) - \mathbf{APPROX}(AU^{n-1},\varepsilon)) \tag{4.9.20}$$

where $\|F - \mathbf{COARSE}(F,\varepsilon)\| \le \varepsilon$ and $\|AU - \mathbf{APPROX}(AU^{n-1},\varepsilon)\| \le \varepsilon$, and the U^n are now finite dimensional vectors supported by adaptive sets of indices Λ^n. The procedure **COARSE**, which simply corresponds to thresholding the data vector F at a level corresponding to accuracy ε, can be practically achieved without the full knowledge of F by using some a-priori bounds on the size of the coefficients $\langle f, \psi_\lambda \rangle$, exploiting the local smoothness of f and the oscillation properties of the wavelets. The procedure **APPROX** is performed by the sparse matrix-vector multiplication algorithm which has been defined in §4.6, assuming that A belongs to a class $\mathcal{M}_{\alpha,\beta}$.

Clearly the modified iteration (4.9.20) satisfies

$$\|U-U^n\| \le \|U-U^{n-1}-\tau(F-AU^{n-1})\|+2\varepsilon \le \rho\|U-U^{n-1}\|+2\tau\varepsilon, \tag{4.9.21}$$

and therefore ensures a fixed reduction rate until the error $\|U - U^n\|$ is of the order $\frac{2\tau}{1-\rho}\varepsilon$. In a similar way, the residual satisfies

$$\|F - AU^n\| \le \rho\|F - AU^{n-1}\| + 2\tau\|A\|\varepsilon, \tag{4.9.22}$$

and is therefore reduced until it reaches the order $\frac{2\tau\|A\|}{1-\rho}\varepsilon$. In order to obtain a converging algorithm, a natural idea is therefore to dynamically update the tolerance ε according to the following procedure: first set it to 1 and divide it by 2 each time the error (measured by the approximate residual $\mathbf{COARSE}(F,\varepsilon) - \mathbf{APPROX}(AU^{n-1},\varepsilon)$) has reached a prescribed order. This algorithm reads as follows:

1. Start with $U^0 := 0$, $\Lambda^0 := \emptyset$ and $\varepsilon_0 := 1$.
2. At stage j, set $U^{j,0} = U^j$ and $\varepsilon_j = 2^{-j}$.
3. For $n = 0, 1, \cdots$, apply

   $U^{j,n+1} := U^{j,n-1} + \tau(\mathbf{COARSE}(F,\varepsilon_j) - \mathbf{APPROX}(AU^{j,n-1},\varepsilon_j))$.

   Continue while

   $$\|\mathbf{COARSE}(F,\varepsilon_j) - \mathbf{APPROX}(AU^{j,n},\varepsilon_j)\| \geq (\frac{2\tau\|A\|}{1-\rho} + 3)\varepsilon_j.$$

4. Once it is found that

   $$\|\mathbf{COARSE}(F,\varepsilon_j) - \mathbf{APPROX}(AU^{j,n},\varepsilon_j)\| \leq (\frac{2\tau\|A\|}{1-\rho} + 3)\varepsilon_j.$$

   set $U^{j+1} := U^{j,n}$ and Λ^{j+1} its support (the set of its non-zero coefficients). Then go to step 2 with $j := j + 1$.

According to (4.9.22), it is always ensured that the exact residual will go below the threshold $(\frac{2\tau\|A\|}{1-\rho} + 1)\varepsilon_j$ after a number of steps in the inner loop $n \leq M$ for some fixed M, and therefore that the approximate residual will go below the threshold $(\frac{2\tau\|A\|}{1-\rho} + 3)\varepsilon_j$ after at most the same number of steps. We therefore obtain a converging adaptive strategy, so far without information about its convergence rate. In particular we would like to know if the optimal rate

$$\|U - U^j\| \leq \#(\Lambda_j)^{-t}, \tag{4.9.23}$$

is achieved when $U \in \ell^p_w$ with $1/p = 1/2 + t$. Since the error $\|U - U^j\|$ can be estimated by

$$\begin{aligned}
\|U - U^j\| &\leq \|A^{-1}\|\,\|F - AU^j\| \\
&\leq \|A^{-1}\|\,(\|\mathbf{COARSE}(F,\varepsilon_j) - \mathbf{APPROX}(AU^{j,n},\varepsilon_j)\| + 2\varepsilon) \\
&\leq (\|A^{-1}\|\tfrac{2\tau\|A\|}{1-\rho} + 5)\varepsilon_j \\
&\lesssim \varepsilon_j = 2^{-j},
\end{aligned}$$

the optimal rate (4.9.23) is equivalent to the following growth condition on the support of U^j,

$$\#(\Lambda^j) \lesssim 2^{j/t}. \tag{4.9.24}$$

It remains an open problem to determine under which circumstances the optimal bound (4.9.24) is achieved by the above algorithm. However, we shall see that (4.9.24) can be proved if we append to each step of the outer loop an additional *coarsening step* which consists of applying a thresholding procedure on the current numerical solution. The role of thresholding in ensuring an optimal complexity of the approximate solution is expressed by the following result from COHEN, DAHMEN and DEVORE [2000].

Theorem 4.9.1 *Let $U \in \ell^p_w$ for some $p < 2$ and let $W \in \ell^2$ be such that $\|U - V\| \leq \varepsilon$. Let $a > 1$ and W be the smallest subvector of V such that $\|W - V\| \leq a\varepsilon$. Then we have*

$$\#(W) \lesssim \|U\|_{\ell^p_w}^{1/t} \varepsilon^{-1/t}, \tag{4.9.25}$$

with $t = 1/p - 1/2$ and therefore

$$\|U - W\| \leq (1 + a)\varepsilon \lesssim \|U\|_{\ell^p_w} (\#(W))^{-t}. \tag{4.9.26}$$

Moreover, we have

$$\|W\|_{\ell^p_w} \lesssim \|U\|_{\ell^p_w}. \tag{4.9.27}$$

Proof Let N be the smallest number such that

$$\|U - U_N\| \leq (a - 1)\varepsilon, \tag{4.9.28}$$

where U_N is the best N-term approximation of U. We know that

$$N \lesssim \|U\|_{\ell^p_w}^{1/t} \varepsilon^{-1/t}. \tag{4.9.29}$$

Denoting by Λ_N the set of indices λ associated with the N largest $|u_\lambda|$, and R_{Λ_N} the operator of restriction of this set, we have

$$\begin{aligned} \|V - V_{\Lambda_N}\| &\leq \|(I - R_{\Lambda_N})(U - V)\| + \|U - U_N\| \\ &\leq \|U - V\| + \|U - U_N\| \leq a\varepsilon. \end{aligned}$$

Therefore

$$\#(W) \leq N \lesssim \|U\|_{\ell^p_w}^{1/t} \varepsilon^{-1/t}, \tag{4.9.30}$$

which proves (4.9.25).

In order to prove (4.9.27), we proceed in a similar way as in the proof of (4.6.31) in Corollary 4.6.4: we define by $U_{\#(W)}$ the best approximation of U with $N = \#(W)$ terms, and write

$$\begin{aligned} \|W\|_{\ell^p_w} &= \|U_{\#(W)}\|_{\ell^p_w} + \|U_{\#(W)} - W\|_{\ell^p_w} \\ &\leq \|U\|_{\ell^p_w} + \|U_{\#(W)} - W\|_{\ell^p} \\ &\leq \|U\|_{\ell^p_w} + (2\#(W))^t \|U_{\#(W)} - W\| \\ &\leq \|U\|_{\ell^p_w} + (2\#(W))^t (\|U_{\#(W)} - U\| + \|U - W\|) \\ &\lesssim \|U\|_{\ell^p_w}, \end{aligned}$$

where we have successively used Hölder's inequality and the approximation properties of $U_{\#(W)}$ and W. ◇.

Theorem 4.9.1 shows that thresholding the numerical solution at the appropriate level will help maintain the optimal bound (4.9.24) and in turn the optimal complexity of the algorithm. The modified algorithm therefore simply amounts to a modification of step 5 in the previous algorithm according to:

5. Once it is found that

$$\|\mathbf{COARSE}(F,\varepsilon_j) - \mathbf{APPROX}(AU^{j,n},\varepsilon_j)\| \geq (\frac{2\tau\|A\|}{1-\rho}+3)\varepsilon_j,$$

set

$$U^{j+1} := \mathbf{COARSE}(U^{j,n}, a(\|A^{-1}\|\frac{2\tau\|A\|}{1-\rho}+5)\varepsilon_j), \tag{4.9.31}$$

and Λ^{j+1} its support. Then go to step 2 with $j := j+1$.

According to Theorem 4.9.1 a should be fixed strictly larger than 1, for example $a = 2$. With this modification, the new algorithm now achieves the ultimate goal, according to the following result which was established in COHEN, DAHMEN and DEVORE [2002].

Theorem 4.9.2 *Assume that $A \in \mathcal{M}_{\alpha,\beta}$ and that $U \in \ell^p_w$ with the restriction $t = 1/p - 1/2 < \min\{\alpha/d, \beta/d\}$. Then the algorithm produces approximate solutions U^j supported on finite sets Λ^j such that*

$$\#(\Lambda^j) \lesssim \|U\|_{\ell^p_w}^{1/t} 2^{j/t}. \tag{4.9.32}$$

Therefore, the algorithm converges at the optimal rate

$$\|U - U^j\| \lesssim 2^{-j} \lesssim \|U\|_{\ell^p_w} \#(\Lambda^j)^{-t}. \tag{4.9.33}$$

The number of arithmetic operations for computing U^j, is also bounded by

$$\mathcal{N}(U^j) \lesssim \|U\|_{\ell^p_w}^{1/t} 2^{j/t}. \tag{4.9.34}$$

Proof The bound (4.9.32) is a direct consequence of (4.9.25) in Theorem 4.9.1. In order to verify that the complexity remains of the same order, we need to examine the intermediate approximations $U^{n,j}$ which are computed inside the inner loop and make sure that they do not ruin the optimal complexity. For this, note that, by Theorem 4.9.1, we also have

$$\|U^j\|_{\ell^p_w} \lesssim \|U\|_{\ell^p_w}. \tag{4.9.35}$$

According to Theorem 4.6.3, we also know that

$$\|F\|_{\ell^p_w} \lesssim \|U\|_{\ell^p_w}. \tag{4.9.36}$$

Using Corollary 4.6.4, we thus obtain that the intermediate interation

$$U^{j,n+1} := U^{j,n-1} + \tau(\mathbf{COARSE}(F, \varepsilon_j) - \mathbf{APPROX}(AU^{j,n-1}, \varepsilon_j)),$$

preserves the bounds

$$\|U^{n,j}\|_{\ell^p_w} \lesssim \|U\|_{\ell^p_w}, \tag{4.9.37}$$

and

$$\#(U^{n,j}) \lesssim \|U\|_{\ell^p_w}^{1/t} 2^{j/t}, \tag{4.9.38}$$

with uniform constant since n stays bounded by a fixed M. We also obtain from Corollary 4.6.4 that the complexity of this iteration remains bounded by $\|U\|_{\ell^p_w}^{1/t} 2^{j/t}$ up to a multiplicative constant, which yields (4.9.34) by summing over all $n \leq M$ and $l \leq j$. $\diamond$.

Remark 4.9.1 *In the above theorem, we have measured the complexity of the algorithm by the total number of arithmetic operations. In addition, the application of the routines* **COARSE** *and* **APPROX** *also requires a sorting procedure of the wavelet coefficients of the numerical solution. This procedure has complexity $N \log N$ with N the number of coefficients in the current solution. In turn, the total complexity of computing U^j behaves like $\mathcal{O}(j2^{j/t})$. We also have left aside the complexity coming from the computations of quantities such as the stiffness matrix entries $\langle A\psi_\lambda, \psi_\mu\rangle$ or right hand side data coefficients $\langle f, \psi_\mu\rangle$, which are here assumed to be accessible at $\mathcal{O}(1)$ cost for each of them. In practice, this cost depends on the type of numerical quadrature which is used in these computations and in some instances (such as integral equations with a singular kernel) might dominate the overall complexity.*

Remark 4.9.2 *The optimality properties of the above adaptive wavelet algorithm have been confirmed by numerical tests in papers such as* COHEN *and* MASSON *[1999] and* BARINKA, BARSCH, CHARTON, COHEN, DAHLKE, DAHMEN *and* URBAN *[2001]. However, the implementation of these tests involves a substantial amount of fine tuning of each step of the algorithm such as coarsening and matrix-vector multiplication. Indeed, while the algorithm proposed in this section is designed in order to ensure the optimal asymptotical behaviour of the approximation error, it is not optimized with respect to the involved multiplicative constants which might be quite large. In particular, the refinement rules imposed by the matrix-vector multiplication often appear to be too severe, in the sense that accuracy ε can be achieved* *with less coefficients than those produced by* **APPROX**(MU, ε). *Another observation is that the coarsening steps are not really needed in the practical implementations of the adaptive wavelet method (for those problems which have been considered so far) which still does exhibit optimal convergence rate. However, it is not known how to prove (4.9.32) and (4.9.34) without these coarsening steps. There seems to be a similar situation in the finite element context: it has recently been proved in* BINEV, DAHMEN *and* DEVORE *[2002] that optimal appoximation rates can be achieved by an adaptive mesh refinement algorithm which incorporates coarsening steps, while these steps are not needed in practice.*

Remark 4.9.3 *An earlier version of Theorem 4.9.2 was obtained in* COHEN, DAHMEN *and* DEVORE *[2000] for a more complicated adaptive wavelet algorithm, closer in spirit to the classical approach: Galerkin solutions u^n are computed on adaptive wavelet spaces V_{Λ^n} which are iteratively refined based on error indicators derived in earlier works such as* BERTOLUZZA *[1995a] and* DAHLKE, DAHMEN, HOCHMUTH *and* SCHNEIDER *[1997]. These error indicators are approximations to the wavelet coefficients of the residual $F - AU^n$ computed by the procedures* **COARSE** *and* **APPROX**. *In the approach that was presented in this section, the approximate residual is directly used to update the numerical solution.*

Remark 4.9.4 *The adaptive strategy that we have detailed for elliptic problem extends to non-elliptic problems such as saddle-point problems, according to the general paradigm: the Richardson iteration will then be replaced by an Uzawa algorithm or a gradient descent applied to the least-square system (see* COHEN, DAHMEN *and* DEVORE *[2002] and* DAHLKE, DAHMEN *and* URBAN *[2002]). A striking fact is that the convergence of the adaptive algorithm in this setting occurs without the need for compatibility conditions such as LBB: one inherits the well-posedness of the continuous problem through the convergence of the infinite dimensional algorithm. It should also be recalled that matrix compressibility also applies in the case of integral operators which have quasi-sparse wavelet discretizations. Therefore several of the obstructions from the classical approach - conditioning, compatibility, dense matrices - have disappeared in the wavelet approach. The extension to nonlinear variational problems, based on infinite dimensional relaxation or Newton iterations is currently being considered. It requires a specific procedure for the application of the nonlinear operator in the wavelet coefficients domain which generalizes the* **APPROX** *procedure.*

4.10 Conclusions

Multiscale decomposition into wavelet bases allow us to generate optimal adaptive approximations of functions by simple thresholding procedures. The accuracy of such nonlinear approximations can be predicted from the smoothness of the function in certain scales of quasi-normed Besov spaces. While wavelet adaptive techniques have been successfully applied to the reduction of parameters in the description of data (e.g. compression, curve and surface design, statistical estimation and denoising), their ability to compute the unknown solution of a physical process in an optimally compressed form is still to be confirmed in many instances.

The potential of these techniques for a given problem can be understood through a fine analysis of the smoothness of solutions in the scales of Besov spaces characterized by the rate of nonlinear approximation. Their practical implementation requires in addition the development and the numerical analysis of efficient resolution schemes. These schemes should benefit from the many positive features of wavelets discretizations that have been described throughout this book: high order and local approximations, fast decomposition and reconstruction algorithms, characterization of function spaces, operator compression and adaptive approximation of functions.

Let us conclude by saying that despite its theoretical success, in the sense of achieving for certain classes of problem the optimal convergence rate with respect to the number of degrees of freedom, the wavelet-based approach to adaptive numerical simulation suffers from several major curses.

The curse of geometry: while the construction of wavelet bases on rectangular domains is fairly simple - one can use tensor product techniques and inherit the simplicity of the univariate construction - it is far less trivial for domains with complicated geometries. Several approaches, including those addressed in chapter 2 of this book, have been proposed to deal with this situation and concrete implementations are nowadays available, but they result in an unavoidable loss of structural simplicity in comparison with the basic prototypes such as the Haar system.

The curse of anisotropy: as explained in this chapter, adaptive wavelet approximation has roughly speaking the same properties as isotropic refinement. However, many instances of singularities such as boundary layers and shock waves have anisotropic features which suggests that the refinement should be more pronounced in one particular direction. Wavelet bases should therefore be reconsidered if one wants to obtain better rates which take some advantage of the geometric smoothness of the curves of discontinuities. On the adaptive finite element side, anisotropic refinement has been considered and practically implemented, yet without a clear theory available for the design of an optimal mesh.

The curse of data structures: encoding and manipulating the adaptive wavelet approximations u_{Λ^n} to the solution means that we need to store both the coefficients and the indices of the adaptive set Λ^n which should be dynamically updated. The same goes for the indices of the matrix A which are used in the matrix-vector multiplication at each step of the algorithm. This dynamic adaptation, which requires appropriate data structures, results in major overheads in the computational costs which are observed in practice: while the wavelet adaptive algorithms indeed exhibit the optimal rate of convergence and might outperforms adaptive finite element algorithms from this perspective, the latter remain significantly more efficient from the point of view of computational time.

The significance of wavelets in numerical analysis remains therefore tied to these curses and future breakthroughs are to be expected once simple and appropriate solutions are proposed in order to deal with them.

4.11 Historical notes

Since the 1960's, the study of nonlinear methods has become a central issue in approximation theory. In his famous monograph on Besov spaces, Peetre mentions the study of nonlinear approximation spaces as a motivation to consider the case $p < 1$ (PEETRE [1974]). In the 1980's, a systematic treatment of nonlinear approximation spaces in relation with interpolation theory was achieved by R. DeVore, V. Popov and their collaborators. In particular, the nonlinear version of Theorem 3.5.2 was given in DEVORE and POPOV [1986] with applications in the context of free knot spline approximation. Results specific to wavelets were firstly proved in DEVORE, JAWERTH and POPOV [1991], including Theorem 4.6.2 with a slightly different proof.

On a more practical point of view, since the pioneering work in the 1970's, e.g. BABUSHKA and RHEINBOLDT [1978], adaptive methods have been intensively developed and analyzed by numericians in the context of finite differences, finite elements and finite volumes. These methods have been documented in numerous contributions (see e.g. VERFURTH [1994] and ERIKSSON, ESTEP, HANSBO and JOHNSON [1995]). A key ingredient is the derivation of *local a-posteriori error indicators* from the currently computed solutions.

The first adaptive wavelet method for PDE's was proposed in MADAY,

PERRIER and RAVEL [1991], in the context of evolution problems, with the idea that a typical indicator for refinement at the next time step is provided by the size of a wavelet coefficient of the current numerical solution. In the same period, the idea of using wavelets in order to sparsify integral operators was introduced in BEYLKIN, COIFMAN and ROKHLIN [1991]. Since then, wavelet adaptive discretizations have been extensively analyzed and implemented in various situations.

References

ABGRALL, R. (1995), Multiresolution analysis on unstructured meshes: application to CFD, in: K.W.MORTON and M.J.BAINES, eds., *Numerical methods for fluid dynamics V*, Oxford Science Publications.

ADAMS, R. (1975), *Sobolev Spaces* (Academic Press, New York).

ALPERT, B. (1993), A class of bases in L^2 for the sparse representation of integral operators, *SIAM J. Math. Anal.* **24**, 246-262.

ANTONINI, M., M. BARLAUD, I. DAUBECHIES and P. MATHIEU (1992), Image coding using wavelet transforms, *IEEE Trans. Image Processing* **1**, 205-220.

AVELLANEDA, M., G. DAVIS and S. MALLAT (1997), Adaptive greedy approximations, *Constructive Approximation* **13**, 57-98.

BABUSHKA, I., J. GAGO, D. KELLY and O. ZIENKIEWICZ (1982) Hierarchical finite element approaches, error estimates and adaptive refinements, in: J.R. WHITEMAN, ed, *The mathematics of finite elements and applications IV* (Academic Press, London).

BABUSHKA, I and W.C. RHEINBOLDT (1978), Error estimates for adaptive finite element computations, *SIAM J. Numer. Anal.* **15**, 736-754.

BAKER, G.A. (1975), *Essentials of Padé Approximants* (Academic Press, New York).

BAKHVALOV, N.S. (1966), On the convergence of a relaxation method with natural constraints on the elliptic operator, *USSR Comp. Math. and Math. Phys.* **6**, 101-135.

BANK, R.E. and T. DUPONT (1981), An optimal order process for solving finite element equations, *Math. Comp.* **36**, 35-51.

BANK, R.E., A.H. SHERMAN and A. WEISER (1983), Refinement algorithms and data structures for regular local mesh refinement, in: R. STEPLEMAN and AL., eds., *Scientific Computing*, Amsterdam: IMACS, 3-17 (Noth Holland, Amsterdam).

BANK, R.E. and A. WEISER (1985), Some a posteriori error estimates for elliptic partial differential equations, *Math. Comp.* **44**, 283-301.

BARINKA, A., T. BARSCH, P. CHARTON, A. COHEN, S. DAHLKE, W. DAHMEN and K. URBAN (2001), Adaptive wavelet schemes for elliptic problems – Implementation and numerical experiments, *SIAM J. Sci. Comp.* **23**, 910-939.

BARRON, A.R. (1994), Approximation and estimation bounds for artificial neural networks, *Machine Learning* **14**, 115-133.

BENNETT, C. and R. SHARPLEY (1988), *Interpolation of operators* (Academic Press, New York).

BERGER, M. and P. COLLELA (1989), Local adaptive mesh refinement for shock hydrodynamics, *J. Comp. Phys.* **82**, 64-84.

BERGER, M. and J. OLIGER (1984), Adaptive mesh refinement for hyperbolic partial differential equations, *J. Comp. Phys.* **53**, 482-512.

BERGH, J. and J. LÖFSTRÖM (1976) *Interpolation spaces* (Springer Verlag, Berlin).

BERTOLUZZA, S. (1995a), A-posteriori error estimates for wavelet Galerkin methods, *Appl. Math. Lett.* **8**, 1-6.

BERTOLUZZA, S. (1995b), Adaptive wavelet collocation for the solution of steady state equation, in: *SPIE Proc. Wavelet Appl. II.* **2491**.

BERTOLUZZA, S. (1997) An adaptive collocation method based on interpolating wavelets, in: W. DAHMEN, A.J. KURDILA and P.OSWALD, eds., *Multiscale Wavelet Methods for PDE's* (Academic Press, New York).

BERTOLUZZA, S. (1998) Stabilization by multiscale decompositions, *Appl. Math. Lett.* **6**, 129-134.

BESOV, O.V. (1959), On a family of functions spaces. Embedding theorems and applications, *Dokl. Akad. Nauk. USSR* **126**, 1163-1165.

BEYLKIN, G., R. COIFMAN and V. ROKHLIN (1991), Fast wavelet transforms and numerical algorithms, *Comm. Pure and Appl. Math.* **44**, 141-183.

BEYLKIN, G. and J.M. KEISER (1997), An adaptive pseudo-wavelet approach for solving nonlinear partial differential equations, in: W. DAHMEN, A.J.KURDILA and P.OSWALD, eds., *Multiscale Wavelet Methods for PDE's* (Academic Press, New York).

BEYLKIN, G. and M. BREWSTER (1995), A multiresolution strategy for numerical homogeneization, *Appl. Comp. Harm. Anal.* **2**, 327-349.

BINEV, P., W. DAHMEN and R. DEVORE (2002), Adaptive finite element methods with convergence rate, preprint IGPM-RWTH Aachen, to appear in Numer. Math.

BIRMAN, M.S. and M. SOLOMJAK (1967), Piecewise polynomial approximation of functions of the class W_p^α, *Math. USSR-Sb.* **2**, 295-317.

DE BOOR, C. (1978), *A practical guide to splines* (Springer Verlag, Berlin).

DE BOOR, C. (1973), The quasi-interpolant as a tool in elementary polynomial spline theory, in: G.G. LORENTZ, ed., *Approximation theory* (Academic Press, New York).

DE BOOR, C., R. DEVORE and A. RON (1993), On the construction of multivariate pre-wavelets, *Constructive Approximation* **9**, 123-166.

DE BOOR, C. and G. FIX (1973a), Approximation from shift-invariant subspaces of $L_2(\mathbb{R}^d)$, *Trans. Amer. Math. Soc.* **341**, 787-806.

DE BOOR, C. and G. FIX (1973b), Spline approximation by quasi-interpolants, *J. Approx. Theory* **8**, 19-45.

BRAESS, D. and W. HACKBUSCH (1983), A new convergence proof for the multigrid method including the V-cycle, *SIAM J. Numer. Anal.* **20**, 967-975.

BRAMBLE, J.H. and S.R. HILBERT (1970), Estimation of linear functionals on Sobolev spaces with applications to Fourier transform and spline interpolation, *SIAM J. Numer. Anal.* **7**, 112-124.

BRAMBLE, J.H. and S.R. HILBERT (1971), Bounds for a class of linear functionals with applications to Hermite interpolation, *Numerische Mathematik* **16**, 362-369.

BRAMBLE, J.H., J.E. PASCIAK and J. XU (1990), Parallel multilevel preconditioners, *Math. Comp.* **55**, 1-22.

BRAMBLE, J.H. (1993), *Multigrid methods* (Longman Scientific and Technical, Harlow, England).

BRANDT, A. (1977), Multilevel adaptive solutions to boundary value problems, *Math. Comp.* **31**, 333-390.

BRENNER, S.C. and R.L. SCOTT (1994), *The mathematical theory of finite element methods* (Springer Verlag, New York).

BREZINSKI, C. and J. VAN ISEGHEM (1994), Padé Approximations, in: P.G. CIARLET and J.L. LIONS eds., *Handbook of Numerical Analysis III* (North Holland, Amsterdam).

BRUDNYI, Y. (1970), Approximation of functions of n-variables by quasi-polynomials, *Math. USSR Izv.* **4**, 568-586.

BURT, P. and E. ADELSON (1983), The laplacian pyramid as a compact image code, *IEEE Trans. Comm.* **31**, 482-540.

BUTZER, P.L. and H. BEHRENS (1967), *Semi-groups of operators and approximation* (Springer Verlag, Berlin).

BUTZER, P.L. and K. SCHERER (1972), Jackson and BErnstein type inequalities for families of commutative operators in Banach spaces, *J. Approx. Theory* **5**, 308-342.

CANUTO, C. and I. CRAVERO (1997), Wavelet-based adaptive methods for advection-diffusion problems, *Math. Mod. Meth. Appl. Sci.* **7**, 265-289.

CANUTO, C., A.TABACCO and K. URBAN (1999), The wavelet element method. Part I: construction and analysis, *Appl. Comp. Harm. Anal.* **6**, 1-52.

CARNICER, J.M., W. DAHMEN and J.M. PEÑA (1996), Local decomposition of refinable spaces and wavelets, *Appl. Comp. Harm. Anal.* **3**, 127-153.

CAVARETTA, A., W. DAHMEN and C.A. MICCHELLI (1991), Stationary Subdivision, *Memoirs of AMS* **453**.

CHAITKIN, G.M. (1974), An algorithm for high speed curve generation, *Comp. Graph. and Image Proc.* **3**, 346-349.

CHUI, C.K. (1988), *Multivariate splines* (SIAM, Philadelphia).

CHUI, C.K. (1992), *An introduction to wavelets* (Academic Press, Boston).

CHUI, C.K. and E. QUAK (1992), Wavelets on a bounded interval, in: D. BRAESS and L. SCHUMAKER, eds., *Numerical methods of approximation theory*, 57-76 (Birkhaüser, Basel).

CHUI, C.K. and Y. WANG (1992), A general framework for compactly supported splines and wavelets, *J. Approx. Theory* **71**, 263-304.

CIARLET, P.G. (1978), *The finite element method for elliptic problems* (North Holland, Amsterdam).

CIARLET, P.G. (1991), Basic error estimate for the finite element method, in: P.G. CIARLET and J.L. LIONS, eds., *Handbook of Numerical Analysis*, vol.2 (Elsevier, Amsterdam).

CIECELSKI, Z. and J. FIGIEL, (1983) *Spline bases in classical function spaces on compact C^∞ manifolds*, Studia Math. 76, 1-58.

CLEMENT, P. (1975), Approximation by finite element functions using local regularizations, *RAIRO Mod. Math. Anal. Num.* **2**, 77-84.

COHEN, A., W. DAHMEN, I. DAUBECHIES and R. DEVORE (2001), Tree-structured approximation and optimal encoding, *App. Comp. Harm. Anal.* **11**, 192-226

COHEN, A., W. DAHMEN and R. DEVORE (2000), Multiscale methods on bounded domains, *Trans. Amer. Math. Soc.* **352**, 3651-3685.

COHEN, A., W. DAHMEN and R. DEVORE (2000), Adaptive wavelet methods for elliptic equations - Convergence rates, *Math. Comp.* **70**, 27-75.

COHEN, A., W. DAHMEN and R. DEVORE (2002), Adaptive wavelet methods for elliptic equations - beyond the elliptic case, *Found. of Comp. Math.* **2**, 203-245.

COHEN, A., I. DAUBECHIES and P. VIAL (1993), Wavelets and fast wavelet transforms on an interval, *Appl. Comp. Harm. Anal.* **1**, 54-81.

COHEN, A. and I. DAUBECHIES (1996), A new technique to estimate the regularity of refinable functions, *Rev. Mat. Iberoamer.* **12**, 527-591.

COHEN, A. and I. DAUBECHIES (1992), A stability criterion for biorthogonal wavelets and their related subband coding schemes, *Duke J. of Math.* **68**, 313-335.

COHEN, A., I. DAUBECHIES and J.C. FEAUVEAU (1992), Biorthogonal bases of compactly supported wavelets, *Comm. Pure and Appl. Math.* **45**, 485-560.

COHEN, A., I. DAUBECHIES and G. PLONKA (1997), Regularity of Refinable Function Vectors, *J. Fourier Anal. Appl.* **3**, 295-324.

COHEN, A., R. DEVORE, P. PETRUSHEV and H. XU (1998), Nonlinear approximation and the space $BV(\mathbb{R}^2)$, *Ame. J. Math* **121**, 587-628.

COHEN, A. and A. EZZINE (1996), Quadratures singulieres et fonctions d'echelle, *CRAS Paris* **323**, *Série I*, 829-834.

COHEN, A., K. GRÖCHENIG and L. VILLEMOES (1999), On the regularity of multivariate refinable functions, *Constructive Approximation* **15**, 241-255.

COHEN, A., S.-M. KABER, S. MUELLER and M. POSTEL (2002), Fully adaptive multiscale methods for conservation laws, *Math. of Comp.* **72**, 183-225.

COHEN, A. and R. MASSON (1999), Wavelet adaptive methods for elliptic problems - preconditioning and adaptivity, *SIAM J. Sci. Comp.* **21**, 1006-1026.

COHEN, A. and R. MASSON (2000), Wavelet methods for elliptic problems - boundary conditions and domain decomposition, *Numerische Mathematik* **86**, 193-238.

COHEN, A. and R. RYAN (1995), *Wavelets and multiscale signal processing* (Chapman and Hall, London).

COHEN, A. and J.M. SCHLENKER (1993), Compactly supported wavelets with hexagonal symmetry, *Constructive Approximation* **9**, 209-236.

COIFMAN, R. and M.V. WICKERHAUSER (1992), Entropy-based algorithms for best basis selection, *IEEE Trans. Inf. Theory* **38**, 713-718.

COIFMAN, R., Y. MEYER, S. QUAKE and M.V. WICKERHAUSER (1993), Signal processing and compression with wavelet packets, in: S. ROQUES and Y. MEYER eds., *Progresses in wavelet analysis and applications* (Frontieres, Gif sur Yvette, France).

DAHLKE, S., W. DAHMEN, R. HOCHMUTH and R. SCHNEIDER (1997), Stable multiscale bases and local error estimation for elliptic problems, *Applied Numerical Mathematics* **23**, 21-48.

DAHLKE, S. and R. DEVORE (1996), Besov regularity for elliptic boundary value problems, *Comm. in PDE's* **22**, 1-16.

DAHLKE, S. (1999), Besov regularity for elliptic boundary value problems on polygonal domains, *Appl. Math. Lett.* **12**, 31-36.

DAHLKE, S. (1998), Besov regularity for elliptic boundary value problems with variable coefficients, *Manuscripta Math.* **95**, 59-77.

DAHMEN, W. (1996), Stability of multiscale transforms, *J. of Fourier Anal. and Appl.* **2**, 341-361.

DAHMEN, W. (1997), Wavelets and multiscale methods for operator equations, *Acta Numerica*, 55-228.

DAHMEN, W. and R. SCHNEIDER (1998), Wavelets with complementary boundary conditions - function spaces on the cube, *Results in Math 34*, 255-293.

DAHMEN, W. and R. SCHNEIDER (1999a), Composite wavelet bases for operator equations, *Math. Comp.* **68**, 1533-1567.

DAHMEN, W. and R. SCHNEIDER (1999b), Wavelets on manifold I. Construction and domain decomposition, *SIAM J. Math. Anal.* **31**, 184-230.

DAHMEN, W. and C.A. MICCHELLI (1984), Subdivision algorithms for the generation of box spline surfaces, Computer Aided Geometric Design **1**, 115-129.

DAHMEN, W. and C.A. MICCHELLI (1993), Using the refinement equation for evaluating integrals of wavelets, *SIAM J. Numer. Anal.* **30**, 507-537.

DAHMEN, W., A. KUNOTH and R. SCHNEIDER (1996), Operator equations, multiscale concepts and complexity, *1995 AMS-SIAM Summer Seminar, Math. of Numer. Anal., Amer. Math. Soc., Lectures in Appl. Math.* **32**, 225-261.

DAHMEN, W., A. KUNOTH and K. URBAN (1996), A wavelet Galerkin method for the Stokes problem, *Computing* **56**, 259-302.

DAHMEN, W., A. KUNOTH and K. URBAN (1999), Biorthogonal spline-wavelets on the interval. Stability and moment conditions, *Appl. Comp. Harm. Anal.* **6**, 132-196.

DAHMEN, W., B. HAN, R.-Q. JIA and A. KUNOTH (2000), Biorthogonal multiwavelets on the interval: cubic hermite splines, *Constr. Approx.* **16**, 221-259.

DAHMEN, W., S. PRÖSSDORF and R. SCHNEIDER (1994), Wavelet approximation methods for pseudodifferential equations I: stability and convergence, *Mathematische Zeitschrift*, **215**, 583-620.

DAHMEN, W., S. PRÖSSDORF and R. SCHNEIDER (1993), Wavelet methods for pseudodifferential equations II: matrix compression and fast solution, *Advances in computational mathematics* **1**, 259-335.

DAHMEN, W. and R. STEVENSON (1999) Element by element construction of wavelets satisfying stability and moment conditions, *SIAM J. Num. Anal.* **37**, 319-352.

DAUBECHIES, I. (1992), *Ten lectures on wavelets* (SIAM, Philadelphia).

DAUBECHIES, I. (1988), Orthonormal bases of compactly supported wavelets, *Comm. Pure Appl. Math.* **41**, 909-996.

DAUBECHIES, I. and J. LAGARIAS (1991), Two scale difference equations I. Existence and global regularity of solutions, *SIAM J. Math. Anal.* **22**, 1388-1410.

DAUBECHIES, I. and J. LAGARIAS (1992), Two scale difference equations II. Local regularity, infinite products of matrices and fractals, *SIAM J. Math. Anal.* **23**, 1031-1079.

DAUBECHIES, I. and W. SWELDENS (1998), Factoring wavelet transform into lifting steps, *J. Four. Ana. Appl.* **4**, 247-269.

DAUGE, M. (1988), *Elliptic boundary value problems on corner domains* (Springer Verlag, Berlin).

DAVID, G. and J.L. JOURNÉ (1984), A boundedness criterion for generalized Calderon-Zygmund operators, *Ann. of Math.* **120**, 371-397.

DENY, J and J.L. LIONS (1954), Les espaces de type de Beppo Levi, *Ann. Inst. Fourier* **5**, 305-370.

DESLAURIERS, G. and S. DUBUC (1987), Interpolation dyadique, in: G. CHERBIT, ed., *Fractals, dimensions non entières et applications* (Masson, Paris).

DEVORE, R. (1998), nonlinear Approximation, *Acta Numerica* **7**, 51-150.

DEVORE, R. and G.G. LORENTZ (1993), *Constructive Approximation* (Springer Verlag, Berlin).

DEVORE, R., B. JAWERTH and V. POPOV (1992), Compression of wavelet decompositions, *American Journal of Math.* **114**, 737-785.

DEVORE, R. and B. LUCIER (1990), High order regularity for conservation laws, *Indiana Math. J.* **39**, 413-430.

DEVORE, R., P. PETRUSHEV and X.M. YU (1992), Wavelet approximation in the space C, in: *Progress in approximation theory* (Springer Verlag, New York).

DEVORE, R. and V. POPOV (1988), Free multivariate splines, *Constructive Approximation* **3**, 239-248.

DEVORE, R. and V. POPOV (1988), Interpolation of Besov spaces, *Trans. of the AMS* **305**, 397-414.

DEVORE, R. and V. POPOV (1988), Interpolation spaces and nonlinear approximation, in: M. CWIKEL, J. PEETRE, Y. SAGHER and H. WALLIN, eds., *Function spaces and applications, Lecture Notes in Math.* **1302**, 191-205 (Springer Verlag, Berlin).

DEVORE, R., S. RIEMENSCHNEIDER and R. SHARPLEY (1979), Weak interpolation in Banach spaces, *J. Functional Analysis* **33**, 58-91.

DEVORE, R.A and R.C. SHARPLEY (1993), Besov spaces on domains in $\mathbb{R}^d$, *Trans. of the AMS* **335**, 843-864.

DeVore, R. and X.M. Yu (1990), Degree of adaptive approximation, *Math. Comp.* **55**, 625-635.

Donoho, D., I. Johnstone, G. Kerkyacharian and D. Picard (1994) Wavelet shrikage: Asymptotia ? (with discussion), *J. Roy. Stat. Soc. Ser. B* **57**, 301-369.

Donovan G., J. Geronimo and D. Hardin (1999) Orthogonal polynomials and the construction of piecewise polynomial smooth wavelets, *SIAM J. Math. Anal.* **30**, 1029-1056.

Doerfler, W. (1996), A convergent adaptive algorithm for Poisson's equation, *SIAM J. Numer. Anal.* **33**, 1106-1124.

Duffin, R.J. and A.C. Schaeffer (1952), *A class of nonharmonic Fourier series*, Trans. Amer. Math. Soc., 72, 341-366.

Dyn, N., J. Gregory and D. Levin (1991), Analysis of uniform binary subdivision schemes for curve design, *Constructive Approximation* **7**, 127-147.

Dyn, N. (1992), Subdivision schemes in computer-aided geometric design, in: W.A. Light, ed., *Advances in Numerical Analysis II, Wavelets, Subdivision algorithms, and Radial Basis Functions* (Clarendon Press, Oxford).

Dyn, N. (2000), A construction of biorthogonal functions to B-splines with multiple knots, *Appl. Comp. Harm. Anal.* **8**, 24-31.

Dyn, N., M. Floater and A. Iske (2002), Adaptive thinning algorithms for bivariate scattered data, *J. Appl. and Comp. Math.*

Eirola, T. (1992), Sobolev characterization of solutions of dilation equations, *SIAM J. Math. Anal.* **23**, 1015-1030.

Eriksson, K., D. Estep, P. Hansbo and C. Johnson (1995), Introduction to adaptive methods for differential equations, *Acta Numerica, Cambridge University Press*, 105-158.

Fedorenko, R.P. (1964), The speed of convergence of one iterative process, *USSR Comp. Math. and Math. Phys.* **4**, 1092-1096.

Frazier, M, B. Jawerth and G. Weiss (1991), *Littlewood-Paley theory and the study of function spaces*, CBMS Conference Lecture Notes 79, (AMS, Providence, RI).

Gersho, A. and R.M. Gray (1992), *Vector quantization and signal compression* (Kluwer, Boston).

Gilbert, A. (1998), A comparison of multiresolution and classical one-dimensional homogeneization schemes, *Appl. Comp. Harm. Anal.* **5**, 1-35.

Greengard, L. and V. Rokhlin (1987), A fast algorithm for particle simulations, *J. Comp. Phys.* **325**.

Grisvard, P. (1983), *Elliptic problems on non-smooth domains* (Pittman, New York).

HACKBUSCH, W. (1985), *Multi-grid methods and applications* (Springer Verlag, New York).

HACKBUSCH, W. and Z.P. NOWAK (1989), On the fast matrix multiplication in the boundary element method by panel clustering, *Numer. Math.* **54**, 463-491.

HARTEN, A. (1993), Discrete multiresolution and generalized wavelets, *J. Appl. Num. Math.* **12**, 153-193.

HARTEN, A. (1995), Multiresolution algorithms for the numerical solution of hyperbolic conservation laws, *Comm. Pure Appl. Math.* **48**, 1305-1342.

HARTEN, A. (1996), Multiresolution representation of data II: generalized framework, *SIAM J. Numer. Anal.* **33**, 1205-1256.

JAFFARD, S. (1991), Pointwise smoothness, two-microlocalization and wavelet coefficients, *Publicacions Matemàtiques* **35**, 155-168.

JAFFARD, S. (1992), Wavelet methods for fast resolution of elliptic problems, *SIAM J. Numer. Anal.* **29**, 965-986.

JIA, R.Q. and B. HAN (1998), Multivariate refinement equations and convergence of subdivision schemes *SIAM J. Math. Anal.* **29**, 1177-1199.

JIA, R.Q. and J.J. LEI (1993), Approximation by multi-integer translates of functions having global support, *J. Approx. Theory* **72**, 2-23.

JIA, R.Q (1999) Smoothness of multivariate refinable functions in Sobolev spaces, *Trans. Amer. Math. Soc.* **351**, 4089-4112.

JIA, R.Q., S. RIEMENSCHNEIDER and D.X. XHOU (1999), Smoothness of multiple refinable functions and multiple wavelets, *SIAM J. Matrix Anal. and Appl.* **21**, 1-28.

JOHNEN, H. and K. SCHERER (1976), On the equivalence of the K-functional and moduli of continuity and some applications, in: *Constructive theory of functions of several variables*, Lecture Notes in Mathematics **571** (Springer Verlag, Berlin).

KAHANE, J.P. (1961), *Teoria Constructiva de Functiones*, Course Notes, University of Buenos Aires.

KOVACEVIC, J. and M. VETTERLI (1995), *Wavelets and subband coding* (Prentice Hall, Englewood Cliff, N.J.).

KUNOTH, A. (1995), Multiscale preconditioning - Appending boundary conditions by Lagrange multipliers, *Advances in Comp. Math.* **4**, 145-170.

KYRIASIS, G (1996), Wavelet coefficients measuring smoothness in $H_p(\mathbb{R}^d)$, *Appl. and Com. Harm. Anal.* **2**, 100-119.

LAX, P. (1972), *Hyperbolic system of conservation laws and the mathematical theory of shock waves* (SIAM, Philadelphia).

LAZAAR, S., J. LIANDRAT and P. TCHAMITCHIAN (1994), Algorithme à base d'ondelettes pour la résolution numérique d'équations aux dérivées partielles à coefficients variables, *CRAS Paris, Série I* **319**, 1101-1107.

LEMARIÉ, P.G. (1997), On the existence of compactly supported dual wavelets, *Appl. Comp. Harm. Anal.* **3**, 117-118.

LEMARIÉ, P.G. (1988), Ondelettes à localisation exponentielle, *J. Math. Pures et Appl.* **67**, 227-236.

LEMARIÉ, P.G. (1992), Analyses multirésolutions non-orthogonales, commutation entre projecteurs et dérivation et ondelettes vecteurs à divergence nulle, *Rev. Mat. Iberoamer.* **8**, 221-236.

LEMARIÉ, P.G. and G. MALGOUYRES (1991), Support des fonctions de base dans une analyse multiresolution, *CRAS Paris Série I* **213**, 377-380.

LEVEQUE, R.J. (1992), *Numerical methods for conservation laws* (Birkhäuser, Basel).

LIONS, J.L and E. MAGENES (1972), *Non-homogeneous boundary value problems and applications* (Springer Verlag, New York)

LIONS, J.L. and J. PEETRE (1964), *Sur une classe d'espaces d'interpolation, Publ. Math. Inst. Hautes Etudes Sci.* **19**, 5-68.

LYCHE, T. (1993), Knot removal for spline curves and surfaces, in: E.W. CHENEY, C.K. CHUI and L. SCHUMAKER, eds., *Approximation Theory VII*, 207-227 (Academic Press, Boston).

MADAY, Y., V. PERRIER and J.C. RAVEL (1991), Adaptativité dynamique sur bases d'ondelettes pour l'approximation d'équations aaux dérivées partielles, *CRAS Paris, Série I* **1**, 405-410.

MALLAT, S. (1989), Multiresolution approximation and wavelet orthonormal bases of $L^2(\mathbb{R})$, *Trans. Amer. Math. Soc.* **315**, 69-88.

MALLAT, S. (1998), *A wavelet tour of signal processing* (Academic Press, New York)

MASSON, R. (1996), Biorthogonal spline wavelets on the interval for the resolution of boundary problems, *M3As* **6**, 749-791.

MEYER, Y. (1990), *Ondelettes et opérateurs* (Hermann, Paris), english translation by D.H. SALINGER (Cambridge University Press, Cambridge, 1992).

MEYER, Y. (1991), Ondelettes sur l'intervalle, *Rev. Mat. Iberoamer.* **7**, 115-134.

MEYER, Y. (1992), *Ondelettes, algorithmes et applications* (Armand Colin, Paris), english translation by R. RYAN (SIAM, Philadelphia, 1994).

MEYER, Y. (1997), Wavelets and fast numerical algorithms, in: P.G. CIARLET and J.L. LIONS, eds., *Handbook of Numerical Analysis*, vol.5 (Elsevier, Amsterdam).

MORIN, P., R. NOCETTO and K. SIEBERT (2000), Data oscillation and convergence of adaptive FEM, *SIAM J. Num. Anal.* **38**, 466-488.

NEWMAN, D.J. (1964), Rational approximation to $|x|$, *Michigan Math. J.* **11**, 11-14.

NGUYEN, V. and G. STRANG (1996), *Wavelets and filter banks*, (Wellesley-Cambridge Press, Cambridge).

OSWALD, P. (1990), On the degree of nonlinear spline approximation in Besov-Sobolev spaces, *J. Approx. Theory* **61**, 131-157.

OSWALD, P. (1994), *Multilevel finit element approximation: theory and applications* (Teubner, Stuttgart).

OSWALD, P. and R. LORENTZ (1996), Multilevel finite element Riesz bases in Sobolev spaces, in: P. BJORSTAD, ed., *Proceeding of the Ninth International Conf. on Domain Decomposition, Bergen.*

OSWALD, P. and E.A. STOROZHENKO (1978), Jackson's theorem in the space $L_p(\mathbb{R}^k)$, $0 < p < 1$, *Siberian Math J.* **19**, 630-639.

PEETRE, J. (1974), *New thoughts on Besov spaces*, (Duke University Math. Series, Duham).

PEETRE, J. and G. SPARR (1972), Interpolation of normed abelian groups, *Ann. Mat. Pura. Appl.* **92**, 217-252.

PELLER, V. (1985), Description of Hankel operators of the class σ_p for $p < 1$, investigation of the order of rational approximation, *Math. USSR Sbornik* **50**, 465-494.

PERRIER, V. and P. MONASSE (1998), Orthogonal wavelet bases adapted for partial differential equations with boundary conditions, *SIAM J. Math. Anal.* **29**, 1040-1065.

PETRUSHEV, P. (1988), Direct and converse theorems for spline and rational approximation and Besov spaces, in: M. CWIKEL, J. PEETRE, Y. SAGHER and H. WALLIN, eds., *Function spaces and applications, Lecture Notes in Math.* **1302**, 363-377 (Springer Verlag, Berlin).

PETRUSHEV, P. (2000), Bases consisting of rational functions of uniformly bounded degrees or more general functions, *J. Funct. Anal.* **174**, 18-75.

PLONKA, G. (1995), Factorization of refinement masks for function vectors, in: C.K. CHUI and L.L. SCHUMAKER, eds., *Wavelets and Multilevel Approximation*, (World Scientific Publ. Co., Singapore).

RIOUL, O. (1992), Simple criteria for subdivision scheme, *SIAM J. Math. Anal.* **23**, 1544-1576.

SAID, A and W.A. PEARLMAN (1996), An image multiresolution representation for lossless and lossy compression, *IEEE Trans. Image. Proc.* **5**, 1303-1310.

SCHOENBERG, I.J. (1946), Contributions to the problem of approximation of equidistant data by analytic functions, *Quarterly Applied Math.* **4**, 45-99 (part A), 112-141 (part B).

SCHUMAKER, L. (1981), *Spline functions: basic theory* (Wiley, New York).

SHAPIRO, J. (1993), Embedded image coding using zerotrees of wavelet coefficients, *IEEE Signal Processing* **41**, 3445-3462.

SHEN, Z. (1998), Refinable function vectors, *SIAM J. Math. Anal.* **29**, 235-250.

STEIN, E.M. (1970), *Singular Integrals and Differentiability Properties of Functions* (Princeton University Press, Princeton).

STEVENSON, R. (1998), Piecewise linear (pre)-wavelets on non-uniform meshes, in: W. HACKBUSCH and G. WITTUM eds., *Multigrid Methods V*, 306-319 (Lecture Notes in Computational Science and Engineering, Vol. 3, Springer-Verlag, Heidelberg).

STEVENSON, R. (1997), Experiments in 3D with a three point hierarchical basis preconditioner, *Applied Numerical Math.* **23**, 159-175.

STRANG, G. and G. FIX (1969), Fourier analysis of the finite element method in Ritz-Galerkin theory, *Stud. Appl. Math.* **48**, 265-273.

STRANG, G. and G. FIX (1973), *An analysis of the finite element method* (Wellesley-Cambridge press, Cambridge).

SWELDENS, W. and R. PIESSENS (1994), Quadrature formulae and asymptotic error expansions for wavelet approximation of smooth functions, *SIAM J. Num. Anal.* **31**, 2140-2164.

SWELDENS, W. (1996), The lifting scheme: a custom design construction of biorthogonal wavelets, *Appl. Comp. Harm. Anal.* **3**, 186-200.

SWELDENS, W. (1998), The lifting scheme: a construction of second generation wavelets, *SIAM J. Math. Anal.* **29**, 511-546.

TAUBMAN, D. (1998), Embedded block coding with optimized truncation, *Signal Processing: Image Communication*, **17**, *Elsevier, Amsterdam.*

TCHAMITCHIAN, P. (1996), Inversion explicite de certains operateurs elliptiques, *SIAM J. Math. Anal.* **27**, 1680-1703.

TCHAMITCHIAN, P. (1996), Wavelets functions and operators, in: G. ERLEBACHER, Y. HUSSAINI and L. JAMESON, eds., *Wavelets, Theory and Applications*, 87-178 (Oxford University Press).

TEMLYAKOV, V. (1998), Best N-term approximation and greedy algorithms, *Adv. Comp. Math.* **8**, 249-265.

TRIEBEL, H. (1983), *Theory of function spaces* (Birkhauser, Basel).

URBAN, K. (1995), On divergence-free wavelets, *Avances in Comp. Math.* **4**, 51-82.

VAIDYANATHAN, P.P. (1992), *Multirate systems and filter banks* (Prentice Hall, Englewood Cliff, N.J.).

VERFURTH, R. (1994), A posteriori error estimation and adaptive mesh refinement techniques, *J. Comp. Appl. Math.* **50**, 67-83.

VILLEMOES, L. (1994), Wavelet analysis of two-scale refinement equations, *SIAM J. Math. Anal.* **25**, 1433-1460.

VOLKMER, H. (1992), On the regularity of wavelets, *IEEE Trans. Inf. Theory.* **38**, 872-876.

von Petersdorff, T. and C. Schwab (1997), Fully discrete multiscale Galerkin BEM, in: W. Dahmen, A.J.Kurdila and P.Oswald, eds., *Multiscale Wavelet Methods for PDE's* (Academic Press, New York).

Young, R.M. (1980), *An introduction to non-harmonic Fourier series* (Academic Press, New York).

Yserantant, H. (1986), On the multi-level splitting of finite element spaces, *Numerische Mathematik* **49**, 379-412.

Yserantant, H. (1993), Old and new convergence proof for multigrid methods, *Acta Numerica, Cambridge University Press*, 285-326.

Index

www.ingramcontent.com/pod-product-compliance
Lightning Source LLC
LaVergne TN
LVHW011910200926
848484LV00002B/582
* 9 7 8 0 4 4 4 5 1 1 2 4 9 *